T5-DHH-233

The
Malcolm Baldrige National Quality Award

A Yardstick for Quality Growth

 ADVANCED FOAM DEVELOPMENT

Engineering Process Improvement Series

John W. Wesner, Ph.D., P.E., Consulting Editor

The Malcolm Baldrige National Quality Award

A Yardstick for Quality Growth

Maureen S. Heaphy
Gregory F. Gruska

ADDISON-WESLEY PUBLISHING COMPANY

Reading, Massachusetts · Menlo Park, California · New York · Don Mills, Ontario
Wokingham, England · Amsterdam · Bonn · Sydney · Singapore · Tokyo
Madrid · San Juan · Paris · Seoul · Milan · Mexico City · Taipei

Many of the designations used by manufacturers and sellers to distinguish their products are claimed as trademarks. Where those designations appear in this book and Addison-Wesley was aware of a trademark claim, the designations have been printed with initial capital letters or all capital letters. The Malcolm Baldrige National Quality Award service mark is used by permission of the National Institute of Standards and Technology. This use does not imply a recommendation of, endorsement by, or affiliation with the National Quality Program managed by NIST.

The publisher offers discounts on this book when ordered in quantity for special sales.

For more information, please contact:

Corporate & Professional Publishing Group
Addison-Wesley Publishing Company
One Jacob Way
Reading, Massachusetts 01867

Library of Congress Cataloging-in-Publication Data
Heaphy, Maureen S.,1950–
    The Malcolm Baldrige National Quality Award : a yardstick for quality growth /
Maureen S. Heaphy, Gregory F. Gruska.
        p.   cm.—(Engineering Process Improvement Series)
    Includes bibliographical references and index.
    ISBN 0-201-63368-X (pbk. : alk. paper)
    1. Malcolm Baldrige National Quality Award. 2. Total quality management—
Awards—United States. I. Gruska, Gregory S., 1943–   . II. Title. III. Series.
HD62. 15.H4 1995
658.5'62'07973—dc20   95-8179
                                                                             CIP

Copyright © 1995 by Addison-Wesley Publishing Company

All rights reserved. No part of this publication may be reproduced, stored in a retrieval system, or transmitted, in any form or by any means, electronic, mechanical, photocopying, recording, or otherwise, without the prior written permission of the publisher. Printed in the United States of America. Published simultaneously in Canada.

Text design by Wilson Graphics & Design (Kenneth J. Wilson)

0-201-63368-X
1 2 3 4 5 6 7 8 9 CRW 98979695
First printing May 1995

DEDICATION

To Dr. W. Edwards Deming

On December 20, 1993, Dr. Deming died at the age of 93.
He is missed by many, not only for his words of wisdom but also
for the guidance he provided to so many aspiring professionals.
He challenged traditional paradigms, making some people
uncomfortable, yet opening new ways of thinking to others.
We were blessed to have been associates of and,
even more so, friends with such a great man.

Engineering Process Improvement Series

Consulting Editor, John W. Wesner, Ph.D., P.E.

Global competitiveness is of paramount concern to the engineering community worldwide. As customers demand ever-higher levels of quality in their products and services, engineers must keep pace by continually improving their processes. For decades, American business and industry have focused their quality efforts on their end products rather than on the processes used in the day-to-day operations that create these products and services. Experts across the country now agree that focusing on continuous improvements of the core business and engineering processes within an organization will lead to the most meaningful, long-term improvements and production of the highest-quality products.

Whether your title is researcher, designer, developer, manufacturer, quality or business manager, process engineer, student, or coach, you are responsible for finding innovative, practical ways to improve your processes and products in order to be successful and remain world-class competitive. The **Engineering Process Improvement Series** takes you beyond the ideas and theories, focusing in on the practical information you can apply to your job for both short-term and long-term results. These publications offer current tools and methods and useful how-to advice. This advice comes from the top names in the field; each book is both written and reviewed by the leaders themselves, and each book has earned the stamp of approval of the series consulting editor, John W. Wesner.

Key innovations by industry leaders in process improvement include work in benchmarking, concurrent engineering, robust design, customer-to-customer cycles, process management, and engineering design. Books in this series will discuss these vital issues in ways that help engineers of all levels of experience become more productive and increase quality significantly.

All of the books in the series share a unique graphic cover design. Viewing the graphic blocks descending, you see random pieces coming together to build a solid structure, signifying the ongoing effort to improve processes and produce quality products most satisfying to the customer. If you view the graphic blocks moving upward, you see them breaking through barriers—just as engineers and companies today must break through traditional, defining roles to operate more effectively with concurrent systems. Our mission for this series is to provide the tools, methods, and practical examples to help you hurdle the obstacles, so that you can perform simultaneous engineering and be successful at process and product improvement.

The series is divided into three categories:

Process management and improvement This includes books that take larger views of the field, including major processes and the end-to-end process for new product development.

Improving Functional Processes These are the specific functional processes that are combined to form the more inclusive processes covered in the first category.

Special process topics and tools These are methods and techniques that are used in support of improving the various processes covered in the first two categories.

CONTENTS

FOREWORD

You may wonder why, with so many books on quality and its variations, you should read this particular one. This book is important because it is written by knowledgeable professionals who have a strong background in industry and consulting. They know the field well and write about it in a clear and interesting way. They not only tell us what the Malcolm Baldrige National Award is all about, but give advice on how to use the Criteria to improve organizational performance and the outflow of products and services.

Why read the book when so much information flows from the Baldrige office? Because authors Heaphy and Gruska give us valuable insights and are good coaches. They take us through the categories of the Award step by step. They give us examples of who does unusually well in each of them. They keep us focused on the reason for continuous quality improvement. All of these steps get us ready to improve our organizations.

One early step is self-assessment and the discipline of writing down what is happening. We are led through the tough job of unbiased evaluation and assignment of scores. Then we're told how to use the results to plan for and implement improvements. One of the nice features is an example of a completed application for the Award, and an example of a feedback report after assessment. If we wonder how our score might compare to others, we have a valuable table by year of the number of companies in each of several scoring ranges. We're given the estimated time spent by the Award winners in preparing the application. But, despite giving us all of this good practical information about the Award itself, the authors do not stray from their goal of helping organizations to improve through use of the Baldrige Criteria.

As an added feature we have a comparison of ISO 9001 with Baldrige Criteria. The misleading statements made by some advocates of only the ISO 9000 series are exposed to the daylight of facts.

Who should read a book like this? Business and government executives as well as those in not-for-profit organizations would benefit. Those who have or are starting internal Baldrige-type assessments would find the book of value. Those who want to see how good they are would benefit by the road map for assessment. Finally, academics would find this a state-of-the-art companion text in business and engineering schools.

Take good notes if you want to get the maximum value from it. Helpful ideas occur in every chapter.

William A.J. Golomski
President, W.A. Golomski & Associates
Senior Lecturer in Business Policy and Quality Management,
Graduate School of Business, University of Chicago
Honorary Member, American Society for Quality Control

PREFACE

The Malcolm Baldrige National Quality Award is not a quality standard.

Now that we have your attention, let us explain what we mean. When we speak of standards, we refer to documented requirements such as those from the American National Standards Institute (ANSI) or the International Organization for Standardization (ISO). They typically identify a minimum set of requirements that must be met in order to achieve a stated objective. These documents prescribe what an organization must do in order to be in compliance with the standard. On the other hand, the Criteria, which are the basis for the Malcolm Baldrige National Quality Award process, are 24 basic, interrelated, results-oriented requirements. They are nonprescriptive in that they do not dictate *how* the results are to be obtained. In fact, the objectives of the Award process include to identify and communicate the diversity of approaches that can be used by organizations to achieve quality and performance excellence. The Malcolm Baldrige National Quality Award Criteria are considered by many to be the national definition of quality.

The Malcolm Baldrige National Quality Award (MBNQA, or simply Baldrige Award) was developed by a public-private partnership administered by the National Institute of Standards and Technology (NIST) in response to Public Law 100-107, The Malcolm Baldrige National Quality Improvement Act of 1987. Congress wanted to stem the eroding leadership of American products and services as well as assist in the improvement of the nation's productivity growth. Legislators realize that improving the quality of products and services is one of the key ways for Americans to regain our competitive position. Improved quality leads to less waste, higher customer satisfaction, and improved profitability.

Although Public Law 100-107 establishes a national quality award for the recognition of excellent companies, it also establishes that the central purpose of the MBNQA is educational—to create an evolving body of knowledge on how organizations are able to change their cultures and achieve eminence; and to provide a means for other companies to learn and use this information.

As part of the Award process, the recipients are required to share information on their successful quality strategies with other U.S. companies.[1] Since the Award's inception, the Award recipients have held thousands of sharing sessions—far exceeding the minimal requirements and NIST's expectations. An examination of the Award recipients' processes shows the diversity of approaches

---

[1] Award recipients are not expected to share proprietary information, even if such information was part of their application.

and implementation strategies used to achieve excellence. Unlike a standard, which prescribes a specific set of requirements, the Baldrige Criteria are nonprescriptive. They do not prescribe any specific tools, techniques, technologies, systems, or starting points. The intent is to foster innovation. The Criteria are, instead, a road map to assist companies on their journey to excellence. They contain well-defined core values, concepts, and a framework, and also detail expectations derived from these. In 1995, the Criteria focus on 24 key requirements (Items) among seven Categories. These Items are further detailed by Areas to Address and extensive notes. These provide users with many insights and ideas for improving their practices.

The MBNQA process is also a yardstick. The review and scoring systems aid in the development of diagnostic feedback that can be used for planning and implementation of appropriate improvement activities. It is here where the rubber hits the road.

We realize that few organizations will seek to win the Award itself. In fact, at the time of publication of this book, some organizations are ineligible to submit applications—including health care organizations, educational institutions, government agencies, and nonprofit organizations. But these organizations also have market constraints and customer requirements. They have the same need to learn how well they are doing and how they could get better. The Malcolm Baldrige National Quality Award Criteria are used by many companies to guide their quality improvement journey. They use the Criteria to aid self-assessment and improvement. This satisfies one of the intents of Public Law 100-107.

The Criteria alone are not sufficient for self-improvement. Reimann and Hertz, of the Office of Quality Programs at NIST, identify four characteristics of an effective self-assessment instrument as "(1) educational value; (2) completeness—addressing all requirements and how they are deployed; (3) an integrated way to collect information so that it may be meaningfully evaluated; (4) results indicators that address how well requirements are met."[2]

The Criteria and supporting documentation contain the key elements necessary for self-assessment; the emulation of the Award process (review, scoring, and feedback) fills in the rest.

---

[2] C. Reimann and H. Hertz, "The Malcolm Baldrige National Quality Award and ISO 9000 Registration: Understanding Their Many Important Differences," National Institute of Standards and Technology, Gaithersburg, MD, 1994.

> The aim of this book is to provide guidance and direction to the readers on their quality journey by utilizing the MBNQA Criteria and process as a *road map* and *yardstick*.

DESCRIPTION OF CHAPTERS

This book offers insights into the Baldrige Criteria through explanations, examples from the Award recipients and other organizations, and a complete case study. The focus of the book is on using the MBNQA process for self-improvement, not applying for a national award. Consequently, the book goes beyond the development of a written response to the Criteria.[3] The review process deliverables, including feedback report and site visit issues, are presented to assist your organization in performing a self-assessment. We follow the learning paradigm that you first should understand "what" before you learn "how."

The first two chapters are a basic introduction to the Criteria and the key concepts. It is intended for those readers who have had little or no exposure to the MBNQA or who wish a quick refresher.

Chapter 3 starts us on our quality journey by discussing the use of the Criteria for self-improvement. This includes a management grid, which can be used to do a macro assessment of your quality journey. The PDSA[4] (Plan-Do-Study-Act) cycle is introduced to show *improvement* and *no improvement* models. Although this chapter does not focus on scoring, the concepts of Approach, Deployment, and Results (ADR) are introduced. Chapter 3 also discusses how various states and municipalities are using the MBNQA as a prototype for their state quality awards to promote quality excellence within their own region.

The Criteria are explained in detail in Chapter 4. Understanding of the elements of this chapter is key in the self-improvement process. Each of the seven Categories is addressed with examples from recipients and other organizations. A graphical summary of the Category and key excellence indicators is included.

Chapter 5 discusses the Award process. This may seem out of place in a book focused on self-improvement, but to be successful, self-improvement needs an effective review, evaluation, and feedback process as well as the response development activities. Multiple-unit organizations may decide to set up a structure

---

[3] The response would be the MBNQA application if the organization decided to enter the Award process.

[4] Although some people use PDCA (C for check), Dr. Deming was insistent on PDSA.

similar to that of the Baldrige process; that is, establish an internal Board of Examiners who would review, score, and provide feedback to a written response from the organization's divisions, plants, and/or business units. AT&T's Chairman's Quality Award is an example of such an internal process. Such a structure may not be appropriate for single-unit organizations. But even these organizations need independent review, evaluation, and feedback, whether it is internal or external. The Eaton Quality Award (EQA) process of the Eaton Corporation is described and discussed as an example of a self-improvement activity.

Chapter 5 also contains the list of MBNQA recipients to date and a short description of each.

The assessment process is the focus of Chapter 6. This chapter discusses what is needed for an effective written response development process and provides suggestions to achieve it. The review process and sources of information are also discussed. This is followed by an example of a written response—Chapter 7, the Collins Technology case study.

This book was finalized shortly after the release of the 1995 MBNQA Criteria. To provide you with the most current information, most of the book was updated to reflect the changes to the Criteria—with one exception. The Collins Technology Case Study is based on the 1994 MBNQA Criteria. The reasons for this decision are (1) the time required to make the changes to reflect the 1995 Criteria would have delayed the publication of the book until fall 1995 and (2) the educational purpose of the case study is not put in jeopardy by using the 1994 Criteria. Because of this, Chapter 7 contains a comparison of the 1994 and 1995 Criteria and includes the 1994 Criteria as part of the case study.

The case study can be read simply as an example of a written response. But it can also be used to calibrate your own "yardstick" because the reviews, scoring, and feedback of this written response are given in Chapter 10. But, don't read ahead.

The review, scoring, and site visit processes are described in Chapters 8, 9, and 11 respectively. These chapters are also useful for individuals developing the written response, since the individuals who will review and score the response are customers of the developers of the response. By understanding how these processes work, the response development activities can become more effective. For the reader who is a member of a Board of Examiners, or is developing a structure for self-improvement, these chapters provide many useful guidelines, examples, and insights.

Chapter 9 also contains the score distribution of participants in the MBNQA process and general findings of companies that are doing well and those that need improvement.

The crucial step in the quality improvement journey is the planning subsequent to the written response review. This is where the PDSA (Plan, Do, Study, Act) cycle begins its next turn. Chapter 10 examines the feedback report: its purpose, development, and use. An effective feedback report can assist tremendously in the planning process. On the other hand, the quality of the feedback report is irrelevant if the organization does not take advantage of its information in the development and implementation of improvement activities. The PDSA cycle is used as a model for converting the assessment into action plans. This chapter also contains the feedback report and Item scores for the Collins Technology Case Study response.

Up to Chapter 11, the book is focused on the technical aspects of the self-improvement process. Although this is not an organizational/management transformation or cultural change book, it would be incomplete without some discussion on the organizational and cultural aspects of the self-improvement process based on the MBNQA. Chapter 12, "Getting Started: Organization and Follow-Through," discusses the importance of a supportive attitude, the need to maintain an improvement focus, and barriers to these activities. A model for change, a model for getting started, and discussion on overcoming the barriers are part of this chapter.

In concert with the core value of continual improvement, the MBNQA process is subject to continual improvement. Since its initiation in 1988, the MBNQA process and Criteria have undergone multiple changes, with the Criteria booklet, the primary educational tool of the process, changing yearly. Chapter 14 discusses this evolution and looks into the future of the MBNQA process.

Except for the summary and conclusions given in Chapter 15, our discussion on using the MBNQA process for self-improvement is complete. However, with the globalization of markets, many organizations are being required to register their quality system to one of the ISO 9000 standards. Other organizations, interested in self-improvement, have heard or read about how the ISO 9000 series standards have helped companies improve quality. Some articles even imply that the MBNQA process and ISO 9000 registration are alternate (and equivalent) strategies. To the casual observer, there must be something to these statements. After all, the number of ISO 9000 ads and articles far exceed those for MBNQA. In fact, although the MBNQA and ISO 9000 series share some expectations for the quality system, they differ in intent, focus, and scope. Chapter 13 was included to clarify the interrelationships between the MBNQA Criteria and ISO 9000. It includes tables cross-referencing the Criteria Items with the ISO item numbers.

To assist the readers, we use the key icon **O⊶** to indicate those paragraphs we feel are key to the aim of the book. We also use the following icons to identify the specific focus of individual sections:

Introductory: These sections are intended for those readers who have had little or no exposure to the MBNQA or who wish a quick refresher.

MBNQA: These sections address and discuss the MBNQA Criteria and process.

Self-Improvement: These sections focus on the use of the MBNQA Criteria and process in organizational self-improvement.

Sections that are not tagged provide background and supporting information including examples and results.

In conclusion, as can be noted by the diverse list of Baldrige Recipients to date (see Chapter 5) the Baldrige Criteria can be applied to large companies (Motorola), small companies (Marlow Industries), high-tech firms (Solectron, Zytec), low-tech firms (Granite Rock), manufacturing (Milliken & Co.), and service organizations (Federal Express). All that is necessary is an organization with a desire to delight the customer, a vision that supports this desire, and a road map and yardstick for implementation. This book cannot give you the desire to delight the customer, but it can help you see the vision and will help you understand the Baldrige Criteria as a road map and yardstick for your quality improvement journey.

ACKNOWLEDGMENT

We gratefully acknowledge our appreciation to the people who made this book possible:

- To our parents, who preserved our joy of learning

- To our colleagues, clients, and seminar participants, whose sharing and insight provided new depth

- To our teachers, who guided us along the way

- To the reviewers, for their suggestions and focus

- To Jennifer Joss at Addison-Wesley for her immediate and continual interest in the project.

CHAPTER

1| Background

INTRODUCTION

Some organizations think they cannot afford to provide a quality product or service. Actually, quality is an investment that pays off. An organization with $10 million in sales could be paying $2 million for scrap, warranty costs, lost productivity, and other losses related to product or service problems. Add to this something for customer dissatisfaction, which according to the late Dr. W. Edwards Deming is unknown and unknowable, and soon we ask ourselves if a fraction of these expenses could be used to prevent the problems. According to one Baldrige Award recipient, "Despite what seemed to be an enormous initial investment in quality, Globe [Metallurgical] had gained a $40 return for every dollar spent on quality." [1]

Improving the quality of U.S. products and services is essential to regaining our competitive position. In the mid-1980s many industry and public leaders realized this. As part of the national quality improvement efforts, industry and government joined to establish an annual U.S. National Award for quality—the Malcolm Baldrige National Quality Award (MBNQA). The Malcolm Baldrige National Quality Improvement Act was signed into law on August 20, 1987. The Award was developed to help U.S. companies reach new levels of quality and customer satisfaction.

The intent of the Award is to promote:

- Awareness of quality and its impact on competitiveness

- Understanding of the requirements for excellence in quality

- Sharing of information on successful strategies and on benefits derived.

---

[1] "Baldrige Winner Gains Market Share," *Quality in Manufacturing*, May/June 1993.

The Award is named for Malcolm Baldrige, who was Secretary of Commerce from 1981 until his accidental death in 1987. Baldrige was a proponent of quality management as a key to this country's prosperity and long-term strength. He helped draft one of the early versions of the Quality Improvement Act. In recognition of his contributions, Congress named the Award in his honor.

The Criteria defined in the MBNQA are not prescriptive. They do not state *how* to achieve a desired outcome. Rather they state the desired outcome—they are descriptive rather than prescriptive. Twenty different companies may take 20 different approaches to meet the Criteria. There is no one right way. Among the objectives of the Baldrige Award is the recognition of these various successful quality strategies and the sharing of information about them.

 ## KEY CONCEPTS OF THE AWARD CRITERIA:

Although the Criteria are descriptive, they are built on a core set of values and paradigms, described as follows:

Customer-driven quality: The ultimate goal of quality efforts is to delight the customer. This is a strategic concept that requires a company to pleasantly surprise and delight the customer, not just prevent defects.

Leadership involvement: The senior executives of a company establish the company culture by their words and actions. Personal involvement in the performance initiative is required to demonstrate commitment, understanding, and the company's values.

Continual improvement and learning: Throughout the Criteria, the last area to address is often "How do you evaluate and improve (the process just described)?" Ongoing improvement of key processes is essential to achieve or maintain a leading position in the market.

Employee participation and development: Customer delight is the goal defined by the Baldrige Criteria. One of the ways to achieve this is to develop the full potential of the workforce. This includes employee involvement, training, recognition, safety, and satisfaction.

Fast response: Success in the competitive marketplace requires efficient and effective processes. Reducing the time it takes to bring a new product or service to the marketplace is now a recognized objective. Likewise, time to respond to a customer inquiry or complaint or time to respond to an employee suggestion often differentiates the high-performance organizations from the bureaucratic ones.

Design quality: Products and services that are designed to be insensitive to expected variation (robust) give a company an advantage. Success will be achieved even when conditions are not optimal. Moving the focus of quality from manufacturing or delivery to the design stage allows the biggest quality impact on the overall product or service.

Prevention: Intuitively it is accepted that prevention is more effective than fire fighting. But it is hard to talk to someone about fire prevention when their house is on fire. If we don't make time for prevention activities, the fires will continue.

Partnership development: Partnerships between labor and management, between suppliers and the company, and between the company and customers offer win–win opportunities. Partnerships with educational institutions to develop the workforce and partnerships with other companies for joint developments are examples of other win–win opportunities.

Long-range outlook: Developing partnerships with suppliers, developing the full potential of the workforce, and creating the future for our customers all require a long-range outlook. At least 3 to 5 years, and in some industries 10 to 15 years, should be the planning horizon.

Management by fact: Facts combined with product or process knowledge are very powerful. Trends, projections, and cause-and-effect relationships should be studied with an understanding of variation.

Good corporate citizen: A company's quality goals and objectives should address corporate responsibility and citizenship such as public health, public safety, and the environment. Giving back to the community by working with educational institutions or local government are included in the Criteria.

Results orientation: Internal results should be correlated with field results and financial indicators. Performance indicators are a means of communicating and monitoring progress. These include customer satisfaction and retention, market share, product and service quality, human resource performance and development, financial indicators, and public responsibility.

USEFUL CRITERIA?

How do we know the Baldrige Criteria form a useful set of guidelines? A study was done of 20 companies that scored high in the Baldrige review process. Common results[2] were as follows:

[2] "Management Practices by General Accounting Office," GAO/NSIAD-91-190, May 1991. Call 202/275-6241 to order.

Employees

- increased job satisfaction, improved attendance, reduced turnover

Quality

- improved quality, reduced cost, increased reliability, increased on-time delivery, fewer errors, reduced lead time

Customers

- improved satisfaction, fewer complaints, higher customer retention rate

Profitability

- improved market share, improved financial indicators

These are the types of results that convince people that quality is an investment.

INTERNAL USE OF CRITERIA

Many companies and organizations are viewing the Criteria as the national definition of Total Quality Leadership (TQL). The Criteria are being used for more than just applying for the Award. Companies are using them to review and guide their performance initiatives. Companies such as Eaton Corporation, Baxter Healthcare, AT&T, and University of Michigan Hospitals use the Criteria for self-assessment.[3] Findings from the assessment should be used as input to the business planning process to build on the strengths and close the gaps where improvements are needed.

According to Mr. Ron Brown, secretary of commerce under President Bill Clinton, "Since the beginning of the Award program over 1 million copies of the Criteria have been distributed."[4] Additionally, hundreds of thousands of copies are made by others for internal company use, for state award processes, etc.

As noted earlier by Globe Metallurgical, a 1988 Award recipient, quality is an investment that pays large dividends.

[3] "Internal Award Programs: Benchmarking the Baldrige to Improve Corporate Quality," *Quality Digest*, May 1992, and Michigan Quality Leadership Award.

[4] Presentation at Quality Forum IX, October 18, 1993, Washington, DC.

2 | Criteria Overview

INTRODUCTION

The Malcolm Baldrige National Quality Award Criteria were developed as an assessment strategy. The Criteria are defined by seven major thrusts called *Categories*:

1.0 Leadership

2.0 Information and Analysis

3.0 Strategic Planning

4.0 Human Resource Development and Management

5.0 Process Management

6.0 Business Results

7.0 Customer Focus and Satisfaction.

The seven Categories are further defined by 24 Items. The Categories and Items are summarized in this chapter. Category and Item details and process flowcharts are given in Chapter 4. A relationship matrix identifying the interdependency of Items is a significant feature of this chapter. These items are further amplified by 54 Areas to Address.

 ## CATEGORIES

The Criteria are grouped into a framework, as shown in Figure 2.1. Leadership, which is Category 1.0 of the Criteria, is considered to be the driving force. Active, visible leadership is needed to achieve world-class levels of customer satisfaction and organizational performance excellence. The overall operating systems to achieve these outcomes are defined by Categories 2.0 through 5.0. The internal measures of progress such as quality, on-time delivery, response time, and economic results are reported in Category 6.0. Approaches and deployment of

Figure 2.1

Criteria Framework

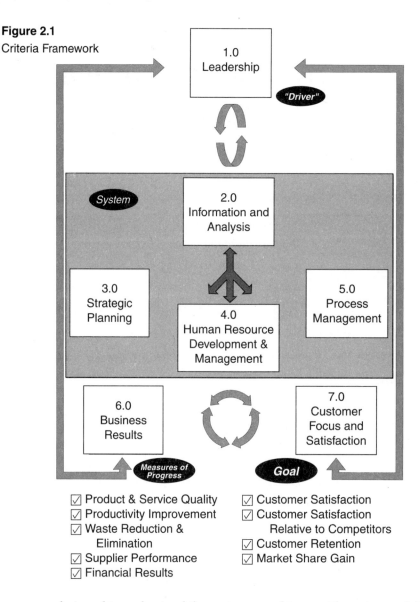

☑ Product & Service Quality
☑ Productivity Improvement
☑ Waste Reduction & Elimination
☑ Supplier Performance
☑ Financial Results

☑ Customer Satisfaction
☑ Customer Satisfaction Relative to Competitors
☑ Customer Retention
☑ Market Share Gain

processes designed to understand the customer and to provide customer delight and measurements of the results of customer satisfaction are reported in Category 7.0.

The seven Categories, detailed by the 24 Items, are not presented in the order of implementation. For example, the design activity, which involves translating the voice of the customer is Item 5.1. But getting the voice of the customer is not covered until Item 7.1. This example also demonstrates the interdependency of the Items. The activity of Item 7.1, getting the voice of the customer, is used as input to Item 5.1, translating the voice of the customer (see also Figure 2.2, later in this chapter). The Criteria are grouped as shown to display the overall framework.

The major focus of each of the seven Categories is given here in an excerpt from NIST material.

1.0 Leadership

The senior executives' success in creating and sustaining a quality culture and customer focus

2.0 Information and Analysis

The effectiveness of data collection, analysis, and benchmarking for quality planning and improvement

3.0 Strategic Planning

The integration of performance requirements into the business plan and the establishment of strategies to assure the quality of products and services

4.0 Human Resource Development and Management

The success of efforts to develop the full potential of the workforce to meet the company's quality and performance objectives

5.0 Process Management

The effectiveness of systems and practices to assure the quality of all operations including suppliers

6.0 Business Results

The measures of progress demonstrating quality achievement and quality improvement through internal quantitative measures

7.0 Customer Focus and Satisfaction

The effectiveness of systems to focus on the customer, define his requirements and successfully meet them, and to provide external results.

 CATEGORIES AND ITEMS

In 1995 the seven Categories are detailed with 24 examination Items and 54 Areas to Address. Although the *customer* has a whole Category (Category 7.0), there is frequent mention of the customer throughout.[1] For example, Item 1.2, Leadership

---

[1] Although many organizations have recognized that there are internal and external customers of their processes, the term "customer" in the MBNQA documentation refers to only the external customer. The Criteria include the interrelationships among departments and processes, but reserve the term "customer" to refer to the external dealers, retailers, and/or end users of the organization's products and services unless stated otherwise.

System and Organization, considers how the leadership system focuses on the customers' objectives. This theme of "customer" is present in all of the Categories. A summary of the Categories and their examination Items follows.

1995 Award Examination Criteria

1.0 *Leadership*

"The *Leadership* Category examines senior executives' personal leadership and involvement in creating and sustaining a customer focus, clear values and expectations, and a leadership system that promotes performance excellence. Also examined is how the values and expectations are integrated into the company's management system, including how the company addresses its public responsibilities and corporate citizenship."[2]

Item 1.1 Senior Executive Leadership: senior executives' personal involvement and visibility in creating an environment for quality.

Item 1.2 Leadership System and Organization: integrate customer focus and performance objectives into the company's organization.

Item 1.3 Public Responsibility and Corporate Citizenship: include public responsibility in quality policies and be a good corporate citizen.

2.0 *Information and Analysis*

"The *Information and Analysis Category* examines the management and effectiveness of the use of data and information to support customer-driven performance excellence and marketplace success."[3]

Item 2.1 Management of Information and Data: breadth and depth of data, including customer, product performance, internal operations, support areas, supplier, employee, and financial, used to guide quality activities.

Item 2.2 Competitive Comparisons and Benchmarking: scope and use of information from outside the company including competitive comparisons and benchmarking (best practices) information from outside the industry.

Item 2.3 Analysis and Use of Company-Level Data: the conversion of the data, described in Item 2.1, into useful information, and the correlation between the indicators presented, such as internal measures, and customer

---

2 *1995 Malcolm Baldrige National Quality Award Criteria*, National Institute of Standards and Technology, Gaithersburg, MD.

3 *Ibid.*

satisfaction indicators. Also included is the linkage between quality improvement and financial improvement.

3.0 *Strategic Planning*

O⟶ "The *Strategic Planning* Category examines how the company sets strategic directions, and how it determines key plan requirements. Also examined is how the plan requirements are translated into an effective performance management system."[4]

Item 3.1 Strategy Development: the planning process for the short and longer term. The process to develop key business drivers to use for deployment is required.

Item 3.2 Strategy Deployment: how the plans developed in Item 3.1 are translated into action plans and deployed throughout the organization, and how the company's performance projects into the future.

4.0 *Human Resource Development and Management*

O⟶ "The *Human Resource Development and Management* Category examines how the workforce is enabled to develop and utilize its full potential, aligned with the company's performance objectives. Also examined are the company's efforts to build and maintain an environment conducive to performance excellence, full participation, and personal and organizational growth."[5]

Item 4.1 Human Resource Planning and Evaluation: overall human resource plans such as labor and management cooperation, redesign of work organization, employee surveys, establishing networks with educational institutions.

Item 4.2 High Performance Work Systems: how work is organized to create opportunities for employee self-direction and effective communication; how compensation and recognition reinforce the work and job design.

Item 4.3 Employee Education, Training, and Development: use of training to develop employee capabilities including training in quality awareness, teamwork, process analysis, problem solving, error proofing, and for developing new skills.

Item 4.4 Employee Well-Being and Satisfaction: how employee health, safety, and ergonomics are included in the improvement plans; how employee satisfaction is determined.

---

4 *Ibid.*
5 *Ibid.*

5.0 *Process Management*

"The *Process Management* Category examines the key aspects of process management, including customer-focused design, product and service delivery processes, support services and supply management involving all work units, including research and development. The Category examines how key processes are designed, effectively managed, and improved to achieve higher performance."[6]

> **Item 5.1** Design and Introduction of Products and Services: translating the voice of the customer into product and process requirements for new or modified products and services.
>
> **Item 5.2** Process Management: Product and Service Production and Delivery: process control and improvement in the production and delivery of products or services.
>
> **Item 5.3** Process Management: Support Services: process control and improvement in support services, such as finance, sales, marketing, public relations, legal services, information services, and administration.
>
> **Item 5.4** Management of Supplier Performance: communication of requirements and assurance of supplier quality through audits, certifications, or reviews.

6.0 *Business Results*

"The *Business Results* Category examines the company's performance and improvement in key business areas—product and service quality, productivity and operational effectiveness, supply quality, and financial performance indicators linked to these areas. Also examined are performance levels relative to competitors."[7]

> **Item 6.1** Product and Service Quality Results: demonstrated leadership position in improving accuracy, reliability, timeliness, and performance of key product or service features. Field performance is now included in this Item as well.
>
> **Item 6.2** Company Operational and Financial Results: quantitative success in reducing labor hours, material and energy consumption, cycle time, and financial waste; human resource indicators such as employee safety,

6 *Ibid.*

7 *Ibid.*

absenteeism, turnover, and satisfaction; improvements in accuracy and cost effectiveness, and reducing cycle time in support areas such as finance, sales, marketing, public relations, legal services, information services, and administration; and indicators of public responsibility such as environmental improvements.

Item 6.3 Supplier Performance Results: indicators of success in supplier quality, delivery, and price broken down by major groupings of suppliers or supplies.

7.0 *Customer Focus and Satisfaction*

"The *Customer Focus and Satisfaction* Category examines the company's systems for customer learning and for building and maintaining customer relationships. Also examined are levels and trends in key measures of business success—customer satisfaction and retention, market share, and satisfaction relative to competitors."[8]

Item 7.1 Customer and Market Knowledge: getting the voice of the customer for short and longer term requirements. Creating the future for customers, not just reacting to customers' requests.

Item 7.2 Customer Relationship Management: service standards, easy access and follow-up with the customers—going beyond providing a product by servicing the customer.

Item 7.3 Customer Satisfaction Determination: methods used to determine customer satisfaction of all customers: dealers, retailers, and end users; linking that to likely future market behavior.

Item 7.4 Customer Satisfaction Results: data obtained by the methods described in Item 7.3, plus customer assessment of products or services, customer awards, and customer dissatisfaction.

Item 7.5 Customer Satisfaction Comparison: the company's established position on customer satisfaction and market share relative to competitors.

INTERDEPENDENCY OF ITEMS

The Criteria Items are interwoven and have many links. For example, consider a company that uses concurrent engineering to design its products. A full description of this activity should be given by the organization in Item 5.1, Design and Introduction of Products and Services. A brief mention of concurrent engineering

---

[8] *Ibid.*

could be made in Item 4.2 as one of the ways the company has improved communication across departments. Reference would be made in Item 4.2 to Item 5.1 for the full explanation.

Some links are dependent on the company's activities, such as the example just given. Other links are an integral part of the Criteria. In Item 5.1 the design process is described and in Item 5.2 the production and delivery processes are described. The resulting data that are used to track progress in these two Items are reported in Item 6.1, Product and Service Quality Results. This is an example of a link that is an integral part of the Criteria.

Figure 2.2
Relationship Matrix

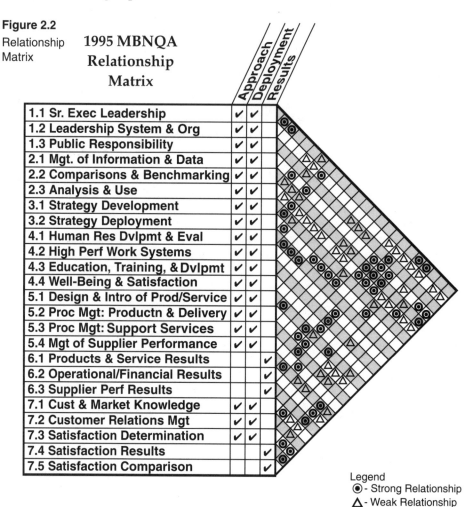

1995 MBNQA Relationship Matrix

Legend
◉ - Strong Relationship
△ - Weak Relationship

The interdependencies within the Criteria are numerous. The idea is to look at the company as an overall system, rather than as independent activities. Figure 2.2 shows the relationships between the Items. The figure indicates if the correlation is strong (⊙) or weak (∆). A strong correlation would be one in which the answer given in one Item has a major impact on the answer in another Item. As an example of reading the table, consider Item 5.4 shown in Figure 2.3. Reading down the diagonal it shows that Item 5.4 has a strong relation with Item 6.3. Reading up the Item 5.4 diagonal, a strong correlation is shown to Item 3.2 and a weak relation with Item 3.1.

The methods described in Item 5.4 about the relationship with suppliers and how their performance is determined should be reflected in the data reported in Item 6.3 so there is a strong correlation shown between these two Items. Item 3.2, Strategy Deployment, includes translating the company's strategic plan to requirements for the supplier so a strong relation is shown between Items 5.4 and 3.2. In Item 3.1, Strategy Development, a description of how supplier capabilities are considered in the planning process is requested. This is one element among many requested so a weak correlation is shown between Items 5.4 and 3.1.

Figure 2.3
Relationship
Matrix Example

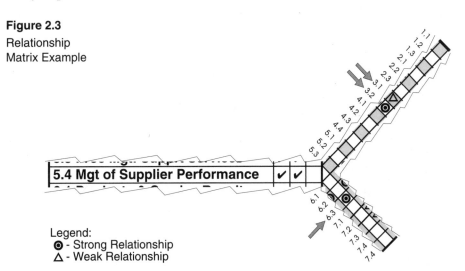

5.4 Mgt of Supplier Performance

Legend:
⊙ - Strong Relationship
∆ - Weak Relationship

If an organization is using the Criteria for self-evaluation and improvement, then this figure may be helpful to those in the organization who are writing the written response to the Criteria. The figure will help to keep a system focus by linking Items together. The same would be true if a company were planning to apply for the Baldrige Award.

SUMMARY

The Criteria were presented as a framework with Category 1.0, Leadership, as the driving force. The senior executives establish the company culture that elevates the customers' needs. The overall operating systems are defined by Categories 2.0 through 5.0. These Categories include the information systems, people systems, process control, and process improvement throughout the organization. The internal measures of progress that the company knows are correlated to customer satisfaction are reported in Category 6.0, including product and service quality, productivity improvements, waste reduction, and improved supplier quality. All of the foregoing lead to the goal of customer delight, Category 7.0. This Category covers the relationship with the customers, customer satisfaction, satisfaction relative to competitors, customer retention, and market share gain.

3 | Using the Criteria for Self-Improvement

INTRODUCTION

This chapter focuses on using the Criteria for self-improvement and as a guide for the quality journey. General findings are presented for companies that are doing well and those not doing well, as evaluated by the Malcolm Baldrige National Quality Award (MBNQA) Criteria. An implementation model with a common pitfall (the "no improvement" model) and the desired model (the "improvement" model) are shown. A management summary of the Baldrige Criteria is the highlight of this chapter. This summary is presented as steps on the quality journey from traditional, through awareness and improving, to outstanding. This information is presented in a grid format against the seven Categories. The management summary grid can be used to facilitate a discussion within the company and lead to a macro assessment of the quality processes, extent of use (deployment), and resulting levels of progress. Effective use of the Criteria and advantages of a written response to the Criteria are presented before closing with some recommendations on preparing the written response for internal use.

◯ QUALITY JOURNEY

In an effort to regain or maintain their competitive position, many organizations are looking to make changes. But what changes? Inefficiencies, rework, customer complaints, and slow bureaucratic processes have been the norm in many organizations. Customers have increased their demands for high-quality services and products at a price that represents value to them. These demands have spiraled to the point that the organizations can no longer continue to do business as usual. To satisfy their customers' needs, a team of employees—leaders and workers alike—could sit down and brainstorm what a high-performance organization should look like. Or they could pick up the MBNQA Criteria and determine if their mental model of the business world (i.e., their organizational paradigm) is consistent with the Criteria. The Criteria could be used to generate a lively discussion of the journey that lies ahead for them. Answers to questions such as the following can help people understand where the organization is headed:

- Do we want to empower employees?

- Is it necessary to collect, analyze, and use data in areas other than our key product or service?

- Does the leadership need to be involved or can they just say they support it?

- Do we want to know how we compare with industry leaders and industry levels of excellence?

If your answers to any of the preceding questions are no, then the Criteria may take you down a path you would otherwise have not selected on your own.

These types of discussions may help an organization to determine their guiding principles—a prerequisite for a successful journey. "Correct principles are like compasses: they are always pointing the way. And if we know how to read them, we won't get lost, confused, or fooled by conflicting voices and values."[1]

There should be a clear decision made that the Baldrige Criteria do or do not capture the direction in which the company wants to head. Don't start using the Criteria just because a customer or competitor is using them. And do not avoid using the Criteria simply because you are not interested in competing for the Award. Some organizations that are not eligible to apply for the Award have used the Criteria for self-evaluation and improvement because they feel strongly that the Criteria do give proper guidance for their journey.

The organizations using the Criteria for self-assessment and improvement look at themselves, compare themselves to the Criteria, and then identify areas needing improvement. Because the Criteria are comprehensive and look at the whole organization, this review may identify a lengthy list of needed improvements.

The reasons these improvements are needed can be divided into two general categories: local and systemic sources of variation. Organizations have typically attacked the local issues first simply because they can be handled by a single department, team, or person. Local issues tend to be fairly well defined and can be resolved

---

[1] Stephen R. Covey, *Principle-Centered Leadership,* Summit Books, NY, 1991, p. 19.

without much disturbance to the status quo. System issues, on the other hand, usually require the involvement of several groups of people, can be time consuming, and
the resolution will cause a change in the way the organization acts and interacts. Yet,
more than 85% of an organization's problems can be attributed to system issues.[2]

An organization should review the list of needed improvements, prioritize their
impact, and pick a small number (six or eight) of the system issues on which to
work. The local issues should be left primarily to employee problem-solving teams.

Developing and implementing an improvement plan that is integrated into
the business plan for these system issues is crucial. The review and improvement
cycle needs to become a normal part of the business, not an extra activity that only
gets done if time permits. If an organization has a business plan and an independent quality plan, the message sent to all employees is that "quality is not part of
our normal business activities." By integrating the review and improvement
process into the business plan, the organization is helping itself to align its various
activities and focus on common goals. "Problems of the future command first and
foremost constancy of purpose and dedication to improvement of competitive
position to keep the company alive and to provide jobs for their employees."[3]

This cycle of activities is described in the model pictured in Figure 3.1.
This model uses the Plan-Do-Study-Act (PDSA) cycle[4] to categorize the steps.
Beginning with the MBNQA assessment, first a plan (P) to conduct the assessment
is developed, next do (D) the assessment, and study (S) the findings so that
strengths and areas needing improvement are identified. The action (A) is to prioritize issues to work on from the identified areas that need improvement. This
feeds into a second PDSA cycle involving the business plan. Now a plan (P) is
developed to close the gap on the areas needing improvement. The do (D) portion
of this cycle is to carry out the plans, then study (S) the results to see if the desired
improvements were realized and act (A) to reinforce progress or revise the plan as
needed. Once this business improvement cycle is completed, note the arrow that
goes back to the MBNQA assessment cycle. Typically the assessment is repeated
once a year. In one large company the authors know about, they conduct assessments twice a year at all of the facilities. It can be difficult for an organization to
define and implement system changes in such a short period of time so an annual
assessment is more common.

[2] Dr. Joseph Juran uses 85% and the late Dr. W. Edwards Deming stated it may be as high as 95%.

[3] Dr. W. Edwards Deming, *Out of the Crisis,* The MIT Press, Cambridge, MA, 1982, p. 25.

[4] The PDSA cycle is sometimes called the Shewhart cycle or the Deming cycle (see Chapter 10).

Figure 3.1

PDSA—Improvement Model

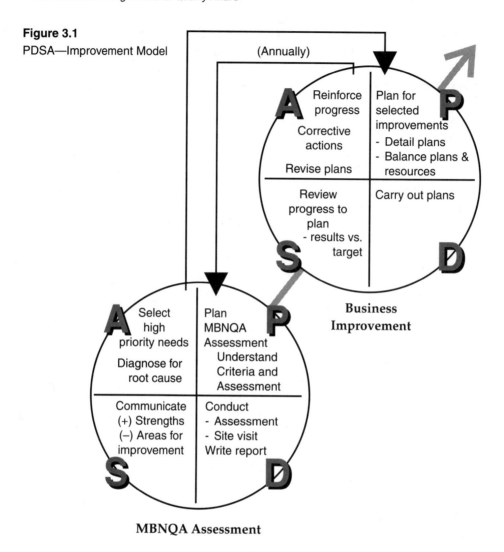

MBNQA Assessment

Have some organizations attempted to use the Criteria without success? Sure, but they typically were going through the motions and not buying into the evaluation and improvement process. Figure 3.2 shows how that can happen. The PDSA cycle is used again to show the various steps. In this case the plan (P) is a mandated order that the organization must be assessed. The assessments are conducted (D), usually by people who are not part of the system. The strengths and areas needing improvement are identified (S) and promises of improvement and change obtained. Awards are given out (to the survivors) in the act (A) step of the cycle. There is no link to the business plan. People completed the assessment because it was required but no benefit was obtained because it was business as usual after the assessment was completed.

Figure 3.2
PDSA—No Improvement Model

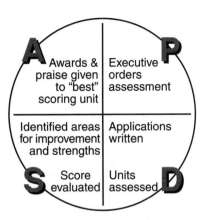

O⊤ GENERAL FINDINGS

Because the Criteria are descriptive, not prescriptive, there are many ways to implement the various processes and achieve outstanding success. But there are common threads seen at companies doing well and at companies needing improvement. Some of the characteristics of organizations doing well are shown in Figure 3.3. In organizations needing improvement we see quite a different picture. The common characteristics seen there are also given in Figure 3.3.

Figure 3.3 Organizational Characteristics	*Companies Doing Well*	*Companies Needing Improvement*
	Leadership creates the quality culture	Management delegates the quality responsibility
	Entire organization is customer driven	Often it is unclear what quality means to the organization
	There are aggressive quality goals and strategies	There is a lack of alignment, no common goals or vision
	Strong information systems are accessible and timely	Weak information systems Lack of measures, indicators, or benchmarks
	Involvement of customers and suppliers leads to win–win situations	Failure to use all avenues to get feedback from customers and employees
	Comprehensive training is available to all	Lack of awareness of best
	Employees act with authority	Employees are "hired hands"
	Quality is evident in all operations	There are partial quality systems in place
	Prevention and improvement drives activities	Reactive systems

The key concepts in the Criteria (described in Chapter 1), such as customer-driven quality, continuous improvement, fast response, and prevention, seem reasonable enough. But if your house is on fire, it is hard to get you involved in fire prevention. Many organizations find themselves in this position. The Criteria define activities they want implemented, but right now their house is on fire. Unfortunately, if time is not spent on prevention, there will still be fires next week, next month, and next year. The cycle will not be broken.

USING THE CRITERIA

Although the focus of this book is on self-improvement, it is useful to note multiple uses of the Criteria. When the first set of Criteria was released in 1988, the intention was to get this material in the hands of industry leaders, university professors, government organizations, and others looking for guidance on defining the elements of total quality leadership. With more than a million copies of the Criteria requested by 1993, that objective apparently has been met. The government never intended to have thousands of companies applying for the Award, but rather to have them use the Criteria for self-improvement. One use of the Award Criteria that the Award Office of NIST probably did not anticipate is all of the spin-off awards at state and community levels. Organizations saw the same benefits of using the Criteria at the local level as the national level, for individual business units as well as for "full" companies, and for not-for-profit organizations as well as for-profit businesses. Some of these benefits are shown in the following list.

Benefits of Using the Baldrige Criteria

- Provides framework for an organization's total quality leadership system

- Helps achieve consensus on what needs to be done

- Helps maintain direction over time

- Eliminates need to internally develop definition of total quality

- Focuses improvement where it is most needed

More than 20 states have brought the Baldrige Criteria to the state level.[5] Most of the state quality award processes use the national model with little or no modification. One example is Michigan, which has a Michigan Quality Leadership Award that uses the Baldrige Criteria. The intention of the Award process at the state and community level is to expose more organizations to the Criteria. The eligibility restrictions for the national Award are typically removed at the state and community level. For example, for the national Award, an organization must be a for-profit company and be a complete business in the sense that it must have significant responsibility in the design, manufacturing, and delivery of the product or service. To contrast that with a state Award, in Michigan, as an example, the Award is open to any Michigan-based organization, including both for-profit and not-for-profit organizations. "There are six eligibility categories in Michigan: manufacturing, service, health care, education, public sector, and small enterprise (less than 100 employees)."[6] If an organization was a distributor of service manuals for a computer company, they might not be eligible for the national Award but could apply for the local Award. The same is true for a plant or single facility belonging to a national or multinational organization. Eligibility and implementation may vary from state to state.

Some industry-specific awards based on the Baldrige Criteria have also been established. The National Housing Quality Award is one example. The Category headings are similar to Baldrige, namely:

- Leadership
- Strategic Quality Planning
- Customer Focus and Satisfaction
- Management of Process Quality
- Human Resource Development and Management
- Information and Analysis
- Quality and Operational Results
- Product Quality.

Their Criteria are less rigorous than Baldrige. There are no Items and Areas to Address, just overall Categories.

[5] See "Designing and Implementing a State Quality Award," published by the National Governors' Association, for guidelines and examples of this. To order, call 800/553-6847 and request document PB 93-154458.

[6] State of Michigan, "*Quality Leadership Award, 1995 Application Forms and Criteria.*" NIST (301/975-2036) can provide the contact name and number for all state Awards that participate in the national network.

The Award Criteria also facilitate sharing and networking among multiple units of a company[7] or among various organizations. For instance, a group of businesses may form a roundtable to discuss lessons learned and best practices in trying to implement a plan to meet the Criteria. Others are using the Criteria for training and education. A committee of one professional society used the Criteria to train all of their major committees on the topic of total quality.

Another use of the Criteria centers on the customer and supplier interaction. Customers, such as Motorola, may request or require their suppliers to use the Criteria to evaluate themselves. Motorola feels so strongly about the benefits of using the Criteria that they want all of their suppliers to apply for the MBNQA. The additional benefit gained in requiring this is that the supplier will get a feedback report.

Business school education is being affected by the Award Criteria. People may complain that the business schools produce MBA graduates who focus only on dollars and have no awareness of the role the customer and quality play. However, some business school professors have volunteered to become MBNQA Examiners so they could better understand the process and the Criteria. These professors gain an awareness of quality as a major business factor and can return to their schools to influence the curriculum and classroom activities.

Some government agencies are using the Criteria to guide their journey. They are not eligible to apply for the Award[8] but they feel the Criteria supply the best road map to follow. Numerous countries have also developed Awards modeled after the Baldrige. The 1995 European Quality Award has nine Categories, which closely parallel the Baldrige:

- Leadership

- Policy and Strategy

- People Management

- Resources

- Processes

---

[7] See Chapter 5 for an example of one company's internal application of the MBNQA process.

[8] There is a Presidential Award for Quality administered by the Federal Quality Institute (FQI) for which federal government agencies may apply. Contact FQI at 202/376-3765 for more information.

- Customer Satisfaction

- People Satisfaction

- Impact on Society

- Business Results.

A potential use of the MBNQA Criteria that has had limited implementation to date is the harmonization of quality requirements within an industry. For example, the automotive industry could decide to use the Baldrige Criteria as the basis for their supplier assessments. This would replace the present practice of each company (General Motors, Ford, Chrysler, Honda, Toyota, Mazda, etc.) having its own somewhat unique audit. A single audit of a specific supplier could be done by one of the automotive companies or by a third party and be recognized by the other customers. The suppliers that provide materials and services to multiple companies would benefit by not having multiple audits in one year and each customer would benefit by not having to conduct all of the assessments. This notion of harmonization is not unique to MBNQA. An industry could select any set of criteria or a standard and agree to use that for supplier audits. This is occurring with the ISO 9000 standards series.[9]

From these many uses it is clear that there is as much, if not more, benefit if the focus is on using the Criteria for improvement rather than for winning an award.

MANAGEMENT GRID

Because the MBNQA Criteria consider the impact and involvement of all the elements of a business, trying to grasp the total picture can seem to be overwhelming at first. Understanding the core values and key concepts that are the foundation of the Criteria can help. The application of these values and concepts to a specific organization may not be very straightforward—especially if the various elements of the organization are at different stages in the quality journey. A management summary of the Baldrige Criteria was developed by the authors to show the phases in the quality journey from *traditional*, through *awareness* and *improving*, to *outstanding*.[10] This summary, shown in Figure 3.4 and presented against the seven Categories, can be used to facilitate a discussion within the company leading to a macro assessment of the quality processes, extent of use, and resulting levels of progress.

---

9 GM, Ford, and Chrysler released QS-9000, an expansion of ISO 9001, in 1994 as their common quality system requirement. Contact the Automotive Industry Action Group to obtain a copy at 810/358-3570.

10 Initially printed in *Industrial Engineering*, November 1993.

Figure 3.4
Malcolm Baldrige National Quality Award (MBNQA)—Management Grid

Characteristics of Progress on Quality Journey

MBNQA Category	Traditional	Awareness	Improving	Outstanding
1.0 Leadership	• State lofty quality goals • Motivating people is solution • Delegate quality • Business units compete with each other	+ Mission/vision statement includes all stakeholders + Recognize that procedures and practices cause most variation	+ Open to new ideas from others including other companies + Consider the whole system + Remove barriers that rob people of pride in work	+ Leadership highly visible and active in quality activities + Leadership acts as coach + Good corporate citizen + Ongoing communication of quality values inside and outside the company + Leadership considers total system
2.0 Information and Analysis	• Fight fires • Most data regarding quality are for key products and services • A lot of data, little information	+ Root cause analysis prevents reoccurrence • Some useful visual displays of data • Switch from percent to opportunities per 1000	+ Value-added reports and graphs + Data in design, and support areas (accounting, purchasing, personnel, computer services) + Track opportunities per million + Aware of best practices globally	+ Extensive depth and breadth of information + Ready access to information throughout organization + Positive correlation between internal measures and customer satisfaction + Understand why best practices have worked at other companies
3.0 Strategic Planning	• There is a business plan (no quality plan) • Optimization occurs within a department	+ Progress to the business plan is reviewed + Business plan is updated annually • There is a quality plan	+ Business plan and quality plan are one + Plan includes quality goals + Plan covers 1- to 2-year horizon + Cross-functional input to plan but limited in scope	+ Business plan includes quality goals and stategies + Customers, suppliers, and employees provide input to plan + Resources provided to achieve goals + Short- and long-term views + Plan includes products and services
4.0 Human Resource Development and Management	• Training is offered to employees • There is a suggestion box	+ Employees on problem-solving teams • Increased focus on getting more suggestions • Mass training	+ Employees are told they are empowered + Training is just in time + Employees are involved in process analysis and improvement + There are opportunities for teams and individuals to contribute + Employee satisfaction levels show positive trends	+ Total system is optimized + Internal supplier customer relation + Employees feel and act empowered + Culture supports use of new skills + Employees feel their opinions are important + Employee satisfaction, attendance, and turnover are outstanding

Figure 3.4
Malcolm Baldrige National Quality Award (MBNQA)—Management Grid *(continued)*

MBNQA Category	Traditional	Awareness	Improving	Outstanding
5.0 Process Management	• Quality effort is focused on manufacturing of product or delivery of service • Receiving inspection	• Quality is owned by manufacturing • Control charts are used to maintain some processes • End-of-line assessment is the safety net • Skip lot sampling	• Some prevention activities + Key processes throughout business identified, studied, and improved + Motto is "If it ain't broke, improve it" + Process control and improvement activities in support areas + Suppliers are part of the team	+ Prevention + Robust designs + Process control and improvement + Reduced cycle time + Supplier certification + Assessments are done on processes and systems
6.0 Business Results	• Data are available to quantify quality of product or service	• Some data are tracked in support areas	• Data on key processes throughout the organization show mostly positive trends • Some comparisons to industry norms	+ 3- to 5-year trends show continual improvement + One of the leaders according to the data for product, service, and response time throughout the organization
7.0 Customer Focus and Satisfaction	• Take care of customer problems • Tracks customer satisfaction by "looking in the rear view mirror" only (ex. market share) • Assumes silence means satisfaction	• Warranty is typical for industry • There is a customer satisfaction survey	+ Listen to the customer • Satisfy the customer • Have service standards + Customers to have easy access to company + Front-line employees have current information and make things right for the customers + Positive trends in customer satisfaction	+ Lead the customer regarding the art of the possible + Commitments to customers lead the industry + Delight the customer + Customer problems are rare + One of the leaders in customer satisfaction, loyalty, and market share niche
Equivalent MBNQA Score	10–30%	40–50%	60–70%	80–100%

Key: + means this attribute should be carried forward to achieve and sustain outstanding status.

A three-prong view of activities in a company is taken against the Criteria: *approach, deployment* and *results*.[11] *Approach* refers to the processes, methods and practices used to address the Criteria. The approach should be systematic and prevention based, with evaluation and improvement cycles. *Deployment* addresses how broadly or narrowly the approach is applied. For example, all employees are using the approach versus just certain departments or classifications of employees using the approach. The operative words for *results* are levels and trends. When results are requested, the expectation is three to five years of positive trends, and demonstration that the organization is one of the best in the field.

The *traditional* approach is based on the Frederick Taylor model from the late nineteenth century, where there are separate planners and doers. Someone does the planning and then gives the assignments to the workers. People who operate with this paradigm think their problems would go away *"if only they could get good workers."* Contests are held between employees, departments, or divisions to try to motivate people. The destructive nature of this type of extrinsic motivation has long-term effects. "Extrinsic motivators . . . are not only ineffective but corrosive. They eat away at the kind of motivation that *does* produce results."[12] The traditional approach involves fixing problems when customers complain. Leaders in the traditional model have the attitude "if it ain't broke, don't fix it." And it "ain't broke" if no one is actively complaining. Organizations operating like this do not try to prevent problems. They think that silence means satisfaction. However, the unsatisfied consumer may change suppliers rather than complain.

One could argue that some large organizations today operate as described in the traditional approach and have been in existence longer than you or I have been around. This is true—just as the scissors and rollover method in high jumping produced world champions. But when Dick Fosbury came along and developed the Fosbury flop, he changed the event forever. Our old management methods worked well for a long time but a Dick Fosbury–type came along and now the focus has changed—and changed for the better. A traditional approach will result in a 10 to 30% score on the Baldrige scale of 100%.

In the *awareness* stage the organization realizes that the rules of the game have changed and what achieved success in the past will not necessarily provide success in the future. The needs of all of the stakeholders, namely, the customers, employees, shareholders, suppliers, and the community, are considered. Some data are tracked and displayed. Problems are tracked as opportunities per thousand rather than percent. A 95% sounds good but in fact it represents 50,000 problems

[11] See Chapter 9 for details.

[12] Alfie Kohn, *No Contest*, Houghton Mifflin, Boston, 1986, p. 60.

per million. Some process control techniques, generally control charts, are used in manufacturing but the role of engineering in developing designs that are not sensitive to variation is not recognized yet at a company in the awareness stage. The battle cry of companies in this stage is *"if only suppliers and manufacturing would do their job we wouldn't have these problems."* Organizations operating in this stage know that they need customer, employee, and supplier input but they may not be good at getting it or using it. The typical North American company would be in the awareness stage with a score of 40 to 50% on the Baldrige scale.

Organizations in the *improving* stage are doing very well. They recognize that one of the roles of leaders is to remove the roadblocks that keep people from doing a good job. New ideas are sought and the culture allows and encourages new skills learned to be used immediately. Quality-related data are tracked and used throughout the organization, including in the support areas, to make improvements. There is good correlation between the improvement of internal processes and customer satisfaction. Input from the customers is obtained through multiple channels and the customers feel they have easy access to the organization. There are positive trends and comparisons to industry averages or leaders. On the Baldrige scale, this type of organization would score 60 to 70%, which may be high enough to be a contender for the national Award. Baldrige recipients are not perfect, but they are a giant step ahead of others, so a score of 60 to 70% could reflect the score of some recipients. The percent is converted to a scale of 1000 so this would be 600 to 700 points.

The *outstanding* stage reflects excellence in everything the company does. The leadership is visible and actively involved in quality activities. The depth and breadth of the data reflect a maturity of the tracking systems. Data are being collected and used because they provide useful information. People are working on the right things as reflected by the strong correlation between internal improvements and customer delight. The total system is optimized. Cooperation among employees and departments is valued. Quality is designed into the product by developing designs that are not sensitive to variation. The cycle time, that is, the time from start to finish, is continually reduced. Whether it is the time it takes to go from concept to market with a new product or service, the time to respond to a customer inquiry, or the time it takes to provide feedback to an employee suggestion, cycle time is continually reduced by streamlining processes. The three to five years of data show ongoing improvements, and comparisons to best levels indicate the organization is one of the leaders. The customers' future requirements are anticipated and created by the company. You and I did not ask for a fax machine or antilock brakes, but proactive companies developed them and created a new need. Problems occur only rarely and are tracked in opportunities per million or billion. Yes, some companies are already tracking in opportunities per billion.

EFFECTIVE USE

Over the years, organizations have made comments to the effect that they are going to use Dr. Juran's trilogy or Dr. Deming's 14 points for management but then quickly add an exception. For example, points 4 and 11 from Dr. Deming state: (point 4) "End the practice of awarding business on the basis of price tag alone. Instead, minimize total cost by working with a single supplier" and (point 11) "Eliminate numerical quotas for the workforce and numerical goals for management." If an organization wants to use only 12 of the 14 points by ignoring points 4 and 11, they may find contradictions in some of their practices. In a similar manner, if an organization wants to use the Baldrige Criteria to guide its quality initiative, it should use them without modifications.

Experience by the authors has shown that changing the Criteria causes more confusion than good. This includes renumbering or renaming the Categories. Although this personalizes the Criteria to the organization, it causes confusion and miscommunication within the organization. This is especially prevalent when individuals discuss the process, or attempt to get more information by reading books and periodicals that discuss the MBNQA process, or use any of the training material and case studies that follow the Criteria. Something as simple as changing the order of the Criteria, for example, making Customer Focus and Satisfaction into Category 1.0, causes a translation problem. Now when people are discussing the Criteria the question needs to be asked over and over, are we talking about our Category 1.0 or the national Category 1.0?

For effectiveness, we recommend that an organization initially pilot the use of the Criteria. There is a school of thought that says go full speed ahead without any pilot. This is a decision an organization needs to make based on its particular situation. Our recommendation is to pilot the use of the Criteria so lessons learned can be incorporated in subsequent activities. On the basis of what occurred in the pilot, the strategy developed for the pilot assessment may need modifications or people may need to be trained in the understanding of the Criteria more extensively for future assessments.

 A written response to the Criteria should be developed, i.e., an application. *The term **application** is being used even though it is recognized that it may not be submitted into the Award process.* Rather, the term is used to indicate that written responses have been developed for all of the Items in the Criteria. One company thought they could shorten the cycle by having an auditor come into the organization, conduct verbal interviews, and than write a Feedback Report. The Feedback Report did not have much credibility with the organization, even though it may have been 100% accurate. So, it was dismissed with various rationalizations. More will be said about the role of a written response to the Criteria for internal use in the next section.

Once the response to the Criteria is written it should be reviewed by a team of trained Examiners, who may be employees of the organization. *This* team should write a feedback report that identifies the strengths and areas for improvement.[13] The organization then needs to define a plan of action. Why collect the data (in this case, do an assessment) unless you are going to use the data? The improvement plan based on the assessment findings should be linked into the business plan so the organization is not being pulled in different ways as described earlier in this chapter when discussing the quality journey.

Another point regarding the effective use of the Criteria is the need to keep the review process focused on improvement. Don't let it become a numbers game. If there is *fear* in the organization, the written response to the Criteria will typically report the good news and hide the bad news. This certainly is not helpful in the long run. One way to avoid having units play the numbers game is to minimize the adverse effects of a low score in the early years of using the Criteria. Because the rules of the game are being changed now through use of the Baldrige Criteria as a road map and a yardstick for the journey, it seems reasonable to give the individual units time to make improvements. It would seem that a unit which acknowledges that its area needs improvement and has an improvement plan in place would be looked on more favorably than a unit that may have scored higher but thinks that no improvement is needed. The attitude and culture displayed by the first unit would lead to higher levels of achievement in the future than that of the second unit where status quo is thought to be good enough.

Keeping the executives and line managers involved in the review and improvement process is the final point in the effective use of the Criteria. The better the Criteria are understood, the more likely they will be used to guide the ongoing journey. It is frustrating and confusing for employees to hear one month that the MBNQA Criteria are going to be used and then the next month words and actions indicate either business as usual, or worse yet, a change of direction.

---

[13] These observations should be based solely on the written response. If the Examiners "know" of other strengths or areas for improvement, the response development process should be improved.

Employees often refer to this as working at "Baskin-Robbins" because the company has a flavor of the month.[14]

 ## ADVANTAGES OF WRITING A RESPONSE TO THE CRITERIA

Even if a company does not apply for the Baldrige Award, there are numerous benefits to having written the response to the Criteria—an application. By having the leadership work with the Criteria and develop (and sometimes struggle) with answers, they learn the comprehensive nature of quality. There is a learning process that occurs in putting the document together. The more the unit struggles in answering the Criteria, the clearer it will become that their processes are not well defined or that the data are not readily available. Otherwise the unit would not be struggling for the answers. So, the unit observes how well it is doing and obtains insight into its own processes and performance by writing its response to the Criteria.

Additionally, the process of writing the response facilitates teamwork. A work process generally cuts across many departments, so a team effort is required to write the complete answer. Writing a complete response to all of the Criteria forces the review process to be slower. This allows the unit to focus on improvement rather than the assessment itself.

In the spirit of cooperation and a win–win strategy, it will be easier for units to share success strategies when their response to the Criteria is written. In fact, the documentation developed for the response should be useful to teams working on improving their process, new employee training, or explaining the interface between two departments.

A written response makes the assessment traceable. The "strengths" and "areas for improvement" are more credible because the unit can refer back to the basis of the conclusions. This also helps to build a relationship of trust with your customers. With the written response and backup documentation, your customer will gain confidence that the excellence in your products and services is "because of the system" not "in spite of the system"; that the products and services you will be supplying in the future will have the same or better value than those of today.

As mentioned in Chapter 2, there is an extensive amount of interdependency in the Criteria. What is said in the response to one Item influences other responses. If design cycle time is one of the key indices listed in Category 2.0 then it should be tracked and reported in Category 6.0. By having a written response, the

---

[14] Please note that this reference deals only with the popular slogan of the Baskin-Robbins company, and no inference should be made regarding the company's organization or strategies.

Examiners can review the entire application and look for this type of consistency. By looking at the whole system, i.e., the entire written response, more meaningful feedback can be provided. This would be lost in a case where the entire assessment is done by interviewing managers and employees.

By being able to consider the whole system, the Examiners have an opportunity to form a consolidated opinion. When the assessment is verbal, the examining teams are usually focused on a single Category and can overlook some necessary interrelationships. With a written response to the Criteria, an Examiner typically spends 20 hours reviewing an application and writing comments with a system's focus. This time estimate is prior to a consensus meeting or site visit. The working team writing the response should also take a system view and consider how various activities fit together.

Since the self-assessment should be done periodically, for example, once a year, and not as a one-time event, subsequent assessments will be easier to perform when the response is written. Updating last year's material is easier than writing a response from scratch.

The benefits of writing the response to the Criteria are summarized in the following list.

Advantages of a Written Response

- Management learns comprehensive nature of quality.
- Unit observes how well they are doing.
- Process facilitates teamwork.
- Slower process encourages focus on improvement.
- Easier to share information.
- Assessment is traceable and therefore more credible.
- Examiners can cross-reference Items.
- Multiple Examiners can respond.
- Examiners have time to think.
- Reviewing progress on journey is easier in the future.

PREPARATION OF THE WRITTEN RESPONSE

Because the Criteria were written to have universal applicability to all types and sizes of companies, they are written in general terms. They are not industry specific and generally do not use words that have recently gained popularity ("buzz words" or jargon). Due to the need to keep the language of the Criteria generic enough to have widespread applicability (i.e., not using industry specific terminology) and due to the comprehensive nature of the Criteria, some people are overwhelmed when they first read them.

It is recommended that you first read the Criteria from start to finish once. Then go back and highlight key points in each *Item* and *Area to Address*. Review the purpose of each Item to understand the distinction between them.[15] Each Item is classified as *approach*, *deployment*, or *results*. Approach and deployment address the process and how extensively the process is used. If the Item is classified as results then data in a graph or table format are required.

In writing an application, respond to an Item by providing information on all of the Areas to Address. Indicate 1.1a, 1.1b, etc., with the answer. This will help you keep things straight and will be useful to the Examiners and other readers. (Also, this is required if an application is being submitted for the MBNQA.)

Be honest but modest in claiming how much has been done. If only one simulation was done, don't lead the reader to think that simulation is a commonly used tool. The depth and breadth of the quality initiative should be evident. For example, if the data that are tracked are mainly related to the product,[16] the response indicates a limited breadth. To demonstrate breadth, other data need to be included such as response time to customer inquiries, time to close the books in accounting, accuracy of invoices, and lines of error-free computer code. The depth of the data is reflected by the maturity of the data—in the ways they are collected, tracked, and reported. For example, a customer survey where the data had been collected earlier as a "yes" or "no" answer and are now collected using a five-point scale from "very satisfied" to "very dissatisfied" shows maturity in the way the data are collected.

Use your own terminology in writing the response. The Criteria need to be generic but your written response does not. If your employee teams are called Eagle Teams, then refer to them as such. It is useful to include a glossary with

---

[15] See Chapter 4 for a summary of the Criteria.

[16] For example, dimensions, reliability, price, technology, etc.

Figure 3.5

Possible Allocation of Pages
by Item Based on 1995 Criteria

Points per Item	Items	Pages per Item
10		
15	2.2	1
20	1.3, 2.1, 3.2, 4.1	1
25	1.2, 4.4	2
30	5.3, 5.4, 7.1, 7.2, 7.3	2
35	3.1	2
40	2.3, 5.1, 5.2	3
45	1.1, 4.2, 6.3	3
50	4.3	4
55		
60	7.5	4
65		
70		
75	6.1	5
100	7.4	7
130	6.2	9
		68 pages 70 allowed

acronyms and jargon used. Remember that your readers, including the Examiners, will be basing their evaluations of your quality systems only on information in your written response to the Criteria.

The 1995 restriction on the number of pages for an application being submitted for the MBNQA is 70 single-sided pages. Even if the written response is not being submitted, it is a good idea to limit the number of pages. This forces people to think about the processes they are describing and to determine the key points. Since a team of people is usually writing the material, each will need to know their page limitation.[17] A rule of thumb is to allocate the pages in accordance with points. A 15- or 20-point Item would be allocated one page, a 25-, 30-, or 35-point Item two pages, and so on. Figure 3.5 shows a suggested allocation.

Because the Criteria are interlinked, it is suggested that reference be made to other Items as needed. For instance, concurrent engineering would be detailed in Item 5.1 when the design process is described, but it could also be mentioned briefly in Item 4.2 for communication across functions. Along the same line of thinking, it is important to show the overall integration of quality activities. Is there a master plan or just a flurry of activities?

[17] See Chapter 6 for ideas about teams.

Avoid anecdotal information. An example may be useful in clarifying a process just described, but an example by itself is anecdotal.

Be concise, factual, and quantitative. Pay attention to approach, deployment, and results. Demonstrate a defined strategy, prevention-based system, and continuity. Avoid empty narrative in your answers. Two partial answers are given below for special services made available to employees, which is part b of Item 4.4, Employee Well-Being and Satisfaction. The first answer is an example of empty narrative.

> b.) Great Northern (GN) has embraced a prevention-based wellness program that focuses on health as a lifestyle. We encourage all of our associates to exercise and eat healthy meals. The quality of work life has a known impact on associates' attitudes and performance so GN provides an environment that is clean and safe. Recognizing the importance of families, some of our celebrations include children and spouses such as our annual picnic. Over a dozen of our associates run in the 5K turkey trot, which is a charitable marathon held at Thanksgiving time. Not only does this improve the health of the associates that run but also contributes to a good cause. Responding to associates' suggestion, last year a nonsmoking policy in all GN operational facilities was implemented, creating a smoke-free environment for all associates. Tuition refund is offered to associates wanting to pursue a college degree.

The following is more typical of the response as it would be written by an outstanding company.

> b'.) "Great Northern (GN) has embraced a prevention-based wellness program that focuses on health as a lifestyle. A fully equipped fitness center is available at headquarters, and GN will assist our field service office associates in paying for a health club membership in their location. Associates and their spouses are encouraged to use our center, and child care is available. The Wellness Team meets twice a month to consider all aspects of associates' wellness. The team sponsors a yearly Health Fair and 5K Run, and offers incentives for weight loss. The Wellness Team is a cosponsor of the Treat Yourself Right program which features low-calorie dishes in our cafeterias at discount prices. The team was responsible for implementation of a nonsmoking policy in all GN operational facilities, creating a smoke-free environment for all associates. Other programs include . . . ."[18]

It is advisable to compare the draft of the written response to the Award guidelines, not someone's interpretation of those guidelines. Think of what happens when a copy of a copy is made: the quality degrades. The same thing happens when an application is compared to an interpretation of the Criteria instead of the original Criteria.

---

[18] The 1994 Great Northern Case Study is available from the American Society for Quality Control, 800/248-1946, item number T551. This is the case study used in Examiner preparation by NIST in 1994.

SUMMARY

Using the Criteria for self-improvement and as a guide on the quality journey was the focus of this chapter. Although there are many ways to achieve success, general findings common to the recipients were presented and summarized in Figure 3.3. The need to capture comments from the feedback report and use them as input to the planning process was presented and reinforced through two figures, Figures 3.1 and 3.2. A management summary of the Baldrige Criteria was the highlight of the chapter. The management summary, presented as steps on the quality journey, led the reader through the stages of traditional, awareness, improving, and outstanding levels. A grid format was used against the seven Categories to present behavior exhibited at the different stages (Figure 3.4). Effective use of the Criteria and advantages of writing the response to the Criteria were presented before closing with some recommendations on preparing an application for self-assessment.

4 1995 Malcolm Baldrige National Quality Award Criteria

INTRODUCTION

The key concepts of the Criteria described in Chapter 1 are customer-driven quality, leadership involvement, continual improvement, employee participation and development, fast response, design quality, prevention, partnership development, long-range outlook, good corporate citizenship, and results orientation. The Award Criteria ask how the key concepts are being integrated into the organization's strategies and activities. Equal emphasis is placed on the actualization of the quality strategies; i.e., quality achievement and quality improvement as demonstrated through the quantitative information furnished by the applicant. As mentioned in Chapter 2, to add order to the response, the Criteria are divided into seven Categories—each focusing on a major facet of the total business process:

1.0 Leadership

2.0 Information and Analysis

3.0 Strategic Planning

4.0 Human Resource Development and Management

5.0 Process Management

6.0 Business Results

7.0 Customer Focus and Satisfaction.

These Categories are defined by examination *Items*, which in turn are supported by *Areas to Address*. In this chapter, the Criteria are presented in a graphical summary format and include details that build on the foundation of Chapter 2. The graphical presentation of a Category is a flowchart with each box containing one

Item and the associated Areas to Address summarized by bulleted points outside the box. A reprint of the Category from the *1995 Malcolm Baldrige National Quality Award Criteria* booklet is included with examples for each Item from selected previous recipients and other sources. At the end of each section there is a narrative description of the key excellence indicators for the particular Category.

CATEGORY 1.0: LEADERSHIP

 Description and Examples

> "The *Leadership* Category examines senior executives' personal leadership and involvement in creating and sustaining a customer focus, clear values and expectations, and a leadership system that promotes performance excellence. Also examined is how the values and expectations are integrated into the company's management system, including how the company addresses its public responsibilities and corporate citizenship."[1]

The first Category, Leadership, is the driver of the overall system. The focus is on the role of the senior executives in creating a successful quality culture that takes into account all of the stakeholders' interests. In this context, the stakeholders are the customers, employees, stockholders, suppliers, and the community. Senior executives must be personally and publicly involved. This may include creating and communicating the company's values, becoming educated in quality topics themselves, training others in the quality culture, recognizing employee contributions, and reviewing the progress of quality plans. If senior executives analyze the Criteria, they will see that changing an old procedure or adding a new one is not enough. They must make fundamental changes in the way they do business to achieve performance excellence.

The Criteria recognize that no organization is a totally independent entity, but exists within social, cultural, and economic communities. For total quality excellence, the leaders' basic values and commitment need to include areas of public responsibility and corporate citizenship. They must be active participants in the creation of strategies, systems, and practices for achieving excellence. Whether

---

[1] *1995 Malcolm Baldrige National Quality Award Criteria,* National Institute of Standards and Technology, Gaithersburg, MD.

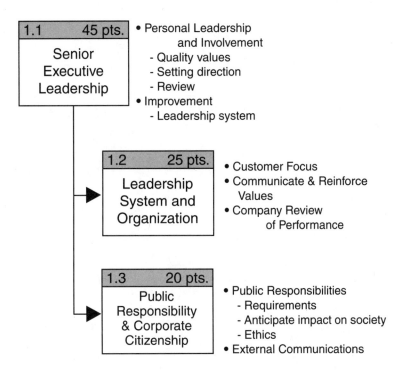

Figure 4.1
Leadership Category

1.0 Leadership

1.1   45 pts.
Senior Executive Leadership
	• Personal Leadership and Involvement
	  - Quality values
	  - Setting direction
	  - Review
	• Improvement
	  - Leadership system

1.2 25 pts. Leadership System and Organization	• Customer Focus • Communicate & Reinforce Values • Company Review of Performance

1.3   20 pts.
Public Responsibility & Corporate Citizenship
	• Public Responsibilities
	  - Requirements
	  - Anticipate impact on society
	  - Ethics
	• External Communications

Maximum points - 90

they want to or not, senior leaders serve as role models, reinforcing the company's values, not only in their own organization, but for the community at large.

Figure 4.1 is a graphical presentation of the Category 1.0 Criteria. Each box contains one Item with the Areas to Address summarized as bulleted points outside the box. The points associated with each Item are shown but not discussed here.[2]

---

[2] See Chapter 9 for information on scoring the application.

Throughout this chapter, the complete listing of the Criteria for each Item in each Category, including the notes, is provided here. This information has been obtained from the *1995 MBNQA Criteria* booklet developed under the direction of the National Institute of Standards and Technology.

Item 1.1: Criteria

1.1 Senior Executive Leadership (45 pts.)

Describe senior executives' leadership and personal involvement in setting directions and in developing and maintaining a leadership system for performance excellence.

☑ Approach
☑ Deployment
☐ Results

Areas to Address

(a) how senior executives provide effective leadership and direction in building and improving company competitiveness, performance, and capabilities. Describe executives' roles in: (1) creating and reinforcing values and expectations throughout the company's leadership system; (2) setting directions and performance excellence goals through strategic and business planning; and (3) reviewing overall company performance, including customer-related and operational performance.

(b) how senior executives evaluate and improve the effectiveness of the company's leadership system and organization to pursue performance excellence goals.

Notes:

1.   "Senior executives" means the applicant's highest-ranking official and those reporting directly to that official.

2.   Values and expectations [1.1a(1)] should take into account all stakeholders—customers, employees, stockholders, suppliers and partners, the community, and the public.

3.   Activities of senior executives appropriate for inclusion in 1.1a might also include customer, employee, and supplier interactions, mentoring other executives, benchmarking, and employee recognition.

4.   Review of company performance is addressed in 1.2c. Responses to 1.1a(3) should reflect senior executives' personal leadership of and involvement in such reviews, and their use of the reviews to focus on key business objectives.

5.   Evaluation of the company's leadership system might include assessment of executives by peers, direct reports, or a board of directors. It might also

Item 1.1: *Senior Executive Leadership*

Category 1.0 is composed of three Items. The first—Senior Executive Leadership—addresses the senior executives' personal involvement in the quality journey. The term *senior executive* as used here refers to "... the applicant's highest ranking official and those reporting directly to that official."[3]

This Item recognizes that, in an effort to do a good job and regardless of their level, most employees will do what they *think* their boss wants them to do. It is not sufficient for leaders to just support and provide the budget for the quality initiative. The leaders must create and reinforce the company's values and set high expectations. They must also be personally committed to communicating and "living" the quality message.

Peter Senge, in his book *The Fifth Discipline*, discusses the need for commitment in the building of shared vision within an organization. He also differentiates between commitment and compliance. "Compliant followers go along with a vision. They do what is expected of them. They support the vision, to some degree." Individuals committed to the vision want it to happen, and they will do whatever can be done to make it happen.

To show the leadership of the senior executives, the company's mission, vision, and/or values should be described in this Item as well as the executives' roles in creating and reinforcing the values. Personal involvement, which might be demonstrated by describing how the executives are involved in planning for quality, reviewing progress of the plans, recognizing employees, giving or getting training, and/or meeting with suppliers and customers to discuss quality, is requested in the Item. Further, as part of the continual improvement in all elements of the leadership system, the senior executives must evaluate and improve the system.

The personal involvement of the executives is to be described, *not* the business planning process or the employee recognition activities. Those descriptions should be given in Categories 3.0 and 4.0, respectively.

As with many of the Items, the last Area to Address basically asks "how do you evaluate and improve what you just described?" In Item 1.1, the last Area requests the organization to explain how the senior executives evaluate and improve the effectiveness of the leadership system. In many organizations, this evaluation and

---

3 *1995 Malcolm Baldrige National Quality Award Criteria*, National Institute of Standards and Technology, Gaithersburg, MD.

improvement is often accomplished through employee surveys or focus groups. In a non-threatening environment, employees are asked questions such as the following:

- Does management walk the talk?

- Do you get cooperation from other departments?

These questions could also be asked in connection with other questions regarding training, employee involvement, and recognition, which are addressed in Category 4.0. Other sources of input that could be used for evaluating and improving the leadership system are assessment by direct reports or by the board of directors.

Examples for Item 1.1

The senior executives describe their personal involvement and visibility in creating an environment for quality.

At **Texas Instrument Defense Systems & Electronics Group** (TI-DSEG) the executives are members of the Quality Improvement Team. This team dedicates one-half day every week to addressing quality issues. They have attended quality-related training including the Six Sigma classes at the Motorola University. The vice presidents at TI-DSEG are the champions of their five thrusts:

- Customer satisfaction

- Stretch goals

- Teamwork and empowerment

- Benchmarking

- Integrated total quality.

While attending a Quest for Excellence Conference[4] and hearing about Cadillac's skip level meetings, the executives at TI got an idea to improve their communication. Today, at TI-DSEG, an open exchange is held at rotating locations. These one-hour meetings are scheduled weekly and each is attended by one of the senior executives. Employees come in and ask questions or make comments. They were doing this before Ross Perot made such town meetings a popular format.

Included in the senior executive involvement are customer visits, which are tracked and posted in the office area. Each executive, assigned five key customers and one supplier, talks to the executives at the assigned company. This allows the TI-DSEG leadership to stay in touch with the supplier and customer community.

---

[4] This conference is the first public announcement by the Baldrige recipients about their success strategies and is scheduled every February.

The leadership at **GTE Directories Corporation** (GTE) conducts training in quality topics. Every member of the company's management board taught quality tools and techniques classes to other managers.

At **AT&T Network Systems Group—Transmission Systems Business Unit** (AT&T Transmission) the senior executives are members of the Quality Council. This group is the umbrella for involvement and leadership of the quality effort, development of the mission and values, and the conducting of business through teams. The senior executives have attended various training classes including classes on the topics of Dr. Deming's management philosophy, the Baldrige Criteria, policy development, quality improvement storyboards, and break-through technology. The leadership and more than 1700 employees attended four-day seminars taught by Dr. Deming for AT&T. Each evening during these seminars, the leadership conducted workshops on how to implement Dr. Deming's philosophy. In 1991 (the year before they received the MBNQA) the executives completed 364 visits with customers to discuss quality. The effectiveness of their leadership style is evaluated through employee surveys.

Each senior executive monitors customers' calls for two hours per month at **AT&T Universal Card Services** (AT&T Card). This helps to keep them in touch with what the customer is saying—both the good news and the bad news. Quarterly meetings are held with key suppliers to keep those communication lines open. To communicate with the employees, a quarterly "all associates" meeting is held. Additionally, the executives participate in benchmarking visits and each is a champion for one of the Baldrige Categories.

The Senior Quality Management Team at the **Ritz-Carlton Hotel Company** is composed of the top 13 executives of the company. They meet weekly to review quality issues. To see things from the employees' view, each executive works as an employee during the startup of a new facility.

At **Granite Rock** the executives are the facilitators of the 10 corporate quality teams to help align the quality efforts across the corporation. They also teach a course, the Front-line Leadership Series, following the cascade pattern of being trained first then presenting the material to others in the organization.

Part of the executives' compensation is determined from a "customer value-added" index at **AT&T Consumer Communications Services** (AT&T Communications). This index is based on several customer satisfaction measures.

Item 1.2: Criteria

1.2 Leadership System and Organization (25 pts.)

Describe how the company's customer focus and performance expectations are integrated into the company's leadership system and organization.

☑ Approach

☑ Deployment

☐ Results

Areas to Address

(a) how the company's leadership system, management, and organization focus on customers' and high performance objectives.

(b) how the company effectively communicates and reinforces its values, expectations, and directions throughout the entire work force.

(c) how overall company and work unit performance are reviewed and how the reviews are used to improve performance. Describe the types, frequency, and content of reviews and who conducts them.

Note:

1. Reviews described in 1.2c should utilize information from business and customer-related results Items—6.1, 6.2, 6.3, 7.4, and 7.5—and also may draw upon evaluations described in other Items and upon analysis (Item 2.3). The descriptions should address key measures and/or indicators used to track performance, including performance related to the company's public responsibilities. Reviews could also incorporate results of process assessments and address regulatory, contractual, or other requirements, including review of required documentation.

Item 1.2: *Leadership System and Organization*

The second Item in the Leadership Category is Leadership System and Organization. Here, the integration of the customer focus into the day-to-day activities is investigated. Communication and reinforcement of the company's values to the entire workforce are described. The overall review of performance to the quality plans should also be discussed. This includes types, frequency, content, and use of the reviews. This is intended to be a macro review, not a "product"[5] audit, which is addressed in Item 5.2 of the 1995 criteria.

Examples for Item 1.2

The integration of customer focus and quality values into day-to-day business. Process assessments and reviews.

The review of quality and operational performance at **AT&T Transmission** occurs as part of its policy deployment process. This is a cascading process that involves employees, customers, and shareholders. Through policy deployment, projects are prioritized. This is the link the Quality Council uses to track progress. Customer feedback is used in the policy deployment process to measure progress, while the quality improvement story method assures progress. A self-assessment is conducted each year at **AT&T Transmission** and **AT&T Card** using the Baldrige Criteria.

Cadillac uses four key initiatives to achieve its missions and integrate the customer focus into day-to-day business:

- Business planning process
- Quality Network, a partnership between labor and management
- Simultaneous engineering (concurrent engineering)
- Human resource people strategy.

This last activity is composed of cross-functional teams of hourly and salaried employees. They have the responsibility to research, design, recommend, and implement people processes.

The company's values and customer focus are communicated in a variety of ways at **AT&T Communications**. This includes field visits by executives, publications, roundtable discussions, and face-to-face meetings with employees.

---

[5] In this context, product audit refers to a review of what the customers receive, whether it is a service or physical merchandise.

Item 1.3: Criteria

1.3 Public Responsibility and Corporate Citizenship (20 pts.)

Describe how the company includes its responsibilities to the public in its performance improvement practices. Describe also how the company leads and contributes as a corporate citizen in its key communities.

☑ Approach
☑ Deployment
☐ Results

Areas to Address

(a) how the company integrates its public responsibilities into its performance improvement efforts. Describe: (1) the risks and regulatory and other legal requirements addressed in planning and in setting operational requirements and targets; (2) how the company looks ahead to anticipate public concerns and to assess possible impacts on society of its products, services, and operations; and (3) how the company promotes legal and ethical conduct in all that it does.

(b) how the company leads as a corporate citizen in its key communities. Include a brief summary of the types of leadership and involvement the company emphasizes.

Notes:

1. The public responsibility issues addressed in 1.3a relate to the company's impacts and possible impacts on society associated with its products, services, and company operations. They include environment, health, safety, and emergency preparedness as they relate to any aspect of risk or adverse effect, whether or not these are covered under law or regulation. Health and safety of employees are not included in Item 1.3. Employee health and safety are covered in Item 4.4.

2. Major public responsibility or impact areas should be addressed in planning (Item 3.1) and in the appropriate process management Items of Category 5.0. Key results, such as environmental improvements, should be reported in Item 6.2.

3. If the company has received sanctions under law, regulation, or contract during the past three years, briefly describe the incident(s) and its current status. If settlements have been negotiated in lieu of potential sanctions, give explanation. If no sanctions have been received, so indicate.

4. The corporate citizenship issues appropriate for inclusion in 1.3b relate to efforts by the company to strengthen community services, education, health care, environment, and practices of trade or business associations. Such leadership and involvement depend upon the company's size and resources. However, smaller companies might take part in cooperative activities with other organizations.

Item 1.3: *Public Responsibility and Corporate Citizenship*

Public responsibility issues include business ethics, environment, and public safety. In this Item, the organization is to describe how these issues are included in the quality initiative. Since the Criteria are interested in prevention, the response should also include a discussion on how risks and requirements involving the environment, public safety, etc., are considered and anticipated.

Communication of the quality values and strategies outside the organization by employees is also covered in this Item. This can include presentations to government forums, health care organizations, educational institutions, or professional organizations as well as membership or active participation in these groups.

Examples for Item 1.3

Public responsibility in quality policies, and the organization as a good corporate citizen.

Texas Instruments Defense Systems & Electronics Group is a member of the Six Sigma Research Institute at Motorola. They are a founding member of a benchmarking clearinghouse.

Texas, like many other states, has brought the Baldrige Criteria to the state level, allowing and encouraging all companies and business entities to use the Criteria. TI helped in this effort by having key employees play active roles as volunteers for "Quality Texas." Additionally, the company hosts a symposium on quality.

Wainwright Industries, Inc. (Wainwright), includes public responsibility and environmental stewardship as one of their core values. By redesigning processes, using organic coolants and lubricants, and developing water-based cleaning systems, they have removed all of the hazardous waste from their facilities.

Key Excellence Indicators

Because the Criteria are descriptive, not prescriptive, there is no one right answer. However, in companies that are doing well, some common elements can be observed. With respect to Category 1.0, Leadership, in excellent companies the executives are highly visible. They are the missionaries preaching and living the quality message. The various leadership styles all include the characteristic of the manager acting as a coach and removing the roadblocks that keep employees from taking pride in their work. The company values are clear, easily remembered, and influence decision-making.

A flatter organization—fewer levels of management—is a common characteristic because of the coaching, not controlling style. In companies doing well, leaders have a strong sense of cycle time, always looking at the total system, trying to shorten the time from start to finish. This may be the time it takes to go from concept to market, the time to respond to an employee suggestion, or the time to respond to a customer. There is an attitude that the company can learn from everyone, rather than the "not invented here" syndrome. Consequently, the leadership maintains contact with suppliers, customers, and employees.

David Garvin developed a litmus test for leadership. He described the *symbolic* involvement of the leadership. Elements include:

- Making quality important to employees

- Elevating quality above finance and efficiency goals

- Positioning the CEO as the quality hero.

The desired level of involvement is *active* involvement. This requires the leadership to participate in activities such as:

- Teaching quality classes

- Personally reviewing quality systems

- Showing a commitment of time and energy beyond slogans and lip service.

A review of log books and executive calendars demonstrates this level of commitment. Listening to employees, customers, and suppliers is also characteristic of active involvement.[6]

---

[6] David Garvin, "How the Baldrige Award Really Works," *Harvard Business Review*, Vol. 69, No. 6 (Nov. - Dec. 1991): 80–93.

CATEGORY 2.0: INFORMATION AND ANALYSIS

 ### Description and Examples

> "The *Information and Analysis* Category examines the management and effectiveness of the use of data and information to support customer-driven performance excellence and marketplace success."[7]

Categories 2.0 through 5.0 are considered to be the overall systems framework in the Baldrige model (see Figure 2.1 in Chapter 2). Category 2.0, Information and Analysis, examines the scope and use of data as well as the benchmarking activities of the company. The right data must be chosen for collection and analysis and then used effectively and efficiently to guide decisions. Measurements need to be derived from the company's strategy and encompass all key processes.

Facts and data needed for quality improvement are of many types. They include *customer, product and service, performance, operations, market, competitive comparisons, supplier, employee related,* and *financial data.* Quality-related data that are limited only to the product or service provided by the company show a lack of maturity of the company's quality efforts. Every department should be tracking and using key indicators that are linked to customer satisfaction—for both the internal and external customer.

Having an overwhelming amount of data is not the key. Having the right data is! That is, we need data that are correlated to internal or external customer satisfaction. There should be processes in place that convert these data into useful information, and make the results easily accessible in a timely manner so that they can help the decision-making process.

Quality and performance data can provide information on the progress of the improvement activities. They cannot tell us how far we need to go to become a world leader. For this we need to look outside the organization. Benchmarking, which is examined in this Category, is an approach for learning what other organizations are doing and for getting new ideas. It has become well known through activities at Xerox. Although benchmarking is a way to gain information for stretch targets and ideas for continual improvement, some cautions are offered (see Item 2.2 later). A story was told[8] about a Russian soldier seeing a light bulb for the first time at the end of World War II. He was fascinated by this bulb hanging from the ceiling by an electrical cord—providing light on demand for the whole

---

[7] *1995 Malcolm Baldrige National Quality Award Criteria,* National Institute of Standards and Technology, Gaithersburg, MD.

[8] Karl Albrecht, *The Only Thing That Matters,* Harper Business, NY, 1992, p. 44.

Figure 4.2
Information and Analysis Category

2.0 Information & Analysis

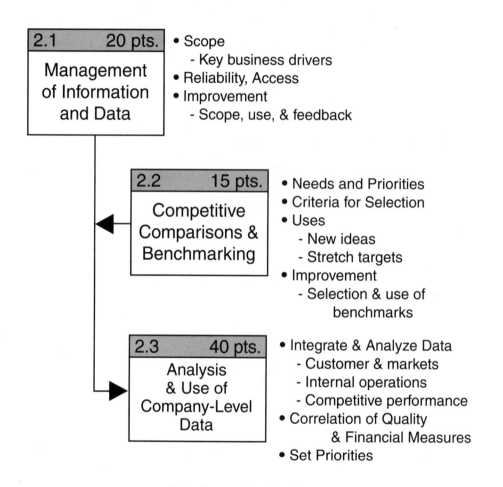

| 2.1 | 20 pts. |
| Management of Information and Data |

- Scope
  - Key business drivers
- Reliability, Access
- Improvement
  - Scope, use, & feedback

| 2.2 | 15 pts. |
| Competitive Comparisons & Benchmarking |

- Needs and Priorities
- Criteria for Selection
- Uses
  - New ideas
  - Stretch targets
- Improvement
  - Selection & use of benchmarks

| 2.3 | 40 pts. |
| Analysis & Use of Company-Level Data |

- Integrate & Analyze Data
  - Customer & markets
  - Internal operations
  - Competitive performance
- Correlation of Quality & Financial Measures
- Set Priorities

Maximum points - 75

room. The soldier cut the cord, and took it with him. His intention was to take it home and show this magical item to others. His ignorance of how the light bulb worked would not occur to him until he was unable to reproduce the magic. Understanding the process that was benchmarked is needed—that is, answer "Why does this practice work?"

Figure 4.2 is a graphical presentation of the Category 2.0 criteria. Each box contains one Item with the Areas to Address summarized as bulleted points outside the box.

Item 2.1: Criteria

2.1 Management of Information and Data (20 pts.)

Describe the company's selection and management of information and data used for planning, management, and evaluation of overall performance.

☑ Approach
☑ Deployment
☐ Results

Areas to Address

(a) how information and data needed to drive improvement of overall company performance are selected and managed. Describe: (1) the main types of data and information and how each type is related to the key business drivers; and (2) how key requirements such as reliability, rapid access, and rapid update are derived from user needs.

(b) how the company evaluates and improves the selection, analysis, and integration of information and data, aligning them with the company's business priorities. Describe how the evaluation considers: (1) scope of information and data; (2) use and analysis of information and data to support process management and performance improvement; and (3) feedback from users of information and data.

Notes:

1. Reliability [2.1a(2)] includes software used in the information systems.

2. User needs [2.1a(2)] should consider knowledge accumulation such as knowledge about specific customers or customer segments. User needs should also take into account changing patterns of communications associated with changes in process management and/or in job design.

3. Scope of information and data [2.1b(1)] should focus primarily on key business drivers.

4. Feedback from users [2.1b(3)] might entail formal or informal surveys, focus groups, teams, etc. However, evaluations should take into account patterns of communications and information use, as users themselves might not be utilizing well the information and data available. Even though the information and data system should be user friendly, the system should drive better practice. This might require training of users.

Item 2.1: *Management of Information and Data*

The criteria for selecting data and how each is linked to key business drivers are requested in this Item. To demonstrate the breadth and depth of data used, the organization should list the types of data collected and why they are being collected. Headings such as *customer related, product and service performance, operational, support services, employee related, supplier performance, cost,* and *financial* could be used to group the data. Operational data would include productivity, efficiency, and effectiveness measures such as labor hours, material usage, energy consumption, waste reduction, environmental improvement, and cycle time reduction. Support services include any activity within the organization that is not directly involved with the development, production, or delivery of the organization's product or service. This can include finance, accounting, software services, sales, marketing, public relations, information services, personnel, purchasing, legal services, administrative services, and basic research and development.

Collecting data is not enough. The data have to be useful and be used. A description of how the organization assures the reliability and rapid access of the data is a necessary part of any response for this Item. Note that actual data are not reported here. Only a description of the data should be given. The charts, graphs, and tables of the data should be presented in Categories 6.0 and 7.0.

The last Area to Address of Item 2.1 deals with the evaluation and improvement of the data selection and analysis process. That is, how is the organization assuring that the right data are getting to the right people at the right time so that they can make the right business decisions. The scope of the data and the cycle time from data gathering to access should be addressed. Broadening the access to the data and aligning data with the process improvement plan is a necessity.

Examples for Item 2.1

Breadth and depth of data, including customer, product performance, internal operations, support areas, supplier, employee, and financial, used to guide quality activities.

There are 29 databases at **AT&T Card.** They are used to track many customer satisfaction indicators such as the average speed of answer, the abandon rate and courtesy on inbound calls, the accuracy and timeliness for claims processing, and the response time for credit approval.

At the **Ritz-Carlton Hotel Company**, detailed guest preference information is gathered. Daily quality production reports are used as an early warning system by each hotel. These reports are derived from data submitted from each of the 720 work areas in the hotel.

Item 2.2: Criteria

2.2 Competitive Comparisons and Benchmarking (15 pts.)

Describe the company's processes and uses of comparative information and data to support improvement of overall performance.

☑ Approach
☑ Deployment
☐ Results

Areas to Address

(a) how competitive comparisons and benchmarking information and data are selected and used to help drive improvement of overall company performance. Describe: (1) how needs and priorities are determined; (2) criteria for seeking appropriate information and data—from within and outside the company's industry; (3) how the information and data are used within the company to improve understanding of processes and process performance; and (4) how the information and data are used to set stretch targets and/or encourage breakthrough approaches.

(b) how the company evaluates and improves its overall process for selecting and using competitive comparisons and benchmarking information and data to improve planning and overall company performance.

Notes:

1. Benchmarking information and data refer to processes and results that represent best practices and performance.

2. Needs and priorities [2.2a(1)] should show clear linkage to the company's key business drivers.

3. Use of benchmarking information and data within the company [2.2a(3)] might include the expectation that company units maintain awareness of related best-in-class performance to help drive improvement. This could entail education and training efforts to build capabilities.

4. Sources of competitive comparisons and benchmarking information might include: (a) information obtained from other organizations such as customers or suppliers through sharing; (b) information obtained from the open literature; (c) testing and evaluation by the company itself; and (d) testing and evaluation by independent organizations.

Fifty-seven baseline measures are used at **Granite Rock**. They include such diverse measures as surveys, discrepancies, short pay, product quality, load-out time, on-time delivery, safety, efficiency, and profit.

Nontraditional areas of a company can also track quality-related data. For example, in a communication department, the measures could include percent of employees that read the company newsletter and the time it takes to publish the company newsletter. Other possibilities are percent of changes made to the documentation prepared and percent of employees that feel they get timely information. Two measures, which are generic and can be used by any department, are accuracy and response time of the service provided.

A legal department may track the time to complete a copyright form (or other legal document), the reading level of contracts, or the percent of lawsuits that go to court that the company lost. In the government affairs department, measures could include the time to respond to a request for information, the percent of information provided that is proactive, or the number of times the department was aware of legislation that impacts the company.

Item 2.2: *Competitive Comparisons and Benchmarking*

Item 2.2 requests the Criteria and uses of information the organization employs in determining its relative position in the business world. The areas that should be addressed are to be linked to key business drivers such as customer-related, product and service performance, operational, support services, employee-related, supplier performance, cost, and financial drivers. This information could come from customers and suppliers through sharing networks, from public literature, from testing of competitive products by the company, or by an independent organization.

Benchmarking is an activity that had frequent misuse in the initial years of the Baldrige Criteria. First of all, some companies thought benchmarking involved simply acquiring a number that represented excellence and embracing that number as the new goal. This misconception actually generated a cottage industry of companies who would sell "benchmark numbers." Several years ago, however, the Criteria were changed to include the phrase "how competitive comparisons and benchmarking information and data are *used to improve understanding of processes...*"[italics added]. This change was made to clarify the point that there is more to benchmarking than finding new goals. The information for breakthroughs that used to set stretch targets are still included, but this is balanced with the requirement that there is an understanding of the process that produced those initial results.

5. The evaluation (2.2b) may address a variety of factors such as the effectiveness of use of the information, adequacy of information, training in acquisition and use of information, improvement potential in company operations, and estimated rates of improvement by other organizations.

Item 2.3: Criteria

2.3 Analysis and Use of Company-Level Data (40 pts.)

Describe how data related to quality, customers and operational performance, together with relevant financial data, are analyzed to support company-level review, action, and planning.

☑ Approach
☑ Deployment
☐ Results

Areas to Address

(a) how information and data from all parts of the company are integrated and analyzed to support reviews, business decisions, and planning. Describe how analysis is used to gain understanding of: (1) customers and markets; (2) operational performance and company capabilities; and (3) competitive performance.

(b) how the company relates customer and market data, improvements in product/service quality, and improvements in operational performance to changes in financial and/or market indicators of performance. Describe how this information is used to set priorities for improvement actions.

Notes:

1. Item 2.3 focuses primarily on analysis for company-level purposes, such as reviews (1.2c) and strategic planning (Item 3.1). Data for such analysis come from all parts of the company and include results reported in Items 6.1, 6.2, 6.3, 7.4, and 7.5. Other Items call for analyses of specific sets of data for special purposes. For example, the Items of Category 4.0 require analysis to determine effectiveness of training and other human resource practices. Such special-purpose analyses should be part of the overall information base available for use in Item 2.3.

2. Analysis includes trends, projections, cause-effect correlations, and the search for deeper understanding needed to set priorities to use resources more effectively to serve overall business objectives.

3. Examples of analysis appropriate for inclusion in 2.3a(1) are:

If a company has followed the same procedure or practice for years, it may be hard to think of new ways to accomplish the task more efficiently. Looking outside the company, or outside the industry may give the organization some new ideas. Benchmarking is often referred to as learning from the "best of the best." However, a second misuse of benchmarking occurs when the company copies a benchmark process without understanding why it works. One aspect of benchmarking is the metric, sometimes called the measure or quantification, which answers "How do we know it is the best?" The other aspect is the process, often called the best practice, that answers "What was done to achieve this level of success?" Understanding the bench-marked process should also include the theory; that is, the answer to "Why does this practice work?" Duplicating someone's process does not necessarily mean you will have the same success. For example, cultural differences between the companies can make a difference in the success of an activity as will the interrelationship with other practices and procedures. Remember, "No number of examples establishes a theory ... To copy an example of success, without understanding it with the aid of theory, may lead to disaster."[9] At best, an example, without theory, teaches nothing.[10]

This is not to say that comparisons and benchmark information should not be obtained from other organizations through sharing, from the literature, or by test-ing performed by an independent organization. These are all valuable *starting* points, indicating those organizations that can yield useful information. An often overlooked source of information is the customer. A company could ask the cus-tomer, "How well do we perform compared to your favorite supplier?"

The Criteria ask for a summary of how the company evaluates and improves the selection and use of competitive comparisons and benchmarking information. Typically, when a company is in the early stages of its quality journey, the only benchmarking that is done relates to the primary product or service. As the jour-ney continues, information is sought in other areas of the business. You need not restrict yourselves to looking only at your own industry. For example, companies from very diverse areas look to Walt Disney for customer service and facilities management ideas. American Express is considered to be outstanding with regard to billing and collection. 3M is a benchmark for environmental management.[11]

9 Working papers. - Handout at the Deming Study Group of Greater Detroit by Dr. W. E. Deming; 1990; see also W. Edwards Deming, *The New Economics*, The MIT Press, Cambridge, MA, 1993, p. 106.

10 This should not be construed to mean that examples are useless. They can provide us with insight and direction. The scientific method utilizes empirical studies, but always to support or reject hypotheses in the development of a theory.

11 "Copycats," *Industry Week*, November 5, 1990.

- how the company's product and service quality improvement correlates with key customer indicators such as customer satisfaction, customer retention, and market share; and

- cost/revenue implications of customer-related problems and problem resolution effectiveness.

4. Examples of analysis appropriate for inclusion in 2.3a(2) are:

- trends in improvement in key operational indicators such as productivity, cycle time, waste reduction, new product introduction, and defect levels;

- financial benefits from improved employee safety, absenteeism, and turnover;

- benefits and costs associated with education and training;

- how the company's ability to identify and meet employee requirements correlates with employee retention, motivation, and productivity; and

- cost/revenue implications of employee-related problems and problem resolution effectiveness.

5. Examples of analysis appropriate for inclusion in 2.3a(3) are:

- performance trends relative to competitors on key quality attributes; and

- productivity and cost trends relative to competitors.

6. Examples of analysis appropriate for inclusion in 2.3b are:

- relationships between product/service quality and operational performance indicators and overall company financial performance trends as reflected in indicators such as operating costs, revenues, asset utilization, and value added per employee;

- allocation of resources among alternative improvement projects based on cost/revenue implications and improvement potential;

- net earnings derived from quality/operational/human resource performance improvements;

- comparisons among business units showing how quality and operational performance improvement affect financial performance;

- contributions of improvement activities to cash flow and/or shareholder value;

- trends in quality versus market indicators;

- profit impacts of customer retention; and

- market share versus profits.

The benchmarking process, including the scope and use of benchmarking information, is requested in this Item. However, other Items ask for the actual data from the benchmarking activity or ask how benchmarking information is used for process improvement. In total, benchmarking is addressed in Items 2.2, 5.2, 5.3, 6.1, 6.2, and 6.3.[12]

Examples for Item 2.2

Selection and use of information from outside the company including competitive comparisons and benchmarking information outside the industry.

A corporate database tracking 250 activities from more than 100 companies is used by **AT&T Transmission** for their benchmarking information. **TI Defense & Electronics** is a member of a benchmarking clearinghouse and they have a benchmarking champion. The people participating on the benchmark studies need to be the process owners, but having a central collection and coordination point may be necessary for larger companies.

Granite Rock compares their on-time delivery with the results of Domino's Pizza, Inc., delivery located in the same geographic location, fighting the same California traffic. When they perform better than Domino's for the period, a pizza party is held for the employees.

Item 2.3: *Analysis and Use of Company-Level Data*

Simply having a massive amount of data available will not help an organization. The data must be of a form that can be readily understood and used. And it must be available when the user needs it.

This Item addresses how data are aggregated, analyzed, and translated into useful information. This should include a description of how feedback and customer complaint data are aggregated for overall evaluation. All data need to be translated into actionable information. Correlation of customer-related data and internal operations should be documented to ensure that the right things are being improved. The correlation between improvements and financial performance would also be reported here.

The notes for this Item are extensive,[13] far longer than the Item itself. These notes offer several examples of correlation analysis. For example, the following correlations could be made:

[12] Also see Chapter 12 for additional information on benchmarking.

[13] For the full set of notes, see the pages 56 and 58.

- product or service improvements correlated with customer satisfaction

- financial benefits correlated with improved employee safety, absenteeism, and turnover

- quality indicators correlated with market indicators.

As companies continue on their quality journey, changes are needed in what data are collected and what analysis is done. Motorola made the term "Six Sigma" a well-known (or at least often used) goal. It expresses a measure of perfection. In the old system of tracking in percent, 99% conformance sounds good. If that was the grade you got in school you were probably happy. But when tracking conformance in percent, companies mask the magnitude of improvement that is still needed. A 99% conformance is equivalent to 10,000 opportunities (i.e., defects) per million. The average American company has quality levels in the magnitude of 6200 to 67,000 opportunities per million.[14] By Motorola's definition six sigma means 3.4 parts per million (ppm).[15] The true statistical calculation of plus and minus six standard deviations is 2 parts per billion (ppb). When the magnitude of improvement needed is 2000×, it does not matter which definition (3.4 ppm or 2 ppb) is used. In either case the system needs an overhaul not just fine-tuning.

To provide a sense of what six sigma represents, consider a process where there is one transaction every second; for example, a watch. Then 99% conformance would mean 1 defect every 1.7 minutes; i.e., the watch would lose one second every 1.7 minutes or approximately 7 minutes a day. If the process satisfied Motorola's definition, there would be 1 defect every 3.4 days. On the basis of the statistical definition, the process would yield only 1 defect every 16 years.[16]

Examples for Item 2.3

The conversion of the data, described in Item 2.1, into useful information, and the correlation between the indicators presented and market indicators.

At **TI Defense Systems & Electronics** (TI-DSEG) concurrent engineering and the "six steps to six sigma" are used to translate data into action. The six steps are (1) select a key product characteristic, (2) determine elements that contribute to the key

---

[14] L. Dobyns and C. Crawford-Mason, *Quality or Else*, Houghton Mifflin, Boston, 1991, p. 137.

[15] As a technical side note, Motorola's definition is based on plus and minus 4.5 standard deviations. They start with plus and minus 6 standard deviations. They then allow the mean to shift plus and minus 1.5 standard deviations. For further information, contact Motorola directly using the phone number or address in the appendix.

[16] Roger Harnishfeger, Dana Corporation, presented this concept and the calculations in 1991.

characteristic, (3) determine the process steps that impact the characteristic, (4) determine the target and allowable tolerance, (5) determine the capability of the process, and (6) if process is not capable, take corrective action; if it is capable determine what ongoing controls are needed to maintain the process response. Four vital measures are always tracked within TI-DSEG regardless of the team: customer satisfaction, cycle time, defects per unit in parts per million, and training hours per employee.

The expression used by **AT&T Transmission** is that the data should be local, friendly, and dirty. It should be local in that it is relevant to the people tracking it. It should be friendly so it is easy to translate the data into actionable information. It should be dirty because it is being used, not filed in a cabinet drawer.

Customer satisfaction and operational performance data are made available throughout the company at **AT&T Communications**. This provides timely feedback to all employees—workers and management—on meeting quality goals. All process and performance indicators are linked to overall efforts to meet customer requirements and improved satisfaction levels.

One service company said they use failure mode and effects analysis (FMEA) on everything. They took an analysis tool that is often associated with manufacturing companies and applied it to the service processes.

Key Excellence Indicators

Companies doing well in this Category have a strong quantitative orientation—decisions are based on facts and data. Their focus is not on the numbers but on what information the data can provide them to improve their processes. If a twofold improvement was made in a process but there was no noticeable change in accuracy, cycle time, or other measure of importance to the internal or external customer, you should wonder if a different process should have been improved. The measures these companies use are correlated to customer and financial indicators. They have widely deployed data that are easily accessible in a timely manner. This allows everyone in the organization to know they are working on the right things.

Facts from data and analysis support a variety of company purposes. This includes planning, reviewing company performance, improving operations, and comparing quality performance with competitors or with best practices. Excellent companies use the benchmarking process aggressively to seek out new ideas.

CATEGORY 3.0: STRATEGIC PLANNING

 ### Description and Examples

> "The *Strategic Planning* Category examines how the company sets strategic directions, and how it determines key plan requirements. Also examined is how the plan requirements are translated into an effective performance management system."[17]

Among the key concepts that provide the foundation for the Criteria are long-range outlook and management by fact. These address this concern: "Are the achievements presented the results of the overall system or were they produced in spite of the system?" This Category examines the planning process for key quality and operational performance requirements as well as the development of short and longer term plans. This also requires a discussion of the effectiveness of the company's translation of plans into actionable key business drivers. A long-term commitment is necessary to achieve quality and market leadership. The planning process needs to determine or anticipate many types of changes. This includes changes that affect customers' expectations, developments in technology, regulatory requirements, and the actions of competitors (Figure 4.3).

Americans seem to have a predisposition to action. This often evolves into a policy of "Ready, Fire, Aim." Dr. Deming talked about the cycle of Plan-Do-Study-Act with a strong emphasis on the planning phase. Some even recommend: "If you have ten days for a project, use five days for planning."[18]

[17] *1995 Malcolm Baldrige National Quality Award Criteria*, National Institute of Standards and Technology, Gaithersburg, MD.

[18] L. Dobyns and C. Crawford-Mason, *Quality or Else*, Houghton Mifflin, Boston, 1991, p. 280.

Figure 4.3
Strategic Planning Category

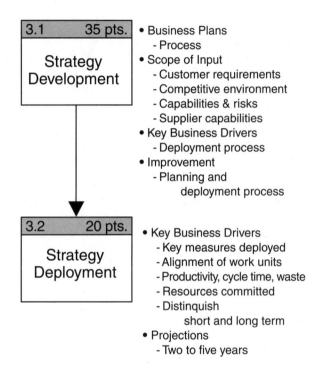

3.0 Strategic Planning

3.1   35 pts.
Strategy Development
	• Business Plans
	  - Process
	• Scope of Input
	  - Customer requirements
	  - Competitive environment
	  - Capabilities & risks
	  - Supplier capabilities
	• Key Business Drivers
	  - Deployment process
	• Improvement
	  - Planning and
	      deployment process

3.2   20 pts.
Strategy Deployment
	• Key Business Drivers
	  - Key measures deployed
	  - Alignment of work units
	  - Productivity, cycle time, waste
	  - Resources committed
	  - Distinquish
	      short and long term
	• Projections
	  - Two to five years

Maximum points - 55

Item 3.1: Criteria

3.1 Strategy Development (35 pts.)

Describe the company's strategic planning process for overall performance and competitive leadership for the short term and the longer term. Describe also how this process leads to the development of key business drivers to serve as the basis for deploying plan requirements throughout the company.

☑ Approach
☑ Deployment
☐ Results

Areas to Address

(a) how the company develops strategies and business plans to strengthen its customer-related, operational, and financial performance and its competitive position. Describe how strategy development considers: (1) customer requirements and expectations and their expected changes; (2) the competitive environment; (3) risks: financial, market, technological, and societal; (4) company capabilities—human resource, technology, research and development and business processes—to seek new market leadership opportunities and/or to prepare for key new requirements; and (5) supplier and/or partner capabilities.

(b) how strategies and plans are translated into actionable key business drivers, which serve as the basis for deploying plan requirements, addressed in Item 3.2.

(c) how the company evaluates and improves its strategic planning and plan deployment processes.

Notes:

1. Item 3.1 addresses overall company strategy and business plans, not specific product and service designs.

2. The sub-parts of 3.1a are intended to serve as an outline of key factors involved in developing a view of the future as a context for strategic planning. Strategy and planning refer to a future-oriented basis for major business decisions, resource allocations, and company-wide management. "Strategy and planning," then, addresses both revenue growth thrusts as well as thrusts related to improving operational performance.

3. Customer requirements and their expected changes [3.1a(1)] might include pricing factors. That is, market success may depend upon achieving cost levels dictated by anticipated price levels rather than setting prices to cover costs.

Item 3.1: *Strategy Development*

Two Items make up this Category: one for the planning process and one for the plans and their deployment. Item 3.1, Strategy Development, asks for a description of the planning process. The sources of input to the planning process should be discussed. This might include customer requirements, projections of competitive environment, capabilities of the company and the suppliers, and risks.

Item 3.1 also covers developing key business drivers. The starting point for developing these drivers could be new product launches, entry into new markets, or a new R&D focus. The company should present a unified front for its improvement efforts, showing that the total system is being considered and optimized.

As with the other Items, the final Area to Address in this Item is the evaluation and improvement of the planning process. This is perhaps the most commonly misunderstood requirement in the Criteria. Often, the answer given is an explanation of how the plans are reviewed and progress determined. What should be addressed is how the planning *process* is improved, not the plans themselves—the plans are addressed in the next Item. Improvements in the planning process might include regrouping those involved in the past year's planning activity to capture lessons learned, providing training on how to increase the effectiveness of the planning process, or learning ideas from other companies about best practices for planning.

Examples for Item 3.1

> *The company's planning process—how all key quality and operationad performance requirements are integrated into the overall planning process and how plans are translated into actionable key business drivers.*

AT&T Transmission uses a formal structured approach to planning called *policy deployment*. In this approach, the priorities of customers, employees, and shareholders are captured. **AT&T Card** uses a cascade process involving annual goals and strategies linked to key initiatives.

Cadillac Motor Car Company had no difficulty showing the integration of their quality and business plan. Their business plan is the quality plan. All employees and some suppliers and dealers are involved in their planning process.

Eastman Chemical addresses performance improvement at all levels of the organization using their Quality Management Process (QMP) as shown in Figure 4.4.

4. The purposes of projecting the competitive environment [3.1a(2)] are to detect and reduce competitive threats, to improve reaction time, and to identify opportunities. If the company uses modeling, scenario, or other techniques to project the competitive environment, such techniques should be briefly outlined in 3.1a(2).

5. Key business drivers are the areas of performance most critical to the company's success. They include customer-driven quality requirements and operational requirements such as productivity, cycle time, deployment of new technology, strategic alliances, supplier development, employee productivity and development, and research and development. Deployment of plans should include how progress will be tracked such as through the use of key measures.

6. Examples of strategy and business plans that might be the starting points for the development of key business drivers are:

    • new product/service lines;

    • entry into new markets or segments;

    • new manufacturing and/or service delivery approaches such as customization;

    • new or modified competitive thrusts;

    • launch of joint ventures and/or partnerships;

    • new R&D thrusts; and

    • new product and/or process technologies.

7. How the company evaluates and improves its strategic planning and plan deployment process might take into account the results of reviews (1.2c), input from work units, and projection information (3.2b). The evaluation might also take into account how well strategies and requirements are communicated and understood, and how well key measures are aligned.

Item 3.2: Criteria

3.2 Strategy Deployment (20 pts.)

Summarize the company's key business drivers and how they are deployed. Show how the company's performance projects into the future relative to competitors and key benchmarks.

Figure 4.4

Eastman Chemical's Quality Management Process

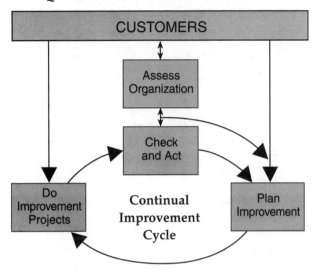

Item 3.2: *Strategy Deployment*

The plans that result from the process described in Item 3.1 are to be addressed in this Item. This should include a discussion of key business drivers. Although the deployment of these plans will affect products and services, you should be careful not to confuse these plans with the *design process for products and services* that is requested in Category 5.0. The plans here may describe how improvements are being made in product quality, response time, employees' capabilities, relationship with suppliers, and so on. They should not describe *how* the product or service is designed.

Discussion of any distinction between the short-term plans (one to three years) and the longer-term plans (three to five years) would be made here. The deployment of the plans, along with the resources committed, is addressed also. The last Area to Address in this Item deals with a two- to five-year projection of key measures or indicators: How will the organization compare with the competition? You are being asked to look into your crystal ball and make a projection—a rational prediction. Certainly the Examiners do not know whether you are right or wrong, but they will get a sense for whether you are looking to the future or reacting to the present.

☑ Approach
☑ Deployment
☐ Results

Areas to Address

(a) summary of the specific key business drivers derived from the company's strategic directions and how these drivers are translated into an action plan. Describe: (1) key performance requirements and associated operational performance measures and/or indicators and how they are deployed; (2) how the company aligns work unit and supplier and/or partner plans and targets; (3) how productivity and cycle time improvement and reduction in waste are included in plans and targets; and (4) the principal resources committed to the accomplishment of plans. Note any important distinctions between short-term plans and longer-term plans.

(b) two-to-five-year projection of key measures and/or indicators of the company's customer-related and operational performance. Describe how product and/or service quality and operational performance might be expected to compare with key competitors and key benchmarks over this time period. Briefly explain the comparisons, including any estimates or assumptions made regarding the projected product and/or service quality and operational performance of competitors or changes in key benchmarks.

Notes:

1. The focus in Item 3.2 is on the translation of the company's strategic plans, resulting from the process described in Item 3.1, to requirements for work units, suppliers, and partners. The main intent of Item 3.2 is alignment of short- and long-term operations with strategic directions. Although the deployment of these plans will affect products and services, design of products and services is not the focus of Item 3.2. Such design is addressed in Item 5.1.

2. Productivity and cycle time improvement and waste reduction [3.2a(3)] might address factors such as inventories, work-in-process, inspection, downtime, changeover time, set-up time, and other examples of utilization of resources—materials, equipment, energy, capital, and labor.

3. Area 3.2b addresses projected progress in improving performance and in gaining advantage relative to competitors. This projection may draw upon analysis (Item 2.3) and data reported in results Items (Category 6.0 and Items 7.4 and 7.5). Such projections are intended to support reviews (1.2c), evaluation of planning (3.1c), and other Items. Another purpose is to take account of the fact that competitors and benchmarks may also be improving over the time period of the projection.

A common mistake made in this Item is to provide a wish list of activities, with no strategy or plan to make it happen. One might want to reduce the time it takes to go from concept to market with a product or service but without a plan, it will just be wishful thinking.

Examples for Item 3.2

The key business drivers that are developed from the process described in Item 3.1.

The Baldrige Criteria are used for an annual self-assessment at **AT&T Transmission**. From this assessment, areas for improvement are identified and prioritized. Perhaps there are 65 areas needing improvement; 8 might be selected for action. The assessment findings are used to drive the improvement plan. At **TI Defense & Electronics**, the goals included a defect reduction six times every year, cycle time reduction of 25%, and an increase in the net revenue per person.

The strategies they used are six sigma training and deployment, business process mapping, benchmarking, an integrated product development process, and the continued deployment of statistical tools. The five strategic indicators at **Wainwright** are safety, internal customer satisfaction, external customer satisfaction, six sigma quality, and business performance. Even as a small company, **Granite Rock** has been developing 10-year plans since 1985. The president and senior executives lead the development of the plans at **AT&T Communications**, which address five key customer requirements: call quality, customer service, billing, reputation, and price.

Key Excellence Indicators

In companies that are doing well, the quality plans and the business plans are integrated; the "arrows" are all aligned. The companies have aggressive goals but also have plans and strategies to achieve these goals. A longer term focus is taken. Decisions are made with a concern that goes beyond that of just looking good for the day, beyond looking good for Wall Street. Plans include not only the products and services provided to customers, but also encompass all areas of the business.

The suppliers and customers are considered as part of the partnership and treated accordingly. Key business drivers are derived from customer requirements and supplier capabilities and ideas are linked with the plans.

CATEGORY 4.0: HUMAN RESOURCE DEVELOPMENT AND MANAGEMENT

 Description and Examples

"The *Human Resource Development and Management* Category examines how the workforce is enabled to develop and utilize its full potential, aligned with the company's performance objectives. Also examined are the company's efforts to build and maintain an environment conducive to performance excellence, full participation, and personal and organizational growth."[19]

The recognition of the workforce as the most important resource a company can have is integrated within the core values of the Criteria. It is also the basis of Category 4.0. Here, the organization can describe how its organization enables all of the workforce to develop and realize its full potential—as persons as well as employees. This Category examines how employee satisfaction and well-being are determined, the opportunities for initiatives and self-direction, training, and recognition for employees (Figure 4.5).

Dr. Juran, Dr. Deming, and others talk about the 85/15 rule as applied to people. The claim is that the system (practices, procedures, methods) causes 85% of the variation and people only cause 15% of the variation. Dr. Deming claims the split is closer to 95/5 but the idea is the same; namely, fix the system (the what), not the people (the who). In line with this, the Criteria ask for information on the company's efforts in building and maintaining "an environment conducive to performance excellence, full participation, and personal and organizational growth."

---

[19] *1995 Malcolm Baldrige National Quality Award Criteria*, National Institute of Standards and Technology, Gaithersburg, MD.

Figure 4.5
Human Resource Development and Management Category

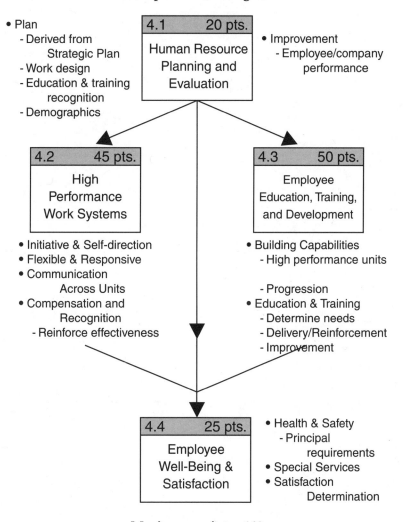

4.0 Human Resource Development & Management

- Plan
  - Derived from Strategic Plan
  - Work design
  - Education & training recognition
  - Demographics

4.1   20 pts.
Human Resource Planning and Evaluation

- Improvement
  - Employee/company performance

4.2   45 pts.
High Performance Work Systems

- Initiative & Self-direction
- Flexible & Responsive
- Communication Across Units
- Compensation and Recognition
  - Reinforce effectiveness

4.3   50 pts.
Employee Education, Training, and Development

- Building Capabilities
  - High performance units
  - Progression
- Education & Training
  - Determine needs
  - Delivery/Reinforcement
  - Improvement

4.4   25 pts.
Employee Well-Being & Satisfaction

- Health & Safety
  - Principal requirements
- Special Services
- Satisfaction Determination

Maximum points - 140

Item 4.1: Criteria

4.1 Human Resource Planning and Evaluation (20 pts.)

Describe how the company's human resource planning and evaluation are aligned with its strategic and business plans and address the development and well-being of the entire workforce.

☑ Approach
☑ Deployment
☐ Results

Areas to Address

(a) how the company translates overall requirements from strategic and business planning (Category 3.0) to specific human resource plans. Summarize key human resource plans in the following areas: (1) changes in work design to improve flexibility, innovation, and rapid response; (2) employee development, education, and training; (3) changes in compensation, recognition, and benefits; and (4) recruitment, including expected or planned changes in demographics of the workforce. Distinguish between the short term and the longer term, as appropriate.

(b) how the company evaluates and improves its human resource planning and practices and the alignment of the plans and practices with the company's strategic and business directions. Include how employee-related data and company performance data are analyzed and used: (1) to assess the development and well-being of all categories and types of employees; (2) to assess the linkage of the human resource practices to key business results; and (3) to ensure that reliable and complete human resource information is available for company strategic and business planning.

Notes:

1. Human resource planning addresses all aspects of designing and managing human systems to meet the needs of both the company and the employees. This Item calls for information on human resource plans. This does not imply that such planning is separate from overall business planning. Examples of human resource plan elements or plan thrusts (4.1a) that might be part(s) of a comprehensive plan are:

    • redesign of work organizations and/or jobs to increase employee responsibility and decision-making;

Item 4.1: *Human Resource Planning and Evaluation*

The overall umbrella for this Category is defined by the first of four Items. Item 4.1, Human Resource Planning and Evaluation, examines the human resource (HR) plans. This plan should be derived from (and be consistent with) the strategic plan described in Item 3.2. The HR plans should address employee development, flexibility, recognition, compensation, and recruitment. These plans may include redesign of work organizations, formation of partnerships with educational institutions, activities to promote labor, and management cooperation.

Improvement of personnel practices, such as recruitment, hiring, and services to employees, are to be described in the last Area to Address of Item 4.1. This also contains the evaluation and improvement of both the HR plans and the alignment of the plans with the overall company strategy. A major challenge in this area is the alignment of human resource plans with the strategic plan.

Examples for Item 4.1

Overall human resource plans such as employee development, labor and management cooperation, establishing networks with educational institutions.

There are four major human resource initiatives at **TI Defense & Electronics**. They are:

- teamwork and empowerment
- people development
- performance and recognition
- employee well-being.

Each initiative is headed by a senior executive and each has an internal facilitator to assist the team.

At the **Ritz-Carlton,** all hotels have a director of human resources and training manager. The company has developed a profile of long-term outstanding employees that is used when recruiting new employees.

Item 4.2: *High Performance Work Systems*

This Item requests a description of how the company promotes high performance through employee contributions as individuals and as part of a team. This involvement might be through problem-solving teams, cross-functional teams,

- initiatives to promote labor-management cooperation, such as partnerships with unions;

- creation or modification of compensation and recognition systems based on building shareholder value and/or customer satisfaction;

- creation or redesign of employee surveys to better assess the factors in the work climate that contribute to or inhibit high performance;

- prioritization of employee problems based upon potential impact on productivity;

- development of hiring criteria;

- creation of opportunities for employees to learn and use skills that go beyond current job assignments through redesign of processes or organizations;

- education and training initiatives, including those that involve developmental assignments;

- formation of partnerships with educational institutions to develop employees or to help ensure the future supply of well-prepared employees;

- establishment of partnerships with other companies and/or networks to share training and/or spread job opportunities;

- introduction of distance learning or other technology-based learning approaches; and

- integration of customer and employee surveys.

2. "Employee-related data" (4.1b) refers to data contained in personnel records as well as data described in Items 4.2, 4.3, and 4.4. This might include employee satisfaction data and data on turnover, absenteeism, safety, grievances, involvement, recognition, training, and information from exit interviews.

3. "Categories of employees" [4.1b(1)] refers to the company's classification system used in its human resource practices and/or work assignments. It also includes factors such as union or bargaining unit membership. "Types of employees" takes into account other factors, such as workforce diversity or demographic makeup. This includes gender, age, minorities, and the disabled.

4. The evaluation in 4.1b might be supported by employee-related data such as satisfaction factors (Item 4.4), absenteeism, turnover, and accidents. It might also be supported by employee feedback and information from exit interviews. Evaluations might also be supported by comparative or benchmarking information.

Evaluation should take into account factors such as employee problem resolution effectiveness, and the extent of deployment of education and training throughout the company.

self-managed work groups, or improvement teams. A description of how the company increases opportunities for flexibility and rapid response to changing requirements is expected. Communication across units is now in this Item.

Alignment of compensation and recognition to reinforce the work design is the last Area to Address. Some companies claim they want employees to work as part of a team, yet continually reward the "lone ranger" who is not a team player. Another example involves having multidiscipline teams across different business units, yet profit sharing is set up based on how good a unit does, not the company. People may do things to make their unit look good even at the expense of damage to another unit.

In the notes for this Item, employee evaluation is mentioned. Students of the late Dr. W. Edwards Deming know that he expressed great concern over this issue. He was opposed to pitting employees against each other by rating and ranking them or by forcing competition through an artificial scarcity of rewards. This portion of the Criteria can be consistent with Deming's philosophy if the focus is on feedback and recognition. Dr. Deming strongly encouraged feedback to employees. It should be frequent, not a once-a-year surprise. There is nothing in the Criteria that states employees must be rated on a scale, such as 1–5 or 1–10. Ongoing feedback can be given. Employees that are doing better than the overall system can provide ideas on lessons learned, whereas employees whose performance is outside the system on the low side are in need of and can be given special attention in a timely manner. So it is possible to meet the intent of this Item and still be consistent with Dr. Deming's philosophy.

Examples for 4.2

Opportunities to have the employees' voices heard as individuals and in groups, including increased opportunities, communication, and responsibility. Alignment of compensation and recognition to reinforce job design.

There are more than 800 teams working on what **AT&T Transmission** calls their Quality Improvement Story (they have 7500 employees). According to the National Association of Suggestion Systems, their response rate is best in class. In addition to these teams, self-directed work groups have been in place for a number of years now. At **AT&T Card** more than 10,000 ideas were submitted in 1992 and the company experienced a $2000/employee gainshare.

At **Wainwright** all associates are involved in the improvement process. On average, each of the 275 associates submitted 1.25 suggestions per week in 1993.

5. Human resource information for company strategic and business planning might include an overall profile of strengths and weaknesses that could affect the company's abilities to fulfill plan requirements. This could result in the identification of specific needs requiring resources or new approaches.

Item 4.2 Criteria

4.2 High Performance Work Systems (45 pts.)

Describe how the company's work and job design and compensation and recognition approaches enable and encourage all employees to contribute effectively to achieving high performance objectives.

☑ Approach
☑ Deployment
☐ Results

Areas to Address

(a) how the company's work and job design promote high performance. Describe how work and job design: (1) create opportunities for initiative and self-directed responsibility; (2) foster flexibility and rapid response to changing requirements; and (3) ensure effective communications across functions or units that need to work together to meet customer and/or operational requirements.

(b) how the company's compensation and recognition approaches for individuals and groups, including managers, reinforce the effectiveness of the work and job design.

Notes:

1. Work and job design refers to how employees are organized and/or organize themselves in formal and informal, temporary or longer-term units. This may include work teams, problem-solving teams, functional units, departments, self-managed or managed by supervisors. In some cases, teams might involve individuals in different locations linked via computers or conferencing technology.

2. Examples of approaches to create flexibility in work design to enhance performance might include simplification of job classifications, cross training, job rotation, work layout, and work locations. It might also entail use of technology and changed flow of information to support local decision–making.

3. Compensation and recognition refer to all aspects of pay and reward, including promotion and bonuses. The company might use a variety of reward and recognition approaches—monetary and non-monetary, formal and informal, and individual and group.

Ritz-Carlton empowers each employee to spend $2000 per incident and sales employees are authorized to spend $5000 per incident to make things right for the customer.

In a 1990 Gallup survey of employees' attitudes several issues relevant to this topic were identified:[20]

1. In companies where quality improvement activities are in place, more than one-third of the employees do not participate.

2. Employees perceive a wide gap between their companies' talk and action on quality principles.

3. Although two-thirds have been asked to be involved in decisions, only 14% feel empowered.

4. Quality programs have the strongest positive impact on job satisfaction, internal communications, and work procedures.

Reasons for the third point are numerous but could include people not having the proper information to make a decision, previous decisions that were over-turned, and punitive action when decisions were counter to what management wanted (but had not expressed).

Teams are a way of life at **AT&T Communications**. Teams implement quality improvements using standardized AT&T developed tools. Process management teams include employees from all levels of the company and, as appropriate, suppliers. Internal contracts are used to clearly assign responsibilities and to establish agreed-on improvement goals.

Some of the recognition provided to employees at **TI Defense & Electronics** includes the President's Award for team excellence, site quality awards, a technical ladder, luncheons, displays, and newsletter articles. The technical ladder offers advancement opportunities for people who want to pursue technical not managerial careers.

AT&T Transmission uses direct reports to assess the supervisors. Some companies now use "360-degree" reviews with input from above, below, and peer level. They have 40 award programs for individuals and teams.

Granite Rock, a company of 400 employees, has an Individual Professional Development Plan (IPDP), which is a proactive, voluntary way to get employees involved in setting goals with their supervisors for ongoing development. An annual recognition day is held at each site.

---

[20] "Quality: Everyone's Job Many Vacancies," 1990 Gallup survey conducted for the American Society for Quality Control.

Compensation and recognition approaches could include profit sharing and compensation based on skill building, use of new skills, and demonstrations of self learning. The approaches could take into account the linkage to customer retention or other performance objectives.

Employee evaluations and reward and recognition approaches might include peer evaluations, including peers in teams and networks.

Item 4.3: Criteria

4.3 Employee Education, Training, and Development (50 pts.)

Describe how the company's education and training address company plans, including building company capabilities and contributing to employee motivation, progression, and development.

☑ Approach
☑ Deployment
☐ Results

Areas to Address

(a) how the company's education and training serve as a key vehicle in building company and employee capabilities. Describe how education and training address: (1) key performance objectives, including those related to enhancing high performance work units; and (2) progression and development of all employees.

(b) how education and training are designed, delivered, reinforced, and evaluated. Include: (1) how employees and line managers contribute to or are involved in determining specific education and training needs and designing education and training; (2) how education and training are delivered; (3) how knowledge and skills are reinforced through on-the-job application; and (4) how education and training are evaluated and improved.

Notes:

1. Education and training address the knowledge and skills employees need to meet their overall work objectives. Education and training might include leadership skills, communications, teamwork, problem solving, interpreting and using data, meeting customer requirements, process analysis, process simplification, waste reduction, cycle time reduction, error-proofing, priority setting based upon cost and benefit data, and other training that affects employee effectiveness, efficiency, and safety. This might include job enrichment and job rotation to

Item 4.3: *Employee Education, Training, and Development*

A company depends on the skill level of its employees. Employee success depends on having meaningful opportunities to learn and apply new skills. In this Item, the training needs, training delivery, and employee input to these are examined. A clear link between the human resource plan and training should be shown. Specific training might pertain to quality awareness, process simplification, problem solving, analyzing and using data, or basic skills such as reading, arithmetic, or language. Companies need to invest in the total development of the workforce.

The evaluation and improvement of quality-related education and training might include links to on-the-job performance and the personal growth within the company of all categories of employees.

Examples for Item 4.3

Training needs determination and training in quality awareness, teamwork, process analysis, problem solving, error-proofing, etc.

TI Defense & Electronics has a job enhancement program in which they spent more than $11 million to improve basic skills in reading, writing, and math. Their supervisory training focuses on coaching and team building. Six sigma training is provided for all and there is a curriculum for every job.

More than 40 hours of training per employee are provided each year at **AT&T Transmission**. Eight weeks of training for new hires is required at **AT&T Card**. **Ritz-Carlton** developed a certification for sales personnel. **Granite Rock** has a try-a-job program. In this program, employees can spend a day in a different job to determine if they would like to train for that job.

A comprehensive training program has been implemented at **GTE**. Most employees involved in the shop floor printing operations are trained to use statistical process control to monitor printing process variation. By acting on these data they are able to maintain and improve product quality, decrease waste, and improve productivity.

Wainwright Industries spends up to 7% of its payroll on training in areas such as basic math and English as well as technical areas.

enhance employees' career opportunities and employability. It might also include basic skills such as reading, writing, language, and arithmetic.

2. Training for customer-contact (frontline) employees should address: (a) key knowledge and skills, including knowledge of products and services; (b) listening to customers; (c) soliciting comments from customers; (d) how to anticipate and handle problems or failures ("recovery"); (e) skills in customer retention; and (f) how to manage expectations.

3. Determining specific education and training needs [4.3b(1)] might include use of company assessment or employee self-assessment to determine and/or compare skill levels for progression within the company or elsewhere. Needs determination should take into account job analysis—the types and levels of skills required—and the timeliness of training.

4. Education and training delivery might occur inside or outside the company and involve on-the-job, classroom, or other types of delivery. This includes the use of developmental assignments within or outside the company.

5. How education and training are evaluated [4.3b(4)] could address: effectiveness of delivery of education and training; impact on work unit performance; and cost effectiveness of education and training alternatives.

Item 4.4: Criteria

4.4 Employee Well-Being and Satisfaction (25 pts.)

Describe how the company maintains a work environment and a work climate conducive to the well-being and development of all employees.

☑ Approach
☑ Deployment
☐ Results

Areas to Address

(a) how the company maintains a safe and healthful work environment. Include: (1) how employee well-being factors such as health, safety, and ergonomics are included in improvement activities; and (2) principal improvement requirements, measures and/or indicators, and targets for each factor relevant and important to the work environment of the company's employees. Note any significant differences based upon differences in work environments among employees or employee groups.

Item 4.4: *Employee Well-Being and Satisfaction*

Because employees are not machines, the last Item for this Category is concerned with employee well-being and satisfaction. In this Item, how the work environment is conducive to the well-being of employees is examined. This includes health, safety, ergonomics, and employee satisfaction. Special services, such as counseling, recreational activities, day care, special leave, and flexible work hours would also be described here. One study[21] found that 69% of employee job satisfaction is tied to the leadership skills of managers.

As with every process, there must be a way to provide feedback to the process. Methods used by the company to determine employee satisfaction, including frequency, are requested. Data (trends and current performance) from this satisfaction determination should be discussed along with other employee well-being measures such as safety, absenteeism, strikes, and employee turnover in Item 6.2.

A common technique used to gather information for this Item is an employee survey. A common complaint by employees about this topic is lack of action after an employee survey. Why was an employee satisfaction survey conducted if no one was going to do anything with the information? Do not collect the data unless you are planning to take action. It is frustrating for employees to share opinions, either positive or negative, and then feel that their opinions were not valued because no apparent action was taken. Employees recognize that all issues will not be resolved. Feedback to employees that tells them what issues were raised and what issues are being pursued will go a long way toward improving employee morale.

Examples for Item 4.4

How the work environment is conducive to employee health, safety, and satisfaction along with a means to determine employee satisfaction and well-being.

At **Granite Rock** an independent survey of employees is conducted to determine satisfaction. The safety record is two times better than industry and they had a $2 million savings in 1991 (the year prior to receiving the Baldrige Award) in workers' compensation claims.

An annual survey for all employees and a monthly opinion survey on a subset of employees is conducted at **AT&T Card**. Additionally, focus group meetings are held in a fashion similar to customer focus groups. Small groups of employees are brought together to voice opinions.

AT&T Transmission conducts an all-employee survey every two years with mini surveys being used throughout the year. **TI Defense & Electronics** also uses an all-employee survey and focus group meetings.

---

[21] Study done by Wilson Learning Corporation, Minneapolis, MN, 1994.

(b) what services, facilities, activities, and opportunities the company makes available to employees to support their overall well-being and satisfaction and/or to enhance their work experience and development potential.

(c) how the company determines employee satisfaction, well-being, and motivation. Include a brief description of methods, frequency, the specific factors used in this determination, and how the information is used to improve satisfaction, well-being, and motivation. Note any important differences in methods or factors used for different categories or types of employees, as appropriate.

Notes:

1. Examples of services, facilities, activities, and opportunities (4.4b) are: personal and career counseling; career development and employability services; recreational or cultural activities; non-work-related education; day care; special leave for family responsibilities and/or for community service; safety off the job; flexible work hours; and outplacement. These services also might include career enhancement activities such as skill assessment, helping employees develop learning objectives and plans, and employability assessment.

2. Examples of specific factors which might affect satisfaction, well-being, and motivation are: effective employee problem or grievance resolution; safety; employee views of leadership and management; employee development and career opportunities; employee preparation for changes in technology or work organization; work environment; workload; cooperation and teamwork; recognition; benefits; communications; job security; compensation; equality of opportunity; and capability to provide required services to customers. An effective determination is one that provides the company with actionable information for use in improvement activities.

3. Measures and/or indicators of satisfaction, well-being, and motivation (4.4c) might include safety, absenteeism, turnover, turnover rate for customer-contact employees, grievances, strikes, worker compensation, as well as results of surveys.

4. How satisfaction, well-being, and motivation information is used might involve developing priorities for addressing employee problems based on impact on productivity.

5. Trends in key measures and/or indicators of well-being and satisfaction should be reported in Item 6.2.

The data at **Ritz-Carlton** regarding employees' perception of cooperation between departments showed an increase from 78% in 1989 to 86% in 1991. **AT&T Communications** conducts an annual workforce survey, conducts employee focus groups, and does site interviews to ascertain employee satisfaction, awareness of company values and goals, and to determine perception of company leadership.

Key Excellence Indicators

O⊤ Employees see themselves as internal customers and suppliers working together to define requirements and ways to achieve them at companies that are doing well. Internal partnerships, including employee development, cross-training, or new work organizations, are common. There is comprehensive training and education to support empowerment and flexible assignments as well as the company's quality initiative. Recognition is balanced between team and individual recognition—and includes both.

O⊤ At companies that are doing well, most and perhaps all employees have contact with the customer. People who are making a product go out to see how it is used and fork truck drivers talk to people in the customer's receiving department to see how handling damage has an impact. People are able to take pride in their work and enjoy having the opportunity to make a difference.

CATEGORY 5.0: PROCESS MANAGEMENT

 ## Description and Examples

> "The *Process Management* Category examines the key aspects of process management, including customer-focused design, product and service delivery processes, support services and supply management involving all work units, including research and development. The Category examines how key processes are designed, effectively managed, and improved to achieve higher performance."[22]

Many examples exist that demonstrate the need for improvements of product quality. Studies have investigated the loss of market share and industry-specific quality levels. One such study was done by David Garvin on air conditioners produced in Japan and the United States.[23] The average defect rate for Japan was 70 times lower than that of the the United States. The *best* American-made units were

---

[22] *1995 Malcolm Baldrige National Quality Award Criteria,* National Institute of Standards and Technology, Gaithersburg, MD.

[23] David Garvin, *"Quality on the Line,"* *Harvard Business Review,* September–October 1983, pp. 65–75.

10% worse than the *worst* Japanese units. More recent studies have indicated this gap is closed now for some products and services.[24]

Category 5.0, Process Management, completes the system framework in the Baldrige model (Categories 2.0 through 5.0 make up the system portion of the Criteria, see Fig. 2.1). It addresses the development and effectiveness of systems and processes for assuring the quality of all operations. Design, production, delivery, support, and operational processes are all considered (Figure 4.6). The relationship between the company and the suppliers is covered. Category 5.0 demonstrates the comprehensive nature of the Criteria since all areas of the business are impacted.

From the organization's response to this Category, it becomes evident if the philosophy of the company is one of prevention or reaction to problems. For example, if a company effectively uses simulations and failure mode analysis during design, selects key characteristics using the loss function, simplifies processes even when there is no crisis, uses process analysis in service areas, and has supplier certification, the company has elements of a prevention-based system. Remember: The Criteria are not prescriptive, so this description is not the only way to demonstrate a prevention philosophy.

The other extreme, a reactive-based system, is demonstrated by a company that relies on the customer to be the final inspector; e.g., it uses prototype phase for design validation and the final customer for design verification. Designs usually have several key characteristics for each part. An arbitrary number is chosen based on past problems which does not consider if the present design is robust. This type of company may feel that they have a customer focus since they react quickly to all customer problems and protect the customer from defects by relying on receiving and final inspection. However, customers are often advised by consumer groups not to buy products during the first year of production—to give the company time to find and correct all the bugs.

Federal Express, and many others, have the motto "If it ain't broke, improve it." The investment of time and energy into simplifying processes, shortening the time it takes to respond and improving the accuracy of the output will pay off. The Federal Quality Institute reported that every dollar invested in quality returns $4 to $5.

There is more overlap between this Category of the Baldrige Criteria and the ISO 9000 series than any other Category (see Chapter 13).

---

[24] "Here Come Japan's Carmakers—Again," *Fortune*, December 13, 1993, pp. 129–137.

Figure 4.6
Process Management
Category

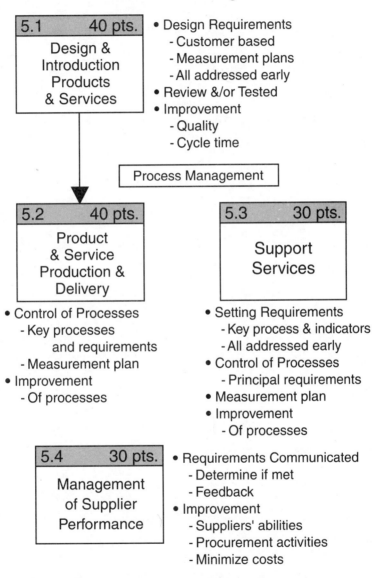

5.0 Process Management

5.1	40 pts.

**Design &
Introduction
Products
& Services**

- Design Requirements
  - Customer based
  - Measurement plans
  - All addressed early
- Review &/or Tested
- Improvement
  - Quality
  - Cycle time

Process Management

5.2	40 pts.

**Product
& Service
Production &
Delivery**

- Control of Processes
  - Key processes
    and requirements
  - Measurement plan
- Improvement
  - Of processes

5.3	30 pts.

**Support
Services**

- Setting Requirements
  - Key process & indicators
  - All addressed early
- Control of Processes
  - Principal requirements
- Measurement plan
- Improvement
  - Of processes

5.4	30 pts.

**Management
of Supplier
Performance**

- Requirements Communicated
  - Determine if met
  - Feedback
- Improvement
  - Suppliers' abilities
  - Procurement activities
  - Minimize costs

Encompasses all processes within the organization

Maximum points - 140

Item 5.1: Criteria

5.1 Design and Introduction of Products and Services (40 pts.)

Describe how new and/or modified products and services are designed and introduced and how key production/delivery processes are designed to meet both key product and service quality requirements and company operational performance requirements.

☑ Approach
☑ Deployment
☐ Results

Areas to Address

(a) how products, services, and production/delivery processes are designed. Describe: (1) how customer requirements are translated into product and service design requirements; (2) how product and service design requirements are translated into efficient and effective production/delivery processes, including an appropriate measurement plan; and (3) how all requirements associated with products, services, and production/delivery processes are addressed early in design by all appropriate company units, suppliers, and partners to ensure integration, coordination, and capability.

(b) how product, service, and production/delivery process designs are reviewed and/or tested in detail to ensure trouble-free launch.

(c) how designs and design processes are evaluated and improved so that introductions of new or modified products and services progressively improve in quality and cycle time.

Notes:

1. Design and introduction might address:

   • modifications and variants of existing products and services;

   • new products and services emerging from research and development or other product/service concept development;

   • new/modified facilities to meet operational performance and/or product and service requirements, and

   • significant redesigns of processes to improve customer focus, productivity, or both.

Item 5.1: *Design and Introduction of Products and Services*

The design process is examined in Item 5.1, Design and Introduction of Products and Services. This process should encompass the translation of customer requirements into product and service requirements and then the translation of these into production and delivery processes. The coordination and integration of the designs, including addressing all requirements early in the process, and review of the designs are also part of this Item. The Item closes by asking how the designs and design processes are evaluated and improved in quality and cycle time.

The renewed emphasis on statistical thinking in the 1970s and early 1980s was focused on manufacturing. Being able to hold the product in our hand made the application of statistical thinking more tangible. The significant role of the design stage and its impact on quality were not always acknowledged. However, companies that are doing well considered the impact of all elements of the total system, including areas such as marketing and engineering, on the quality initiative. For example, in the service industry, a distributor concerned with on-time delivery looked at the entire process, starting with the sales order being completed correctly, rather than just being concerned with the warehouse.

An article in *Business Week*[25] showed that although less than 10% of the cost of a project is incurred in the concept and engineering phase, those two phases impact 90% of the total cost spent. Good manufacturing practices are not going to yield products or services that delight the customer if the marketing analysis or design is flawed.

Control plan development and key characteristic selection should occur in the design phase. In the design stage, control plans address "what should the process be doing" and "what can go wrong" using such tools as process flowcharts, FMEA, and characteristic matrices. Ideas to simplify the process and reduce response time should be implemented early in the development/implementation cycle—before the product or service is introduced.

For some engineers, key characteristics are a sensitive subject. Responses such as "all characteristics are important" reflect the old thinking that key characteristics are those characteristics that must be held within specification or design tolerances. Contemporary thinking maintains that all characteristics must be within

[25] "A Smart Way to Manufacture," *Business Week*, April 1990.

Design approaches could differ appreciably depending upon the nature of the products/services—entirely new, variants, major or minor process changes, etc. If many design projects are carried out in parallel, responses to Item 5.1 should reflect how coordination of resources among projects is carried out.

2. Applicants' responses should reflect the key requirements for their products and services. Factors that might need to be considered in design include: health; safety; long-term performance; environment; measurement capability; process capability; manufacturability; maintainability; supplier capability; and documentation.

3. Service and manufacturing businesses should interpret product and service design requirements to include all product- and service-related requirements at all stages of production, delivery, and use.

4. A measurement plan [5.1a(2)] should spell out what is to be measured, how and when measurements are to be made, and performance levels or standards to ensure that the results of measurements provide information to guide, monitor, control, or improve the process. This may include service standards used in customer-contact processes. The term "measurement plan" may also include decisions about key information to collect from customers and/or employees from service encounters, transactions, etc. The actual measurement plan should not be described in Item 5.1. Such information is requested in Item 5.2.

5. "All appropriate company units" means those units and/or individuals who will take part in production/delivery and whose performance materially affects overall process outcome.

Item 5.2: Criteria

5.2 Process Management: Product and Service Production and Delivery (40 pts.)

Describe how the company's key product and service production/delivery processes are managed to ensure that design requirements are met and that both quality and operational performance are continuously improved.

☑ Approach
☑ Deployment
☐ Results

Areas to Address

(a) how the company maintains the performance of key production/delivery processes to ensure that such processes meet design requirements addressed in Item 5.1. Describe: (1) the key processes and their principal requirements; and (2) the measurement plan and how measurements and/or observations are used to maintain process performance.

Figure 4.7
Loss Functions

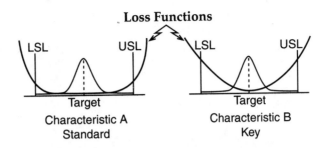

Loss Functions

specification but in some cases it is worth the effort to be on target and to reduce the variation even further. If the internal or external customer is sensitive to the variation within the specification, then it should be flagged as a key characteristic so that effort can be focused on reducing the variation. One might argue that the specification should just be changed to the new desired level. The specifications are usually negotiated based on the customer requirements and process capabilities, so just changing the specification does not result in an improvement because it does not change the customer's sensitivity to the variation. Since the customer requires that all product requirements are to be within the specification limits, this tightening of specification may increase the cost and decrease the value of the product to the customer. This sensitivity to variation can be quantified with the loss function or sensitivity curve (Figure 4.7). Those characteristics with a "steep" curve should be flagged as key characteristics.

Examples for Item 5.1

Translating the voice of the customer into product and process require- ments for new or modified products and services.

Cadillac Motor Car Company has a structured approach for their design process, called the *Four Phase Process*, with activities defined and exit criteria specified for each phase. In the Four Phase Process, simultaneous engineering cross-functional teams conceive, develop, and continually improve the products. Cross-functional business management teams are also in place at **AT&T Transmission.** The teams have responsibility for the entire product life cycle.

The concepts are similar at **TI Defense & Electronics**, which calls their approach *concurrent engineering*. The phases are concept, demonstration, engi- neering and manufacturing development, production, support, and maintenance.

A rule of thumb at **Federal Express** is the 1–10–100 rule. This indicates the cost to correct a problem when it occurs, to correct it downstream, or correct it in the field. Prevention is the best medicine but if a problem occurs, fix it early.

(b) how processes are evaluated and improved to achieve better operational performance, including cycle time. Describe how each of the following is used or considered: (1) process analysis and research; (2) benchmarking; (3) use of alternative technology; and (4) information from customers of the processes—within and outside the company.

Notes:

1. Key production/delivery processes are those most directly involved in fulfilling the principal requirements of customers—those that define the products and services.

2. Measurement plan [5.2a(2)] is defined in Item 5.1, Note (4). Companies with specialized measurement requirements should describe how they ensure measurement effectiveness. For physical, chemical, and engineering measurements, describe briefly how measurements are made traceable to national standards.

3. The focus of 5.2a is on maintenance of process performance using measurements and/or observations to decide whether or not corrective action is needed. The nature of the corrective action depends on the process characteristics and the type of variation observed. Responses should reflect the type of process and the type of variation observed. A description should be given of how basic (root) causes of variation are determined and how corrections are made at the earliest point(s) in processes. Such correction should then minimize the likelihood of recurrence of this type of variation anywhere in the company.

4. Process improvement methods (5.2b) might utilize financial data to evaluate alternatives and to set priorities.

5. Process analysis and research [5.2b(1)] refers to a wide range of possible approaches to improving processes. Examples include process mapping, optimization experiments, basic and applied research, error-proofing, and reviewing critical encounters between employees and customers from the point of view of customers and employees.

6. Information from customers [5.2b(4)] might include information developed as described in Items 7.2, 7.3, and 2.3.

7. Results of improvements in product and service delivery processes should be reported in Items 6.1 and 6.2, as appropriate.

Item 5.2: *Process Management: Product and Service Production and Delivery*

In Item 5.2, the focus moves from the design phase to the production and delivery phase and how these processes are managed to assure that the quality and operational requirements are met and continually improved. It is not enough to have products and services that meet customer requirements on paper—they must consistently satisfy the customer in the field.

In an internal study,[26] Ford Motor Company compared transmissions manufactured in the United States at their Batavia plant to the same design produced in Japan. The same blueprints were followed and the transmissions were being used in the same carline. The offshore transmissions were experiencing only one-third the warranty costs of the domestic transmissions. The difference was not conformance, because all of the Batavia transmissions were built to specification. The individual characteristics on the Japanese units were simply closer to the design target and generally had less variability. The Batavia experience, as it is sometimes called, seems to have been the turning point for the automotive industry; it was the start of the transition from a tolerance philosophy to a target philosophy—a transition from a philosophy that all parts within specification are equally good to one where parts on target are preferred.

This philosophy requires that we understand and document our processes. Building on the development in the design phase, the process control plan should be completed by answering "what is the process doing?"using tools such as capability studies and measurement system analysis. Processes should also be studied to improve quality and reduce the time it takes to complete the product or service. Improvements are still needed in the processes so that the lead time can be reduced. A major issue in the competitive environment is the cycle time from concept to market.

The term *benchmark* is mentioned in numerous Items. In Category 5.0, it is used in the context of getting new ideas for process improvement of production, delivery, or support services. Some people mistakenly think that benchmarking is just getting a metric or number to use as their new goal. They fall into the trap of determining the metric for the "best of the best," say, in inventory turns, and then arbitrarily set their goal at that level. Instead, benchmarking should be used to get new ideas for processes. Copying what worked at one company is not the answer. See Item 2.2 of this chapter for more on the topic of benchmarking.

At one time, companies tracked quality by measuring the AQL, acceptable quality level. Then percent defective was used. Now parts (or opportunities) per

26 "Continuous Improvement—Batavia," Video, Ford Motor Company, Dearborn, MI, 1985.

million or parts per billion is tracked. So 99% converts to 10,000 parts per million; a whole different sense of magnitude. See Item 2.3 of this chapter for more on the topic of tracking progress.

Examples for Item 5.2

Process control and improvement in the production and delivery of a product or service.

AT&T Transmission uses Achieving Process Excellence (APEX) teams, which are breakthrough teams. They are charged with the job of rethinking the way a job is done. It may be the order entry process, the customer field support process, or the employee involvement activity. The International Standards Organization (ISO) 9000 series has been used by AT&T Transmission to guide the development and implementation of procedures. Root cause analysis is done using the Plan-Do-Study-Act cycle.

TI Defense & Electronics uses the six steps to six sigma to link the process and product. The six steps include the selection, analysis, and quantification of key characteristics. The scorecard they use is based on tracking a six sigma level of quality on parts, process, performance, and software.

Granite Rock, a supplier of road construction material, has developed an innovative idea called GraniteXpress. It is based on the same concept as the banking industry's automatic teller machine (ATM) system. A truck driver can go to Granite Rock, 24 hours, seven days a week, and use a card to order material. The driver lines the truck up at a designated chute, indicates the readiness to receive the load, and the truck is filled. There is no operator; everything is automatic. This has greatly reduced the time that the driver is on site and allowed drivers to obtain material when it is convenient for them.

When the **Cadillac** Grand Blanc stamping plant improved their die transition time from 8 to 12 hours to 4 minutes, it was not accomplished by fine-tuning the system, but by rethinking the entire activity. The same is true for **Motorola**'s "Bandit" pager line where production of a pager begins within 17 minutes of the customer placing the order and that pager is on the shipping dock within 44 minutes of the order.[27]

AT&T Communications has seventy-five million transactions with customers each day. This tests the company's goal of perfect customer connection each time. Rapid changes in technology have reshaped the telecommunications field. More than $1 billion is invested each year in technology. This investment in technology has increased the capabilities and reliability of its worldwide telecommunications network.

[27] Presented at Quality Forum IX by R. Soderstrom, October 1993.

Item 5.3: Criteria

5.3 Process Management: Support Services (30 pts.)

Describe how the company's key support service processes are designed and managed so that current requirements are met and that operational performance is continuously improved.

☑ Approach
☑ Deployment
☐ Results

Areas to Address

(a) how key support service processes are designed. Include: (1) how key requirements are determined or set; (2) how these requirements are translated into efficient and effective processes, including operational requirements and an appropriate measurement plan; and (3) how all requirements are addressed early in design by all appropriate company units to ensure integration, coordination, and capability.

(b) how the company maintains the performance of key support service processes to ensure that such processes meet design requirements. Describe: (1) the key processes and their principal requirements; and (2) the measurement plan and how measurements are used to maintain process performance.

(c) how processes are evaluated and improved to achieve better operational performance, including cycle time. Describe how each of the following is used or considered: (1) process analysis and research; (2) benchmarking; (3) use of alternative technology; and (4) information from customers of the processes—within and outside the company.

Notes:

1. Support services are those which support the company's product and/or service delivery but which are not usually designed in detail with the products and services themselves because their requirements do not usually depend a great deal upon product and service characteristics. Support service design requirements usually depend significantly upon internal requirements. Support services might address finance and accounting, software services, sales, marketing, public relations, information services, supplies, personnel, legal services, plant and facilities management, research and development, and secretarial and other administrative services.

Item 5.3: *Process Management: Support Services*

Items 5.2 and 5.3 together encompass all the activities involved with the company's internal process management. Item 5.3, Process Management: Support Services, deals with all the areas of the company not addressed in Item 5.2. This includes areas such as finance and accounting, software services, sales, marketing, information services, purchasing, personnel, legal services, plant and facilities management, basic research and development, and secretarial and other administrative services. At companies where the quality initiative is mature, these areas are delighting their customers, simplifying their processes, and improving the accuracy of their output. Their approach could have been as simple as involving these areas in cross-functional teams and developing control and improvement activities for the support services. Or they may have applied the entire Baldrige Criteria to their area. For example, the legal department of IBM could view themselves as an outside company under contract to IBM. Then they would apply all of the Items of the Baldrige Criteria to their department. This includes executive involvement in the quality activities, getting the voice of their customers, etc. The data they might track could include the time it takes to complete forms for copyright or patent and reading levels of contracts. Be cautious, though, not to play a "numbers game" by focusing on the measures and not the system.[28]

Examples for 5.3

Process control and improvement in support services, such as finance, sales mar-ketimg public relations, legal services, information services, and administration.

At **Federal Express** the accounting department tracks the time it takes to pay C.O.D. (collect on delivery) money collected, accuracy of transactions, and time to close the books. Billing and statements, authorizations, correspondences, payments, and disputes are tracked by **AT&T Card** to measure their improvement in the support areas.

One of the U.S. military units found a common employee complaint to be the time it took to be reimbursed for business travel. This certainly was not going to make or break the military, but it was a sore point with the employees so it became an area that was improved and now the cycle time to pay an employee expense report has been reduced. This is an example of an improvement in a support area.

Die transition time and the number of engineering changes made are tracked at a **Cadillac** stamping plant. **IBM Rochester** includes inventory levels and warehouse space requirements as a measure in the support areas.

---

[28] Dr. Deming has frequently stated that what is most important to our system may be unknown and unknowable.

2. The purpose of Item 5.3 is to permit applicants to highlight separately the improvement activities for processes that support the product and service design, production, and delivery processes addressed in Items 5.1 and 5.2. The support service processes included in Item 5.3 depend on the applicant's type of business and other factors. Thus, this selection should be made by the applicant. Together, Items 5.1, 5.2, 5.3, and 5.4 should cover all operations, processes, and activities of all work units.

3. Process improvement methods (5.3c) might utilize financial data to evaluate alternatives and to set priorities.

4. Process analysis and research [5.3c(1)] refers to a wide range of possible approaches to improving processes. See Item 5.2, Note (5).

5. Information from customers [5.3c(4)] might include information developed as described in Items 7.2, 7.3, and 2.3. However, most of the information for improvement [5.3c(4)] is likely to come from "internal customers"—those within the company who use the support services.

6. Results of improvements in support services should be reported in 6.2.

Item 5.4: Criteria

5.4 Management of Supplier Performance (30 pts.)

Describe how the company assures that materials, components, and services furnished by other businesses meet the company's performance requirements. Describe also the company's actions and plans to improve supplier relationships and performance.

☑ Approach
☑ Deployment
☐ Results

Areas to Address

(a) summary of the company's requirements and how they are communicated to suppliers. Include: (1) a brief summary of the principal requirements for key suppliers, the measures and/or indicators associated with these requirements, and the expected performance levels; (2) how the company determines whether or not its requirements are met by suppliers; and (3) how performance information is fed back to suppliers.

Item 5.4: *Management of Supplier Performance*

Item 5.4 examines management of supplier performance. Taken together, Items 5.1, 5.2, 5.3, and 5.4 should cover all operations and activities of a company or business unit. At many companies, suppliers are now part of the team rather than being treated in the adversarial manner of the past. Their ideas and input are valued in the concept and design stage, not just at the manufacturing stage. Partnerships are demonstrating the win–win concept. In 1986, General Motors published the *Targets for Excellence* manual detailing the quality requirements for their suppliers. If this document is viewed as a management tool, and not simply as a checklist for assessment, many people say it is excellent and provides direction for a company to implement continual improvement. Those supporting this view include GM suppliers. Five areas of business are covered: leadership, quality, cost, delivery, and technology. The requirements are detailed with indicators for clarity. The assessment, augmented with other procedures, is intended to lead to certification, which allows material to go from dock to line without any receiving inspection.

Although receiving inspection is one of the methods a company can use to assure the quality of the supplier, it does not receive high value because it is not prevention based and, in fact, not value added.

Improvement of the company's procurement process is included in this Item. If the applicant is part of a larger company and other units of the company supply goods/services, this would be addressed here.

Examples for Item 5.4

Communication of requirements—quality, delivery, price—and assurance of supplier quality through audits, certifications, or reviews.

AT&T Transmission uses the quality systems guidelines ANSI/ASQC Q94-1987[29] for supplier audits. They hold an annual supplier excellence day. **TI Defense & Electronics** tracks product quality in ppm, on-time delivery, and cost to quantify the progress of their suppliers.

The term *supplier* needs to be defined by a company. In a narrow sense it could be companies supplying a key product or key service. At **AT&T Card** they take a broader view to include suppliers that provide telemarketing, credit card production, product and service literature, maintenance, and availability of systems.

---

[29] This is the U.S. publication of ISO 9004. Note that this is not one of the three standards (ISO 9001-3) that is normally referred to when the ISO 9000 series is used in conjunction with supplier management. See also Chapter 13.

(b) how the company evaluates and improves its management of supplier relationships and performance. Describe current actions and plans: (1) to improve suppliers' abilities to meet requirements; (2) to improve the company's own procurement processes, including feedback sought from suppliers and from other units within the company ("internal customers") and how such feedback is used; and (3) to minimize costs associated with inspection, test, audit, or other approaches used to verify supplier performance.

Notes:

1. The term "supplier" refers to other-company providers of goods and services. The use of these goods and services may occur at any stage in the production, design, delivery, and use of the company's products and services. Thus, suppliers include businesses such as distributors, dealers, warranty repair services, transportation, contractors, and franchises as well as those that provide materials and components. If the applicant is a unit of a larger company, and other units of that company supply goods/services, this should be included as part of Item 5.4.

2. Key suppliers [5.4a(1)] are those which provide the most important products and/or services, taking into account the criticality and volume of products and/or services involved.

3. "Requirements" refers to the principal factors involved in the purchases: quality, delivery, and price.

4. How requirements are communicated and how performance information is fed back might entail ongoing working relationships or partnerships with key suppliers. Such relationships and/or partnerships should be briefly described in responses.

5. Determining how requirements are met [5.4a(2)] might include audits, process reviews, receiving inspection, certification, testing, and rating systems.

6. Actions and plans (5.4b) might include one or more of the following: joint planning, rapid information and data exchanges, use of benchmarking and comparative information, customer–supplier teams, partnerships, training, long-term agreements, incentives, and recognition. They might also include changes in supplier selection, leading to a reduction in the number of suppliers.

7. Efforts to minimize costs might be backed by analyses comparing suppliers based on overall cost, taking into account quality and delivery. Analyses might also address transaction costs associated with alternative approaches to supply management.

A three-stage selection process is used by **IBM Rochester**. They have eliminated receiving inspection on most parts. In theory it might sound like a good idea to eliminate receiving inspection to send a message but unless the quality of the suppliers is high, the problem/disaster will only be moved to a later point in the process. Work close enough with your suppliers to know their capabilities before making such a move.

Supplier evaluation and certification is used by **GTE** to assure high performance. Suppliers are rated on delivery, customer service, documentation, and product/service value.

Key Excellence Indicators

Some of the common threads seen at companies doing well in this Category are quality of design and a focus on response time. There is a linkage between design and process to consider the manufacturing capabilities at the time the design is conceived. The philosophy of continual improvement and prevention is being applied throughout the company, including the support areas such as accounting, purchasing, and information analysis. The attitude is "if it ain't broke, improve it." Leaders in this area have a cooperative relationship with their suppliers—always striving for a win–win relationship.

Excellent companies place strong emphasis on design quality. The concepts of design quality, prevention, and continual improvement emphasize upstream intervention—at early stages of the process. Such upstream intervention needs to consider the company's suppliers. This yields maximum benefits for all the participants.

CATEGORY 6.0: BUSINESS RESULTS

Description and Examples

> "The *Business Results* Category examines the company's performance and improvement in key business areas—product and service quality, productivity and operational effectiveness, supply quality, and financial performance indicators linked to these areas. Also examined are performance levels relative to competitors."[30]

Category 1.0, Leadership, is considered to be the driver in the Baldrige framework. Categories 2.0 through 5.0 are the systems. Category 6.0 captures product and service quality results. The response to this Category will be a summary analysis of the quality and operational data described in Category 2.0 and will reflect all key product and service features discussed in the business overview as well as Category 5.0 and Items 3.2 and 7.1. Also, the measures used to track progress were discussed in Category 5.0 in connection with process control and process improvement activities but the results will not be reported until this Category. These measures and indicators are to be linked to product and process improvements that are predictors of customer satisfaction, and are to cover a wide spectrum of the company. Trends, displayed in graphs or tables, showing three to five years of progress, are expected.

This Category should only contain results. The text should describe the tables and graphics, link the results with other discussions, and/or contain results. If other sections of the written response for self-assessment do not adequately describe the activity producing the results, then improve that section; do not "fix" the discrepancy by adding the description to this Category.

These measures of quality for key products and services must show positive trends. They must also document and substantiate a claim to industry leadership. Inventory levels, energy usage, first-time success rates, lead times, rework rates, and other quality-related business factors must show improvement (Figure 4.8). Sometimes organizations mistakenly report customer satisfaction, customer loyalty, market share, and warranty data in this Category. That type of data, customer reaction, belongs in Category 7.0.

---

[30] *1995 Malcolm Baldrige National Quality Award Criteria,* National Institute of Standards and Technology, Gaithersburg, MD.

Figure 4.8
Business Results Category

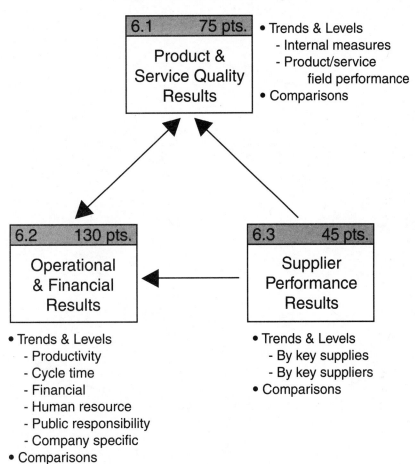

6.0 Business Results

6.1   75 pts.
Product & Service Quality Results

- Trends & Levels
  - Internal measures
  - Product/service
       field performance
- Comparisons

6.2   130 pts.
Operational & Financial Results

- Trends & Levels
  - Productivity
  - Cycle time
  - Financial
  - Human resource
  - Public responsibility
  - Company specific
- Comparisons

6.3   45 pts.
Supplier Performance Results

- Trends & Levels
  - By key supplies
  - By key suppliers
- Comparisons

Maximum points - 250

Item 6.1: Criteria

6.1 Product and Service Quality Results (75 pts.)

Summarize results of improvement efforts using key measures and/or indicators of product and service quality.

☑ Approach
☑ Deployment
☐ Results

Areas to Address

(a) current levels and trends in key measures and/or indicators of product and service quality. Graphs and tables should include appropriate comparative data.

Notes:

1. Results reported in 6.1 should reflect performance relative to specific non-price product and service key quality requirements. Such key quality requirements should relate closely to customer satisfaction and customer retention. These requirements are those described in the Business Overview and addressed in Items 7.1 and 5.1.

2. Data appropriate for inclusion are based upon:

    • internal (company) measurements;

    • field performance;

    • data collected by the company through follow-ups (7.2c) or surveys of customers;

    • data collected or generated by other organizations; and

    • data collected by other organizations on behalf of the company.

3. Product and service quality measures and/or indicators may address requirements such as accuracy, timeliness, reliability, and behavior. Examples include defect levels, repeat services, meeting product or service delivery or response times, availability levels, and complaint levels.

4. Comparative data might include industry best, best competitor, industry average, and appropriate benchmarks. Such data might be derived from independent surveys, studies, laboratory testing, or other sources.

Item 6.1: *Product and Service Quality Results*

Levels and trends for key products and service features are requested in Item 6.1, Product and Service Quality Results. The data, typically three to five years' worth, might be measures of accuracy, reliability, timeliness, performance, delivery, after-sales service, documentation, and/or appearance of key product or service. Data appropriate for Item 6.1 include internal measures made by the company as well as data generated or collected by other organizations. The results reported should reflect all key product and service features described by the organization in the business overview and Items 2.1, 3.2, 5.1, 5.2, and 7.1. Information on trends pertaining to customer complaint response time, effective resolution, and percent of complaints resolved on first contact would be reported here.

Comparisons to industry averages, industry leaders, or benchmarks are required. A company may be making steady progress over the years, yet the current level of performance may be only approaching industry average. This is a completely different story than a company making steady progress and continuing to be the leader in the industry—always setting new challenges for the competition.

Three to five years' worth of data is expected to demonstrate that the organization has had sustained levels of excellent results.

Examples for Item 6.1

Demonstrated leadership position in improving accuracy, reliability, timeliness, and performance of key product or service features.

AT&T Transmission, in the six years prior to receiving the Baldrige Award, achieved a tenfold improvement in equipment product quality and a 50% reduction in cycle time for the last four years. Collectively their efforts have resulted in $400 million cost improvement savings. Their yields are considered to be at best-in-class levels. They resolved 98% of customer technical issues on first contact, which has a strong impact on customer satisfaction.

The application processing time at **AT&T Card** is three days compared to ten, which represents their best competitor.

TI Defense & Electronics is number one for on-time delivery out of 88 defense contractors. They ship 30,000 items per month. For reliability, delivery, and cost there has been an ongoing improvement trend with current levels meeting or exceeding the customer requirement. For example, consider the HARM missile. Its reliability is five times the requirement; has a record of 100% on-time delivery since 1982; and the cost was reduced by one-half in six years.

There are 100 industry-specific measures at **Ritz-Carlton** and they were 10% better than any competition on all measures. They have been named as the benchmark for the industry by an outside firm.

A small business recipient, **Granite Rock**, excels in areas important to their customer. The shrinkage rate of their product, road construction material, is one-half the rate of competitors. They achieved a threefold improvement in batching accuracy in cement content from 1990 to 1992 and reduced the days to resolve problems from 85 in 1990 to 23 in 1992.

In 1993 **AT&T Communications** was recognized for the fourth consecutive year by *Data Communications* magazine with its Best Overall and Best Technology awards for a long-distance carrier. Responding to a cable cut that eliminated service to a business district for nearly half of a day in 1991, the company overhauled its disaster response system. They now have capability to restore service within 10 minutes of a major outage.

The error ratio for **GTE**'s directories published in 1993 is approximately 350 per 1 million listings. This places them at a best-in-class level according to industry studies.

University of Michigan Hospitals, which was one of the first recipients of the Michigan Award based on the Baldrige, reported in *An Agenda for Change*,[31] the following improvements:

- Decreased average admission wait for scheduled patients from 120 minutes to 18 minutes

- Reduced placement days from 2.46 days per patient to 1.26 days, resulting in a savings of nearly $500,000

- Increased throughput in adult operating room of 33% and 10% in pediatrics

- Increased prescribing of eight-hour medications rather than six-hour ones, resulting in a savings of $35,400.

These and other improvements have led to a 720% return on investment of dollars spent on TQM from 1987 to 1991. The return for 1991–1992 was 178%.

---

[31] A document published by University of Michigan Hospitals. For this and other related sources of information, call them at 313/763-9640.

Item 6.2: Criteria

6.2 Company Operational and Financial Results (130 pts.)

Summarize results of improvement efforts using key measures and/or indicators of company operational and financial performance.

☐ Approach
☐ Deployment
☑ Results

Areas to Address

(a) current levels and trends in key measures and/or indicators of company operational and financial performance. Graphs and tables should include appropriate comparative data.

Notes:

1. Key measures and/or indicators of company operational and financial performance should address the following areas:

   • productivity and other indicators of effective use of manpower, materials, energy, capital, and assets;

   • cycle time and responsiveness;

   • financial indicators such as cost reductions, asset utilization, and benefit/cost results from improvement efforts;

   • human resource indicators such as safety, absenteeism, turnover, and satisfaction;

   • public responsibilities such as environmental improvements; and

   • company-specific indicators such as innovation rates and progress in shifting markets or segments.

2. The results reported in Item 6.2 derive primarily from activities described in Items 5.1, 5.2, and 5.3.

3. Comparative data might include industry best, best competitor, industry average, and appropriate benchmarks. For human resource areas such as turnover or absenteeism, local or regional comparative information might also be appropriate.

Item 6.2: *Company Operational and Financial Results*

Trends and current levels for operational results are requested in Item 6.2. These results could include productivity measures such as labor hours, material usage, energy consumption, and cycle time reduction. Trends in financial indicators could be reported here *if* there is a clear connection to the quality efforts. Human resource indicators, described in Item 4.4, such as safety, absenteeism, turnover, and employee satisfaction, would be included. Data representing public responsibility such as safety data pertaining to transporting goods, recycling efforts, and/or reduction or elimination of pollution of local air, water, or land described in Item 1.3 are reported here. Report data from support areas, as described in Item 5.3, including finance and accounting, software services, sales, marketing, information services, purchasing, personnel, legal services, plant and facilities management, basic research and development, and secretarial and other administrative services. Measures tracked might reflect accuracy and timeliness of the activities. Cycle time for design or manufacturing as described in Items 5.1 and 5.2 would be reported here.

Some comparisons to industry average, industry leaders, or benchmarks are needed in order for the organization to demonstrate they are one of the leaders. Because these types of processes have universal application, typical comparisons are to the best practice performance regardless of industry. For employee turnover and absenteeism, local or regional comparisons might be appropriate. A comparison is not needed for every measure and the comparison should not be obtained for the sake of meeting the Baldrige requirement. The purpose of having a comparison is to show that the organization knows its position compared to others.

Examples for Item 6.2

Quantitative success in reducing labor hours, material and energy consumption, financial waste. Measures of progress in improving accuracy and reducing cycle time in support areas. Financial measures linked to quality improvement. Human resource and public responsibility indicators.

The contract services at **TI Defense & Electronics** handle 7800 documents and have a 4000 ppm acceptance rate. They have reduced the cycle time on security clearance from 50 to 20 days. In 1991, they received the energy award from the Association of Energy Engineers and they have eliminated solid hazardous waste from their facilities.

Item 6.3: Criteria

6.3 Supplier Performance Results (45 pts.)

Summarize results of supplier performance improvement efforts using key measures and/or indicators of such performance.

☐ Approach
☐ Deployment
☑ Results

Areas to Address

(a) current levels and trends in key measures and/or indicators of supplier performance. Graphs and tables should include appropriate comparative data.

Notes:

1. The results reported in Item 6.3 derive from activities described in Item 5.4. Results should be broken out by key supplies and/or key suppliers, as appropriate. Data should be presented using the measures and/or indicators described in 5.4a(1).

2. If the company's supplier management efforts include factors such as building supplier partnerships or reducing the number of suppliers, data related to these efforts should be included in responses.

3. Comparative data might be of several types: industry best, best competitor(s), industry average, and appropriate benchmarks.

Ritz-Carlton depends on the accuracy and availability of their information system that has records of the preferences of 120,000 customers. They have continuously reduced the labor hours per occupied room for several years running.

The collection rate before 90 days is 98% or higher for **Granite Rock**, compared to the regional average of 86%. The reduction in gate-to-gate time, that is, the time spent at the facility by a truck driver picking up material, was reduced from 25 minutes to 9 minutes.

GTE has sustained increasing revenue growth while many competitors have had flat or declining revenues.

From 1990 to 1993, accidents decreased by 72%, lost time due to accidents declined by 85%, and lost work days is down by 87% at **Wainwright**. The cycle time to draw up blueprints has gone from 8.75 days to 15 minutes.

Item 6.3: *Supplier Performance Results*

The trend data and current levels for Item 6.3, Supplier Performance Results, might include the number of suppliers, number of certified suppliers, quality levels of products/services provided by suppliers, and on-time delivery. The results should be broken down by major groupings of supplies and suppliers, as appropriate, to demonstrate deployment across many types of supplies and suppliers.

Comparisons are expected in this Item as in all of the Category 6.0 Items. The sources of these comparisons should be explained in Item 2.2 and used throughout Category 6.0.

Examples for Item 6.3

Indicators of success in supplier quality, broken down by major grouping of supplies and/or suppliers.

On-time delivery increased from 60% in 1989 to 100% in 1991 for startup at the **Ritz-Carlton**. The number of days to ship five critical supplies was reduced from 20 days to 1 day.

The circuit suppliers for **AT&T Transmission** demonstrated a twenty-fold improvement in quality over a seven-year period. In a four-year period, a 55% improvement in product quality for external supplies was recognized. Integrated circuit suppliers are now performing at quality levels of 22 ppm.

TI Defense & Electronics has experienced a 46% improvement by their suppliers in on-time delivery since 1988. They reduced the number of suppliers

significantly. It is difficult for a company to have a close working relation with their suppliers if there is a large number of suppliers. A reduction in the number of suppliers allows better communications and less confusion and paperwork. Additionally, this also reduces the sources of variation.

Key Excellence Indicators

Category 6.0, Business Results, contains the quantification of the results of activities described as responses to other Categories. In companies doing well, any data reported in Category 6.0 are linked to a process control or improvement activity described elsewhere (see Figure 2.2, the relationship matrix, in Chapter 2). Leading companies have lower turnover, fewer accidents, and lower absenteeism. The data reported reflect a broad base of improvement trends for products, services, and internal operations. The results show performance excellence when compared to the performance of leaders—inside or outside the industry. Ongoing improvements in supplier quality are shown at the same pace as improvement within the company.

Some tips on how to best display the data graphically include the following:

- The graph should display the results identified as key to the business.

- Both axes need to be clearly and quantitatively labeled.

- Results should be presented for several time periods—generally three to five years.

- Improvements have been quantified; i.e., they are not shown simply as goals.

- Comparisons to industry average, best competitor, and "world class" are clearly shown (see Figure 4.9).

Figure 4.9
Sample Graph

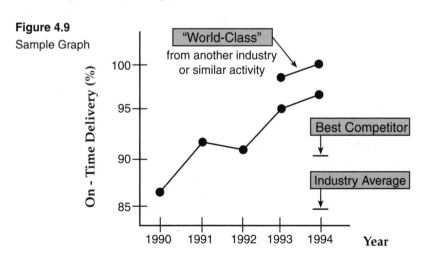

CATEGORY 7.0: CUSTOMER FOCUS AND SATISFACTION

 Description and Examples

> "The *Customer Focus and Satisfaction* category examines the company's systems for customer learning and for building and maintaining customer relationships. Also examined are levels and trends in key measures of business success—customer satisfaction and retention, market share, and satisfaction relative to competitors."[32]

The last Category is Customer Focus and Satisfaction. This Category is the overall goal in the Baldrige framework (see Figure 2.1). Quality is judged by the customer. All product and service features that are perceived by the customer to provide value must be the focus of the company's management system. Consequently, this Category considers the effectiveness of the company's systems to determine customer requirements and the demonstrated success in meeting them.

The organization needs to describe the relationship with the customer and demonstrate its knowledge of the customer. Customers' perceptions of value, preference and satisfaction may be influenced by many factors throughout the overall purchase, ownership, and service experience. Included is the company's relationship with the customer that helps to build trust, confidence, and loyalty. The various activities to determine customer satisfaction and the data related to customer satisfaction should be described. Customer satisfaction, customer loyalty, and market share should be increased through proper identification and fulfillment of customer requirements.

The essence of this Category is to go beyond customer satisfaction to customer delight. A customer may say they are satisfied but may also admit that they are willing to switch suppliers.[33] This seems contradictory at first. Upon further investigation, it is only those customers that say they are delighted or very satisfied that are the loyal customers.

A study done by the Technical Assistance Research Program[34] reported that only 5% of the dissatisfied customers complain to top management, 45% complain to an agent or frontline representative such as the hotel desk clerk, and 50% encounter a problem but never complain. Oftentimes a customer who encounters a problem will go elsewhere for the product or service in the future.

32 *1995 Malcolm Baldrige National Quality Award Criteria,* National Institute of Standards and Technology, Gaithersburg, MD.

33 "Putting the Service-Profit Chain to Work," *Harvard Business Review,* March–April 1994.

34 Presented at Quest for Excellence IV Conference, Washington, DC, February 1992.

"A customer is the most important visitor on our premises. He is not dependent on us; we are dependent on him. He is not an interruption on our work; he is the purpose of it. He is not an outsider of our business; he is part of it. We are not doing him a favor by serving him; he is doing us a favor by giving us an opportunity to do so."[35]

Figure 4.10
Customer Focus and Satisfaction Category

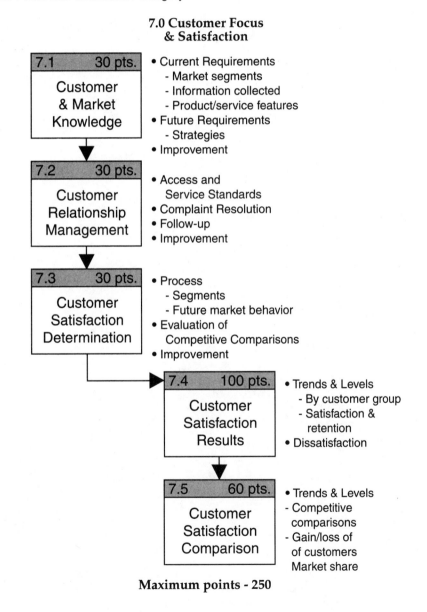

7.0 Customer Focus & Satisfaction

7.1   30 pts.
Customer & Market Knowledge

- Current Requirements
  - Market segments
  - Information collected
  - Product/service features
- Future Requirements
  - Strategies
- Improvement

7.2   30 pts.
Customer Relationship Management

- Access and Service Standards
- Complaint Resolution
- Follow-up
- Improvement

7.3   30 pts.
Customer Satisfaction Determination

- Process
  - Segments
  - Future market behavior
- Evaluation of Competitive Comparisons
- Improvement

7.4   100 pts.
Customer Satisfaction Results

- Trends & Levels
  - By customer group
  - Satisfaction & retention
- Dissatisfaction

7.5   60 pts.
Customer Satisfaction Comparison

- Trends & Levels
- Competitive comparisons
- Gain/loss of of customers Market share

Maximum points - 250

[35] Mahatma Gandhi, reported in "On Q," by the American Society for Quality Control, November 1991.

Item 7.1: Criteria

7.1 Customer and Market Knowledge (30 pts.)

Describe how the company determines near-term and longer-term requirements and expectations of customers and markets, and develops listening and learning strategies to understand and anticipate needs.

☑ Approach
☑ Deployment
☐ Results

Areas to Address

(a) how the company determines current and near-term requirements and expectations of customers. Include: (1) how customer groups and/or market segments are determined and/or selected, including how customers of competitors and other potential customers are considered; (2) how information is collected, including what information is sought, frequency and methods of collection, and how objectivity and validity are ensured; (3) how specific product and service features and the relative importance of these features to customer groups or segments are determined; and (4) how other key information and data such as complaints, gains and losses of customers, and product/service performance are used to support the determination.

(b) how the company addresses future requirements and expectations of customers. Include an outline of key listening and learning strategies used.

(c) how the company evaluates and improves its processes for determining customer requirements and expectations.

Notes:

1. The distinction between near-term and future depends upon many marketplace factors. The applicant's response should reflect these factors for its market(s).

2. The company's products and services might be sold to end users via other businesses such as retail stores or dealers. Thus, "customer groups" should take into account the requirements and expectations of both the end users and these other businesses.

3. Product and service features refer to all important characteristics and to the performance of products and services that customers experience or perceive throughout their overall purchase and ownership. The focus should be primarily on factors that bear upon customer preference and repurchase loyalty—for example, those features that enhance or differentiate products and services from competing offerings.

This Category contains the largest number of Items, five total (Figure 4.10). It also has a significant number of points, 250 points or 25% of the total points. This weighting reflects the level of importance the Criteria place on the core value of customer-driven quality. The weighting is also impacted by the prevalent need by individuals and organizations to "see it work" before they can fully believe. This Category addresses the concern that the excellent strategies described in earlier Categories may be paper systems—theoretically sound but unproven. The company must show that their strategies have yielded fully operational and effective systems.

Item 7.1: *Customer and Market Knowledge*

Customer and Market Knowledge looks at the company's process to determine near-term and future customer requirements. The time horizon depends on many marketplace factors and would be determined by the organization. The customer refers to both the end user and any intermediaries such as dealers or retailer stores. Potential customers, such as customers of competitors, should also be included.

Information on the current and emerging market segments is requested. This should include how the product or service features are selected and their importance determined. Continual process improvement is the focus of the last Area, which asks how the organization evaluates and improves the process for determining customer requirements.

A common problem observed at this point is the tendency for companies to state how they react to customer requests. For example, a customer wants a printed circuit board with certain properties. The organization may describe how they engineer the product. First of all, the engineering description belongs in Item 5.1 and secondly this shows a passive position by the company. Instead, there should be a sense that the company is creating the future, telling the customer about the art of the possible—the various possibilities that exist to meet its future requirements.

Examples for Item 7.1

Getting the voice of the customer for short- and longer-term requirements.

Customer data are obtained from many sources at the **Ritz-Carlton** including the travel industry, focus groups, customer preference information, feedback from current users, and from a guest and travel planner satisfaction system. All of the listening posts are used.

4. Some companies might use similar methods to determine customer requirements/expectations and customer satisfaction (Item 7.3). In such cases, cross-references should be included.

5. Customer groups and market segments (7.1a) might take into account opportunities to select or create groups and segments based upon customer- and market-related information.

6. Examples of listening and learning strategy elements (7.1b) are:

   - relationship strategies, including close integration with customers;
   - rapid innovation and field trials to better link R&D to the market;
   - close monitoring of technological, competitive, societal, environmental, economic, and demographic factors that may bear upon customer requirements, expectations, preferences, or alternatives;
   - focus groups with demanding or leading-edge customers;
   - training of frontline employees in customer listening;
   - use of critical incidents to understand key service attributes from the point of view of customers and frontline employees;
   - interviewing lost customers;
   - won/lost analysis relative to competitors;
   - post-transaction follow-up (see 7.2c); and
   - analysis of major factors affecting key customers.

7. Examples of evaluation and factors appropriate for 7.1c are:

   - the adequacy and timeliness of the customer-related information;
   - improvement of survey design;
   - the best approaches for getting reliable and timely information—surveys, focus groups, customer-contact personnel, etc.;
   - increasing and decreasing importance of product/service features among customer groups or segments; and
   - the most effective listening/learning strategies.

The evaluation might also be supported by company-level analysis addressed in Item 2.3.

TI Defense & Electronics personnel work side by side with Air Force maintenance personnel to learn firsthand how their product is doing in the field. At **AT&T Card** they have four key listening posts:

- customer expectations and needs research

- performance research

- direct customer feedback

- output from the process management activity.

Using this information, cross-functional teams, called Customer Listening Post Teams, aggregate and act on the combined information.

To get input from their customers' customer, **AT&T Transmission** has what they call a Far Side Initiative. They use focus group studies and customer visits to determine customer requirements. **Granite Rock** conducts an annual survey of product users including both customers and noncustomers.

Item 7.2: Criteria

7.2 Customer Relationship Management (30 pts.)

Describe how the company provides effective management of its responses and follow-ups with customers to preserve and build relationships and to increase knowledge about specific customers and about general customer expectations.

☑ Approach
☑ Deployment
☐ Results

Areas to Address

(a) how the company provides information and easy access to enable customers to seek information and assistance, to comment, and to complain. Describe contact management performance measures and service standards and how these requirements are set, deployed, and tracked.

(b) how the company ensures that formal and informal complaints and feedback received by all company units are resolved effectively and promptly. Briefly describe the complaint management process and how it ensures effective recovery of customer confidence, meeting customer requirements for resolution effectiveness, and elimination of the causes of complaints.

(c) how the company follows up with customers on products, services, and recent transactions to determine satisfaction, to resolve problems, to seek feedback for improvement, and to build relationships.

(d) how the company evaluates and improves its customer relationship management. Include: (1) how service standards, including those related to access and complaint management, are improved based upon customer information; (2) aggregation and use of customer comments and complaints throughout the company; and (3) how knowledge about customers is accumulated.

Notes:

1. Customer relationship management refers to a process, not to a company unit. However, some companies may have units which address all or most of the requirements included in this Item. Also, some of these requirements may be included among the responsibilities of frontline employees in processes described in Items 5.2 and 5.3.

2. Performance measures and service standards (7.2a) apply not only to employees providing the responses to customers but also to other units within the company, which make effective responses possible. Deployment needs to take

Item 7.2: *Customer Relationship Management*

The relation the organization has with their customers is examined in Item 7.2, Customer Relationship Management. The concept of quality goes beyond product and service features that meet the customer requirements. It includes characteristics that differentiate the company from their competitors. This could include new offerings, unique services, rapid response, or special relationships.

Does the customer have easy access to seek assistance, make a complaint, or provide a comment? This might be achieved with toll-free telephone lines that are staffed 12 to 16 hours a day to cover customers across various time zones. After the delivery of a product or service, the company should proactively follow up with the customer to determine their satisfaction with their most recent transaction. This is contrasted to the policy of waiting to see if the customer complains. Silence does not mean satisfaction. This type of active follow-up is different from the once-a-year survey to determine overall satisfaction. That would be described in Item 7.4. Instead, its purpose is to close the loop on specific, recent transactions.

The process to handle customer complaints, whether formal or informal, should be described by the applicant. Information on *trends* pertaining to complaint response time, effective resolution, and percentage of complaints resolved on first contact would be reported in Item 6.1.

How service standards are set and tracked should be discussed. Examples of service standards include "all complaints will be resolved within one working day," "the customer will not be on hold for more than 20 seconds," "requests for quotations or product information will be answered within 2 hours." Product or service warranties should *not* be described here; that information is requested in Item 7.3.

How the company evaluates and improves its customer relationship process is the final area. This should include how opportunities to improve the relationship are sought, and how this information leads to improvements in service standards, training, or access.

Examples for Item 7.2

Service standards, easy access, and follow-up with the customers.

The relationship with the customer is enhanced at **AT&T Card** by many features. The toll-free number to call with a question or complaint is printed on the charge card. A call can be handled in any 1 of 140 languages. The toll-free line is staffed 24 hours a day, 365 days a year. There is a free-for-life commitment for all charter members of the AT&T card. A lost or stolen card is replaced within 24 hours. When a cardholder has a disagreement with a merchant regarding a charge, AT&T will draft the letter to the merchant on behalf of the customer. The customer signs the letter and sends it to the merchant.

into account all key points in a response chain. Examples of measures and standards are: telephonic, percentage of resolutions achieved by frontline employees, number of transfers, and resolution response time.

3. Responses to 7.2b and 7.2c might include company processes for addressing customer complaints or comments based upon expressed or implied guarantees and warranties.

4. Elimination of the causes of complaints (7.2b) involves aggregation of complaint information from all sources for evaluation and use throughout the company. The complaint management process might include analysis and priority setting for improvement projects based upon potential cost impact of complaints, taking into account customer retention related to resolution effectiveness. Some of the analysis requirements of Item 7.2 relate to Item 2.3.

5. Improvement of customer relationship management (7.2d) might require training. Training for customer-contact (frontline) employees should address: (a) key knowledge and skills, including knowledge of products and services; (b) listening to customers; (c) soliciting comments from customers; (d) how to anticipate and handle problems or failures ("recovery"); (e) skills in customer retention; and (f) how to manage expectations. Such training should be described in Item 4.3.

6. Information on trends and levels in measures and/or indicators of complaint response time, effective resolution, and percent of complaints resolved on first contact should be reported in Item 6.1.

Item 7.3: Criteria

7.3 Customer Satisfaction Determination (30 pts.)

Describe how the company determines customer satisfaction, customer repurchase intentions, and customer satisfaction relative to competitors; describe how these determination processes are evaluated and improved.

☑ Approach
☑ Deployment
☐ Results

Areas to Address

(a) how the company determines customer satisfaction. Include: (1) a brief description of processes and measurement scales used; frequency of determination; and how objectivity and validity are ensured. Indicate significant differences, if any, in processes and measurement scales for different customer

There is 24-hour technical support available at **AT&T Transmission** and a risk-free trial of new products. **TI Defense & Electronics** has a service standard, which they call Customer Satisfaction Opportunity System (SOS), to respond to customer complaints in one day and resolve the issue in five days.

Service standards at **GTE** address professionalism, time lines, support for sales functions, and complaint resolution.

Granite Rock instituted a policy called short pay. The customer is advised, verbally and also in writing on the invoice, that if they are dissatisfied for any reason they should withhold the portion of the payment representing the product causing dissatisfaction. This is a win–win policy. The initial purpose of short pay was to collect the remaining portion of the money due for products with which the customer was happy, but it turned into a valuable source of customer information and satisfaction. The complaint regarding the unsatisfactory product is investigated and a mutually agreed-on solution is developed.

Federal Express has a money-back guarantee if a delivery is late. They know when they are late so a credit is made to your account automatically. Contrast this with the U.S. postal system, which also has a guarantee but the sender must be aware of the late arrival. Then the sender calls the post office where the letter was sent and files a request for investigation. The sender is told to call back in a week to get the results. With the second phone call, when the sender is told that they are due a refund, they are asked to come to the post office to pick up the refund. Clearly all guarantees are not equal.

groups or segments; and (2) how customer satisfaction measurements capture key information that reflects customers' likely future market behavior, such as repurchase intentions and/or positive referrals.

(b) how customer satisfaction relative to that for competitors is determined. Describe: (1) company-based comparative studies; and (2) comparative studies or evaluations made by independent organizations and/or customers. For (1) and (2), describe how objectivity and validity of studies or evaluations are ensured.

(c) how the company evaluates and improves its overall processes and measurement scales for determining customer satisfaction and customer satisfaction relative to that for competitors. Include how other indicators (such as gains and losses of customers) and customer dissatisfaction indicators (such as complaints) are used in this improvement process. Describe also how the evaluation takes into account the effectiveness of the use of customer satisfaction information and data throughout the company.

Notes:

1. Customer satisfaction measurement might include both a numerical rating scale and descriptors assigned to each unit in the scale. An effective (actionable) customer satisfaction measurement system is one that provides the company with reliable information about customer ratings of specific product and service features and the relationship between these ratings and the customer's likely future market behavior.

2. The company's products and services might be sold to end users via other businesses such as retail stores or dealers. Thus, "customer groups" or segments should take into account these other businesses as well as the end users.

3. Customer dissatisfaction indicators include complaints, claims, refunds, recalls, returns, repeat services, litigation, replacements, downgrades, repairs, warranty work, warranty costs, misshipments, and incomplete orders.

4. Company-based or independent organization comparative studies (7.3b) might take into account one or more indicators of customer dissatisfaction as well as satisfaction. The extent and types of such studies may depend upon factors such as industry and company size.

5. Evaluation (7.3c) might take into account how well the measurement scale relates to actual repurchase and/or customer retention. The evaluation might also address the effectiveness of pre-survey research used in design, and how actionable survey results are—how well survey responses link to key business processes and cost/revenue implications and thus provide a useful basis for improvement.

Item 7.3: *Customer Satisfaction Determination*

This Item, Customer Satisfaction Determination, requires a description of how customer satisfaction is determined. There could be cross references to Item 7.1 if similar methods are used to determine customer requirements. If there are different approaches or measurements for different market segments, the distinctions should be provided. In this context, customers refer to dealers, retail stores, and end users as applicable.

The means of determining competitive comparisons, whether by the organization, an outside agency, or the customer, should be described. In evaluating and improving the process to determine customer satisfaction, other indicators should be considered. This would include gains and losses of customers, and customer dissatisfaction indicators such as complaints and returns. The data for gains and losses of customers, and customer dissatisfaction would not be reported here; it belongs in Item 7.4. Information on how these data are correlated or linked to improvements in determining customer satisfaction is required here.

Examples for Item 7.3

Methods used to determine customer satisfaction of all customers: dealers, retailers, and end users.

An annual survey of customers is conducted by **Granite Rock**. In the survey, customers are asked about the performance and importance of various aspects of the interaction with Granite Rock. They also ask the customers how they compare to other suppliers. This can be done regardless of the industry. A customer could be asked, compared to your favorite supplier, "How do we perform regarding on-time delivery, responsiveness, ease of access, and quality of product?"

TI Defense & Electronics conducts an annual survey of 2000 customers. These surveys were conducted internally starting in 1986 and turned over to a third party in 1991.

6.   Use of data from satisfaction measurement is called for in 5.2b(4) and 5.3c(4). Such data also provide key input to analysis (Item 2.3).

Item 7.4: Criteria

7.4 Customer Satisfaction Results (100 pts.)

Summarize the company's customer satisfaction and customer dissatisfaction results using key measures and/or indicators of these results.

☐ Approach

☐ Deployment

☑ Results

Areas to Address

(a) current levels and trends in key measures and/or indicators of customer satisfaction and customer retention. Results should be segmented by customer group, as appropriate.

(b) current levels and trends in key measures and/or indicators of customer dissatisfaction. Address the most relevant and important indicators for the company's products/services.

Notes:

1.   Results reported in this Item derive from methods described in Items 7.3 and 7.2.

2.   Results data (7.4a) might be supported by customer feedback, customers' overall assessments of products/services, and customer awards.

3.   Indicators of customer dissatisfaction are given in Item 7.3, Note (3).

In addition to following up with a random selection of 3000 customers who had contacted the company, **AT&T Card** also conducts a monthly survey. The monthly survey tracks eight primary product and service satisfiers. It is conducted with 200 AT&T Card customers and 400 customers of competitors.

TI Defense & Electronics completed internal audits and received 23 customer audits in 1991. A process walk-through is used at the **Ritz-Carlton** for internal audits. Independent travel and hospitality rating organizations, such as AAA and Mobil, provide external audits.

Item 7.4: *Customer Satisfaction Results*

After describing how customer satisfaction is determined, the applicant reports the analysis of the data in Item 7.4, Customer Satisfaction Results. Three to five years of customer satisfaction, customer retention, and other related data are requested. Customer awards or other recognition from the customer can be reported in this Item.

Along with the good news, any bad news is also required. Indicators of customer dissatisfaction, such as complaints, returns, claims, recalls, repeat service, warranty work, incomplete orders, or incorrect shipments should be reported.

Item 7.5: *Customer Satisfaction Comparison*

Closely related to the previous Item is Item 7.5, Customer Satisfaction Comparison. In this Item, the organization should compare the company's customer satisfaction results with those of competitors; both domestic and international. The comparisons could be the result of independent studies or surveys done by the customer, if they reflect comparative satisfaction. For example, the J. D. Powers Company conducts surveys to determine customer satisfaction for new cars. Results from the survey, which showed Cadillac to be one of the best in customer satisfaction, was an appropriate use for this Item by Cadillac in their application. Product and service comparisons, such as *Consumer Reports'* test of 10 cars for defects per car, would also be reported here.

Item 7.5: Criteria

7.5 Customer Satisfaction Comparison (60 pts.)

Compare the company's customer satisfaction results with those of competitors.

☐ Approach

☐ Deployment

☑ Results

Areas to Address

(a) current levels and trends in key measures and/or indicators of customer satisfaction relative to competitors. Such indicators may include gains and losses of customers or customer accounts to competitors. Results may include objective information and/or data from independent organizations, including customers. Results should be segmented by customer group, as appropriate.

(b) trends in gaining or losing market share to competitors.

Notes:

1. Results reported in Item 7.5 derive from methods described in Item 7.3.

2. Competitors include domestic and international ones in the company's markets, both domestic and international.

3. Objective information and/or data from independent organizations, including customers (7.5a), might include survey results, competitive awards, recognition, and ratings. Such surveys, competitive awards, recognition, and ratings by independent organizations and customers should reflect comparative satisfaction (and dissatisfaction), not comparative performance of products and services. Information on comparative performance of products and services should be included in 6.1.

Examples for Item 7.4

Quantification determined from previous Item, customer assessment of products or services, customer awards, customer retention, and customer dissatisfaction.

Examples for Item 7.5

The company's established position on customer satisfaction and market share relative to competitors.

Ritz-Carlton exceeds customer expectation in nine key attributes including suitability of room, comfort of furnishings, and caring attitude of staff. The complaints per 100 comments decreased 27% in the three years prior to winning the Baldrige. A variety of numeric indicators show them to be number one in the industry. On the basis of results from a Gallup survey, 94% of the customers were satisfied, compared to 57% at the next highest scoring hotel. They earned 121 awards from various travel industry groups in 1991. The Ritz-Carlton has a high retention of local corporate accounts at all of their facilities.

There was a 25% improvement in customer satisfaction in three years at **AT&T Transmission.** Their international sales doubled in three years and they experienced a tenfold increase in the number of customers. They are the only non-Japanese supplier selling telecommunication products to the Japanese. Also, a 50% reduction in circuit pack returns was experienced in two years.

Granite Rock uses a four-quadrant analysis in which customer satisfaction is compared with importance levels given by the customers. Ideally, a company would want to score high on customer satisfaction for those features identified as being most important. Granite Rock has achieved that result. They increased their market share even though they are in a declining market. On the basis of a five-point customer satisfaction index, they went from 3.5 in 1988 to 4.0 in 1991.

On the basis of surveys conducted by the customer, **TI Defense & Electronics** was rated number one on all attributes surveyed. They are listed on the Department of Navy's "Best Manufacturing Practices" for concurrent engineering, printed wiring board producibility, diamond point optics turning, and their benchmarking process. TI Defense & Electronics earned the best possible rating from the Defense Contract Audit Agency.

AT&T Card knows you need to delight the customers, not just satisfy them. More than 90% of their customers would recommend the AT&T card to a friend. Customers that are satisfied are not necessarily loyal customers. Customers that are *very satisfied* are loyal.

Despite intense competition since deregulation, **AT&T Communications** has maintained a high market share by keeping their focus on the customer. Independent studies show that **GTE** is the preferred directory in 271 of its 274 primary markets. In 1993 **Wainwright** achieved a 95% customer rating on quality, delivery, communication, and service.

Key Excellence Indicators

O┳ Companies doing well have proactive customer systems that use all of the listening posts including customer contact employees, surveys, follow-ups, complaints, and turnover of customers. Quality requirements for different market segments are known for the current and future customers. The frontline employees are empowered, have current information, and act in the customers' best interests. All of this is reflected in high levels of customer satisfaction and a leadership position.

O┳ Customer-focused quality is a strategic business concept. It is directed toward customer retention and market share gain by delighting the customer. A company needs to be sensitive to changing customer and market requirements. Knowledge of developments in technology and of competitors' offerings is constantly monitored. A rapid and flexible response to emerging requirements helps maintain a leadership position.

SUMMARY

The seven Categories of the Criteria were presented in a graphical summary format. The seven Categories are Leadership, Information and Analysis, Strategic Planning, Human Resource Development and Management, Process Management, Business Results, and Customer Focus and Satisfaction. A reprint of the Items from the Award Criteria booklet was included with examples from selected previous recipients and other sources. At the end of each Category, a narrative description of the key excellence indicators for that Category was included.

5 | Award Process

INTRODUCTION

The Malcolm Baldrige National Quality Award (MBNQA) was created by Public Law 100-107, which was signed into law on August 20, 1987. The purpose of this law (Figure 5.1) is to help improve the quality and productivity of American companies by promoting an awareness of quality as an increasingly vital element in achieving a competitive edge and an understanding of the requirements for quality excellence. This is achieved by annually recognizing U.S. companies that excel in *business excellence* and *quality achievement*. These companies are obligated as part of the MBNQA process to share information on their successful quality strategies, if they want to advertise that they were Baldrige Award recipients.

The focus of this book is on the use of the MBNQA Criteria for self-improvement. But self-improvement means more than developing a written response to the Criteria. It requires an unbiased review of the written response, feedback that identifies strengths and areas for improvement, and actions based on the feedback to improve the organization at all levels. Although there are many approaches to the review and feedback processes, the Award process has proven itself to be an effective and robust process, satisfying the needs of all participants. It can provide a model for organizations in establishing a structure for internal review of the written responses of individual units. An awareness of the Award process can also provide a deeper understanding of the Criteria and the intent of the MBNQA.

This chapter examines the MBNQA process and supporting organization. An example of one company's model is described, and the MBNQA recipients are also discussed.

Figure 5.1
Public Law 100-107

The Malcolm Baldrige National Quality Improvement Act of 1987—Public Law 100-107

The Malcolm Baldrige National Quality Award was created by Public Law 100-107, signed into law on August 20, 1987. The Findings and Purposes Section states that:

1. the leadership of the United States in product and process quality has been challenged strongly (and sometimes successfully) by foreign competition, and our Nation's productivity growth has improved less than our competitors' over the last two decades.

2. American business and industry are beginning to understand that poor quality costs companies as much as 20 percent of sales revenues nationally and that improved quality of goods and services goes hand in hand with improved productivity, lower costs, and increased profitability.

3. strategic planning for quality and quality improvement programs, through a commitment to excellence in manufacturing and services, are becoming more and more essential to the well-being of our Nation's economy and our ability to compete effectively in the global marketplace.

4. improved management understanding of the factory floor, worker involvement in quality, and greater emphasis on statistical process control can lead to dramatic improvements in the cost and quality of manufactured products.

5. the concept of quality improvement is directly applicable to small companies as well as large, to service industries as well as manufacturing, and to the public sector as well as private enterprise.

6. in order to be successful, quality improvement programs must be management-led and customer-oriented, and this may require fundamental changes in the way companies and agencies do business.

7. several major industrial nations have successfully coupled rigorous private-sector quality audits with national awards giving special recognition to those enterprises the audits identify as the very best; and

8. a national quality award program of this kind in the United States would help improve quality and productivity by:
   A. helping to stimulate American companies to improve quality and productivity for the pride of recognition while obtaining a competitive edge through increased profits;
   B. recognizing the achievements of those companies that improve the quality of their goods and services and providing an example to others;
   C. establishing guidelines and criteria that can be used by business, industrial, governmental, and other organizations in evaluating their own quality improvement efforts; and
   D. providing specific guidance for other American organizations that wish to learn how to manage for high quality by making available detailed information on how winning organizations were able to change their cultures and achieve eminence.

AWARD ORGANIZATION

The Malcolm Baldrige National Quality Award Act of 1987 led to the creation and continued growth of an active partnership between the private sector and the government (Figure 5.2). Support by the private sector is in the form of funds, volunteer efforts (e.g., as members of the Board of Examiners), and participation in information transfer. Principal support for the program comes from the Foundation for the Malcolm Baldrige National Quality Award, which was established in 1988 to foster the success of the program.

Leaders from U.S. companies serve as foundation trustees to ensure the accomplishment of the foundation's objective—to raise funds to permanently endow the Award program.

Figure 5.2
Award Organization

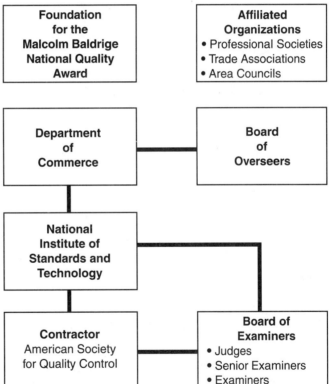

Responsibility for the Award has been assigned to the Department of Commerce. The Award process overseer is the National Institute of Standards and Technology (NIST, formerly known as the National Bureau of Standards or NBS), an agency of the Department of Commerce's Technology Administration. Although the final decision authority lies with the Secretary of Commerce, NIST is responsible for the management and continual improvement of the Award process. This includes the dissemination of information and documentation on the process (see the appendix for contact information); selection, training, and calibration of the Board of Examiners; and eligibility determination of all potential applicants. The selection of NIST to manage the Award process is consistent with this organization's goals to aid U.S. industry through research and services; to contribute to public health, safety, and the environment; and to support the U.S. scientific and engineering research communities. Much of the work performed or sponsored by NIST relates directly to quality and to quality-related requirements in technology development and utilization.

To ensure an independent review and assessment of the Award program, the Secretary of Commerce has appointed a board of overseers consisting of distinguished leaders from all sectors of the U.S. economy (see the appendix for membership). Membership on the Board varies in tenure from one to four years, on a rotating basis. The Board not only evaluates all aspects of the Award program, including the adequacy of the Criteria and processes for evaluating applicants, but also assesses how well the Award is serving the national interest.

The administration for the Award is provided by the American Society for Quality Control (ASQC) under contract to NIST. The ASQC, a professional organization of more than 125,000 individual members, is a world leader in the development, promotion, and application of quality and quality-related technologies for the quality professional, private sector, government, and academia. Its participation in the Award process is just one of its many activities to further its goals: to create a greater awareness of the need for quality, to promote research and the development of standards, and to promote educational opportunities to ensure product and service excellence through improved quality.

The evaluation of Award applications, development of feedback reports, and recommendation of potential Award recipients is the responsibility of the Board of Examiners. The Board of Examiners are business and quality experts, primarily from the private sector, who volunteer to participate in the evaluation process following stated ethical standards of conduct. They do not represent their companies or organizations. Members are selected by NIST through a yearly application process with Judges selected to serve a (rotating) three-year term. Since other membership to the Board is only for one year, every (potential) Examiner must submit an application every year—if they wish to participate for multiple years. In selecting members for the Board of Examiners, NIST considers the applicants' business and geographic demographics as well as their quality expertise. This is done in order to form a Board of Examiners that reflects the diversity of U.S. industries and includes business skills common in such organizations. This also means that individuals from a specific industry or geographic area that has fostered many Examiner applicants will have a lower probability of being selected than an equally qualified person in some other industry or area. In 1995, the Board of Examiners consists of nine Judges and approximately 270 Examiners (approximately 50 are designated as Senior Examiners). All members of the Board must attend a training session for the current year, whether they are new or returning. The purpose of this training is not only to provide the Examiners with information on the Criteria but also to *calibrate* the responses to reduce variability among the Examiners and to build a cooperative team spirit.

 ## REVIEW PROCESS

The review process is initiated when a company submits the eligibility determination form. NIST reviews this form in order to determine if all of the eligibility requirements are met. All potential applicants *must* have their eligibility approved prior to submitting the Award application.

Eligible organizations can submit the application package, which consists of an approved eligibility determination form, with a listing of the applicant's company sites; a completed application form; a business overview, addressing the applicant's key business factors; the application report, responding to all elements of the Award Examination Criteria; and supplemental sections, submitted by applicants whose application includes units that are in different businesses. The applications are reviewed and evaluated by members of the Board of Examiners in a four-stage process. This process results in the recommendation by the Judges of Award recipients to the director of NIST and the Secretary of Commerce and a feedback report to every applicant. Since the feedback report is based on the applicant's responses in the application package, the objective of the applicant should be to provide sufficient information on the management of their products and services and results of improvement processes to permit a complete and rigorous review by the Board of Examiners.

ELIGIBILITY

Application for the Award is limited to for-profit businesses located in the United States or territories. They may be publicly or privately owned, domestic or foreign owned, joint ventures, or, with some restrictions, subsidiaries. In the future, eligibility may change to allow not-for-profit organizations, such as health care and educational institutions, to apply. Government agencies are not eligible, although U.S. government agencies have a similar award, the President's Award.

There are currently three categories of eligibility: manufacturing, service, and small business. The proper classification of companies that perform both manufacturing and service is determined by the larger percentage of sales. Small business is defined as a complete business with 500 employees or less. A small business must be able to document that it functions independently from any other business that is an equity owner.

No more than two Awards will be given in each category and no Award need be given.[1] The maximum number of Awards that can be given in any year is six. The largest number of Awards given was in 1992, which had five Award recipients.

Subsidiaries of companies in the manufacturing or service categories may be eligible for the Award—small businesses must apply as a whole—if they satisfy the following conditions:

- The unit must have existed for at least one year prior to the application.

- The unit must have a clear definition of organization and must function as a complete business unit.

- The unit must have at least 500 employees or 25% of all employees in the worldwide operations of the parent company.

- The entire unit must be included in the application.

Only one subsidiary unit of a company may apply for an Award in the same year in the same category. Further, a subsidiary and its parent company may not both apply for Awards in the same year.

---

[1] In the future, the restriction of two Awards per eligibility category will likely be dropped. In fact, a bill was introduced in the Senate in early 1993 to remove this limitation.

Restrictions on Eligibility

The intent of the MBNQA is achieved by annually recognizing *U.S. companies* that excel in quality management and quality achievement to serve as appropriate models for other U.S. companies. Organizations whose business cannot satisfy the basic requirements of being a full-fledged company and a U.S. company are not eligible—even though they may be a suitable role model in quality management and achievement. Specifically, an organization is not eligible if (1) the major business functions are not in the United States or its territories; (2) more than 50% of the employees or 50% of the physical assets are outside the United States or its territories; (3) more than 50% of a subsidiary's dollar volume of products and services is dedicated to the parent company; (4) it is an individual unit or partial aggregation of units of a chain; or (5) the organization is a company unit with the sole activity of a business support function, such as Sales, Marketing, Distribution, R&D, Purchasing, etc.

If the recipient is a company or a subsidiary unit consisting of more than one-half of the total sales of a company, neither the company nor any of its subsidiary units is eligible to apply for the Award again for a period of five years. A subsidiary with less than one-half of the total sales that receives an Award restricts only its own eligibility for the five years.

Specific questions regarding eligibility should be directed to the MBNQA offices at NIST (see the appendix for contact information).

 ## EVALUATION PROCESS

If qualified, the applicant provides a written application package containing a four-page overview of the applicant's business and a written response to the 24 examination Items. For all applicants, the application is restricted to a maximum of 70 single-sided pages.

The evaluation process (Figure 5.3), always results in a written feedback report to the applicant upon conclusion of the entire review process. Once the submission deadline date is reached, applications are assigned to Examiners, taking into account the nature of the applicants' businesses and the expertise of the Examiners. Assignments are made in accord with strict rules regarding conflict of interest. Each of the five to six Examiners reviewing an application completes an independent assessment. On the basis of the ratings at this first stage, the Judges decide if the application should continue in the review process. If the application is not selected for the second-stage review, a feedback report is generated based on written comments from the independent assessments (see Chapter 10).

The second-stage review consists of the same team of Examiners discussing their ratings from the first stage and reaching a consensus for each examination Item. The Senior Examiner facilitates the discussions. This is the first time in the process that the Senior Examiner has a role different from the other Examiners. The Judges use the consensus ratings to select which applicants will receive a site visit. If an applicant is not selected for a site visit, they will receive a feedback report based on the consensus results.

Figure 5.3

MBNQA Evaluation Process

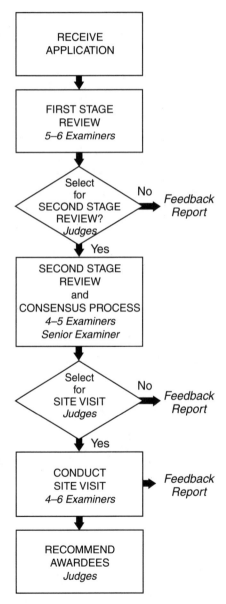

A site visit typically involves the same team of Examiners, including the Senior Examiner, that conducted the initial and consensus reviews. The purpose of the site visit is to verify and clarify specific elements of the application (see Chapter 11). After a site visit, the applicant receives a feedback report based on all of the information that has been reviewed. For the final contenders, this adds up to a report based on about 400 hours of evaluation.

A panel of Judges reviews all of the evaluation reports and makes a recommendation to NIST, the overseer of the Award. NIST presents the Judges' recommendation to the Secretary of Commerce for the final Award decision. The Recipients are then announced and presented in the last quarter of the year. In the past the Awards have been presented by the president or vice president of the Unites States in a special ceremony in Washington, D.C.

The feedback reports for all of the applicants are sent after the Award recipients have been announced.

APPLICATION OF THE AWARD PROCESS TO SELF-IMPROVEMENT

Companies using the Award Criteria as a road map for their quality journey also should look to the Award process as the yardstick for quality improvement. For self-improvement to be successful, an organization must not only develop a written response describing its activities with respect to the Criteria, it must also evaluate how it compares to the Criteria. This evaluation and feedback should be conducted by independent and unbiased reviewers, internal or external to the organization.

For those organizations eligible to participate in the MBNQA process, one possibility would be to submit an application, even though you know your organization is just in the initial stages of its quality journey. For those organizations not eligible for the national process, but who are located in states with a state award, the state process may be able to provide an evaluation and feedback. These choices involve timing constraints and considerations. However, they generally do provide much value for their application fees.

Another obvious approach to the review process is to develop a support structure modeled after the national process. Multiple-unit organizations may decide to take this approach, consolidating the administration at a central site. The Examiners would be employees that have been trained in the Award process. It is advisable to exchange Examiners across units; that is, have an Examiner from unit A review the written response from unit B. A team of Examiners should be reviewing each written response. There is too much variation in the process to have only one Examiner write the comments, do the scoring, and then prepare the feedback report. Teams of Examiners also allow for cross-unit communications of innovative improvement ideas.

Single-unit organizations may be unable to establish this type of structure because of limited personnel. If they are not eligible for the national or state award processes, they can rely on professional organizations for these services. Regardless of the approach, the need for an effective review and feedback process is critical when using the Criteria for self-improvement.

One Company's Model

Eaton Corporation, an international original equipment manufacturer with $5 billion in annual sales, uses the Baldrige Criteria to guide their improvement efforts. In 1990, Eaton, whose headquarters is in Cleveland, Ohio, began emulating the Baldrige process by establishing an internal award called the Eaton Quality Award (EQA). The EQA uses the Baldrige Criteria.

Eaton's award process is similar to the national process. The main difference is the beginning step. At Eaton, an applicant initially does a self-assessment and identifies strengths and areas for improvement based on the Criteria. This is not a full application but only a summary of the strengths and areas for improvement plus a self-evaluation of a score. From this self-assessment, candidates are invited to write a full application. For example, in 1994 there were 83 self-assessments and 29 full applications. From this point on, the EQA evaluation process follows the national model (Figure 5.4).

All organizations in the company are eligible to apply. A plant, division, or staff may apply. For example, a division with seven plants might choose to submit three applications representing three out of the seven plants. This same division may decide next year to submit one application for the entire division covering division offices and the seven plants.

The EQA organization consists of a senior management council, which endorses the EQA activity, selects award recipients, and presents the awards (Figure 5.5). The board of examiners consists of employees that volunteer for a three-year commitment in addition to their regular jobs. The management team at the applicant location is responsible for writing their submission.

The examiner teams review self-assessments, review applications, conduct site visits, and prepare feedback reports. Similar to the national process, several

Figure 5.4
EQA Evaluation Process

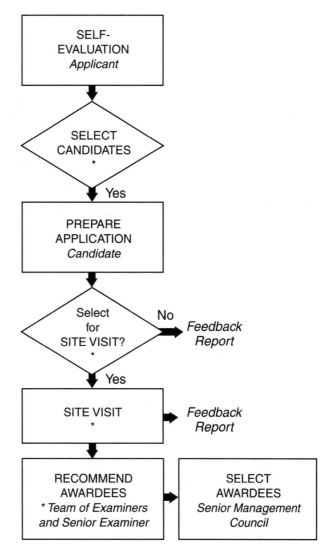

* Review, Evaluation, and Selection by a Team of
EQA Examiners and Senior Examiner

examiners do an independent review of the same application before pooling their findings. Examiners are from around the world including France, England, Spain, and Brazil. Some of the examiners are designated senior examiners and have the additional responsibility of coordinating an examiner team.

Figure 5.5
EQA Structure

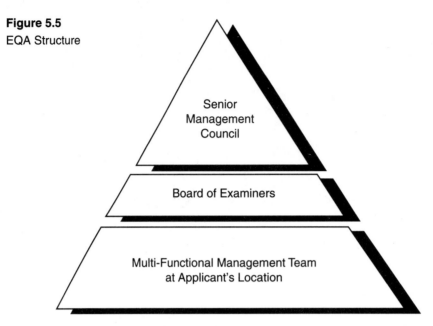

Initially the examiners were from the Quality Departments, but this has changed over the years. In 1990 approximately 90% of the examiners were from Quality Departments. During the next four years this dropped to a low of 25%. Other departments represented include Manufacturing Engineering, Product Engineering, Finance, Operations, and Human Resources. The total number of examiners has grown over the years from 25 in 1990 to more than 60 in 1994.

The timeline for the EQA process is as follows:

Event	*Month*
Information packages distributed	February
Self-assessment submitted	April
Candidates selected to write full application	May
Award applications submitted	July
Finalist selected	September
Site visits	October
Selection of recipients and announcement	November
Award presentation at recipient's location	December
Feedback issued to all candidates	January

In May of each year a 2½-day *new examiner* training course is conducted. As preparation, the participants read a case study. The class consists of an overview of quality, interpreting the Criteria, how to write useful comments, and examples for each of the seven Categories of the Criteria of good practices from other companies. Workshops are held using the case study. In July an *examiner preparation* class is conducted, also 2½-days long. This class is attended by the May participants as well as returning examiners. The topics covered include a review of the Criteria highlighting changes from the previous year's Criteria, explanation of the scoring process, how to write a feedback report, elements of site visits, and proper conduct of examiners regarding issues such as confidentiality. Preparation for this class requires a case study to be read and comments developed.

EQA Criteria training is available to others in the company who are not examiners. This is typically done through Eaton's Quality Institute, an industrial university without walls. The president and vice presidents of the company have completed training on the Criteria.

The EQA process is evaluated and improved each year based on feedback from examiners and applicants. One change that was made based on comments from the applicants is to offer a one-day presentation of the feedback report to the applicant by the examiners so that through discussion the applicant obtains a deeper understanding of the comments.

A Best Practices Day is held annually to facilitate sharing of successful ideas across divisions and staffs. From this, a manual is published and widely distributed within the corporation.

The award process is one of several strategies used by Eaton in their continual improvement efforts. It has been shown to be very useful, not only to improve quality but also the business culture. According to Joy Williams, the award program manager, "in addition to generating business results, this process is strengthening the relationships among departments and divisional groups within Eaton." The EQA process is now in its sixth year and seems to be an integral part of Eaton's strategic business planning process.

 ## MBNQA PROCESS RESULTS

Number of Applicants and Site Visits

The number of applicants and site visits are summarized in Figure 5.6. There may be some growth in the number of companies applying for the national quality award in future years but a large surge is not anticipated. Some companies that are using the Criteria to guide their quality journey make a decision not to apply for the Award because of a concern that it may change employees' focus from improvement to "doing things just to get an award." Other companies using the Criteria are not eligible to apply; for example, individual plants, nonprofit organizations, government agencies, etc. Another factor that limits the number of applicants is other requirements that place constraints on an applicant's time such as ISO 9000 registration. Finally, the soaring number of state and local awards spun off from the MBNQA attract companies to apply at that level in order to highlight success at individual plants or regional areas.

As the list of Award recipients continues to grow each year, it takes away the excuse that "The Criteria don't apply to me," or "We are different." The diversity of the list makes it hard for a company to continue to make these claims. Large and small companies, manufacturing and service companies, high- and low-technology companies now comprise the list of Award recipients. In addition to the diversity of organizations, there is also a diversity of approaches and strategies to achieve quality excellence and customer satisfaction. Each organization selected the path that was suitable to its specific need— they didn't simply try to duplicate that of some other organization. There is one common thread in all these approaches: a focus on the customer and continual improvement.

Figure 5.6
Number of Applicants and Site Visits*

		Category		
Year	Manufacturing	Service	Small Business	Total
1988	45 (10)	9 (2)	12 (1)	66 (13)
1989	23 (9)	6 (2)	11 (0)	40 (10)
1990	45 (6)	16 (3)	34 (3)	97 (12)
1991	39 (9)	21 (5)	47 (5)	106 (19)
1992	31 (7)	15 (5)	44 (5)	90 (17)
1993	32 (4)	13 (5)	31 (4)	76 (13)
1994	23 (6)	18 (5)	30 (3)	71 (14)

* Site visits are shown inside the parentheses.

Award Recipients

A maximum of six Awards can be presented during any given year; only two awards per category for each of the three categories. The Award recipients to date follow:

1994

Consumers Communications Services unit (AT&T)—service

GTE Directories Corporation—service

Wainwright Industries—small business

1993

Eastman Chemical Company (Eastman Kodak)—manufacturing

Ames Rubber Corporation—small business

1992

Defense Systems & Electronics Group (Texas Instruments)—manufacturing

Transmissions Systems unit (AT&T)—manufacturing

Universal Card Services unit (AT&T)—service

Ritz-Carlton Hotel Company—service

Granite Rock Company—small business

1991

Solectron Corporation—manufacturing

Zytec Corporation—manufacturing

Marlow Industries—small business

1990

Rochester IBM—manufacturing

Cadillac Motor Car Division (GM)—manufacturing

Federal Express—service

Wallace Company—small business

1989

Business Products and Systems Group (Xerox)—manufacturing

Milliken & Company—small business

1988

Motorola—manufacturing

(Westinghouse) Commercial Nuclear Fuel Division—manufacturing

Globe Metallurgical—small business

 ## Summary of Award Recipients

The following is a brief description of each of the recipients of the Malcolm Baldrige National Quality Award. A more detailed description is available directly from each recipient (see the appendix for contact information). The diversity of their products and services will become evident. The diversity of their size, as measured by the number of employees is shown in Figure 5.7.

1994—Service

AT&T Consumers Communications Services (AT&T Communications) is the largest of some 20 AT&T business units with 44,000 employees at more than 900 sites throughout the United States. AT&T Communications provides domestic and international long-distance communications services primarily to residential customers. Its customer base includes more than 80 million customers. Despite intense competition since deregulation, the company has maintained a high (about 60%) market share for long-distance services by constantly focusing on the needs of the customer.

They use and integrate a business and quality planning system that focuses on five key customer requirements: call quality, customer service, billing, reputation, and price. Involvement in the planning and deployment encompasses all employees, including the AT&T Communications president and senior executives. To ensure a shared focus and commitment across units, internal contracts are used to assign responsibilities and to establish clear improvement goals for each stage of a process. Process Management Teams include employees from all levels of the company as well as suppliers and other resources, as needed. These teams implement quality improvement using standardized AT&T-developed tools for planning, executing, and evaluating initiatives to improve performance and customer satisfaction.

Figure 5.7

Number of Employees of the MBNQA Award Recipients

Number of Employees	Company Name
160	Marlow Industries
210	Globe Metallurgical
275	Wainwright Industries
280	Wallace Company
400	Granite Rock Company
445	Ames Rubber Corporation
748	Zytec Corporation
2,000	Westinghouse Commercial Nuclear Fuel Division
2,100	Solectron Corporation
2,500	AT&T Universal Card Services
5,150	GTE Directories Corporation
7,500	AT&T Transmission Systems
8,100	IBM Rochester
10,000	Cadillac Motor Car Division (GM)
11,500	Ritz-Carlton Hotel Company
14,300	Milliken & Company
15,000	Texas Instruments—Defense Systems & Electronics Group
17,750	Eastman Chemical Company
44,000	AT&T Consumers Communications Services
50,200	Xerox Business Products and Systems Group
90,000	Federal Express
99,000	Motorola

10 out of 22 recipients have 2500 or fewer employees; only 4 out 22 recipients have 40,000 or more employees

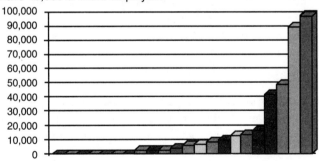

1994—Service

With $1.4 billion in revenue, **GTE Directories Corporation** is one of the world's largest telephone directory companies. Its 5150 employees worldwide produce more than 1200 directory titles in 45 U.S. states and 17 countries. The company's primary focus is on publishing and/or selling advertising for white and yellow pages telephone directories for telephone companies within their official franchise areas, which include all of the Regional Bell Operating Companies.

To meet the company's corporate vision of "100 Percent Customer Satisfaction Through Quality," 14 key customer service requirements have been developed through a Customer Satisfaction Measurement Program. The customer satisfaction measures are linked to a heavy reward system that serves as a powerful motivator for quality service, with a variety of special awards for individuals and teams supporting quality and operational improvement. Team-based management is pervasive throughout the company. It is not unusual to find individuals who have served on ten or more Quality Improvement Teams, even at the manager level. As a result of their efforts, independent studies have shown that GTE Directories is the preferred directory in 271 of its 274 primary markets.

1994—Small Business

Wainwright Industries is a family-owned manufacturer of stamped and machined products for the automotive, aerospace, home security, and information processing industries. With $30 million in annual sales, its major markets are local, national, and increasingly international. All of Wainwright's 275 associates are involved in the company's Continuous Improvement Process. In 1993, associates submitted an average of 54 suggestions, or 1.25 suggestions per week.

Improvement trends in production and delivery areas are positive. For example, reengineering the process to produce drawn housings has led to reductions in lead time from 8.75 days to 15 minutes. All company data are linked to the company's five strategic indicators (listed in order of importance): safety, internal customer satisfaction, external customer satisfaction, six sigma quality, and business performance. Business processes are developed with the safety of associates in mind. As a result, since 1990 accidents have decreased by 72%, lost time due to accidents by 85%, and lost work days by 87%.

1993—Manufacturing

Eastman Chemical Company (a division of Eastman Kodak), founded in 1920, has annual sales of $4 billion and employs 17,750 people. They manufacture more than 400 chemicals, fibers, and plastics for 7000 customers around the world. Eastman is known for its environmental concern, high customer ratings on product quality, product uniformity, supplier integrity, on-time delivery, and reliability. The team structure used involves every employee in the improvement process. Data are widely deployed and used throughout the organization. Eastman averaged 22% of sales from new products commercialized within the last five years. They have a no-fault return policy that is considered to be unique in the industry.

1993—Small Business

Ames Rubber Corporation, based in Hamburg, New Jersey, manufactures rubber rollers used in office equipment such as copiers, printers, and typewriters to transfer and fuse toner to paper. They also provide a specialized component used to protect the transaxle of front-wheel drive vehicles. Ames is in the small business category with 445 employees, who are referred to as "teammates." Every employee is a member of at least one "involvement group" that works on quality improvement. Their defect rate has been reduced to 11 parts per million (ppm). Ames reduced their supplier base and works closely with their suppliers in a win–win environment.

1992—Manufacturing

Defense Systems & Electronics Group (a division of Texas Instruments Inc.) makes precision guided weapons and other advanced defense technology. Employing 15,000 people, this group has become the nation's eighth largest defense electronic contractor. Effective strategic planning, wide use of concurrent engineering methods, and strong relationships with key suppliers have helped the TI Defense Systems & Electronics Group penetrate new defense markets while increasing its share in the company's existing markets.

Teams are used to link all units and levels, from top management to individual work teams. The entire workforce has been trained in design quality, statistical tools, and other quantitative methods. The Defense Systems & Electronics Group had a 21% reduction in production cycle time in 1992, with a 56% reduction in stock-to-production time. An ongoing study by the Navy has identified more best practices in this division than any other company.

1992—Manufacturing

Transmission Systems Business Unit (a division of AT&T) competes in a $15 billion international market to transport data, voice, and images over telecommunication networks. This unit, which employs 7500 people at nine U.S. sites, also has a staff of 3000 at five European facilities. The eight executives on the Quality Council initiate planning and serve as members on any one of the four steering committees. The information system helps executives, managers, and workers track key processes in all phases and at all levels of the business. Design for manufacturability is an essential element of AT&T Transmission's strategy for improvement. Customers use report cards and ten other avenues to evaluate performance against requirements. These actions enabled AT&T Transmission to reduce time to market by 50% in three years. Long-term focus is discussed during forums where AT&T Transmission shares its future technical direction and customers explain their long-term plans and expectations. Twenty-three of these forums were conducted in 1991.

1992—Service

Universal Card Services (a division of AT&T), established in March 1990, was formed around the Baldrige principles. Its strong customer focus helped it to open its one-millionth account 78 days after program launch and to become one of the top ten credit card programs in the nation a month later and the second largest credit card company in just 30 months. AT&T Card's eight broad categories of customer "satisfiers," including price and customer service, are used to define the company's quality focus. These categories are detailed with 125 "satisfiers," each weighted to reflect its relative importance.

Benchmarking activities allowed them to determine what constitutes outstanding performance levels. AT&T Card leads the credit card industry in speed and accuracy of application processing, and customer satisfaction. AT&T Card competes against 6000 national, regional, and local issuers of general-purpose credit cards. A strong culture is displayed for valuing the employees, who are referred to as associates. In recognizing that they are key to delighting the customer, associates are empowered. Additionally, self-directed work teams are responsible for all day-to-day activities and decisions. Employee satisfaction is tracked. This information is used to guide improvements in training, recognition, and other related activities.

1992—Service

Ritz-Carlton Hotels is a management company that develops and operates luxury hotels. At the time of its application, Ritz-Carlton was operating 25 luxury hotels. Participatory executive leadership, thorough information gathering, coordinated planning and execution, and a trained workforce are key to their service quality. The corporate motto for the 11,500 employees is "ladies and gentlemen serving ladies and gentlemen." The aim is to go beyond customer satisfaction, by making each experience a memorable one. Surveys conducted by an independent research firm for Ritz-Carlton indicate 92 to 97% of the guests leave with that impression. Quality reports, derived from data submitted from each of the 720 work areas in the hotel system, serve as an early warning system for identifying problems that might be detrimental to customer satisfaction. Ritz-Carlton has been honored by the travel industry with 121 quality awards from 1991 to 1993 and has received the industry-best ranking by all three major hotel rating organizations.

1992—Small Business

Granite Rock, founded in 1900, is a highly automated, family-owned business that employs 400 people. The company produces rock, sand, gravel aggregates, ready-mix concrete, asphalt, and road materials. Nine corporate objectives, developed from analysis of customers' requirements, are the cornerstone of Granite Rock's quality program. Data for each product line are used to assess performance relative to competitors on key products and service attributes, ranked according to customer priorities.

As part of the effort to reduce process variability and increase product reliability, many employees are trained in statistical process control, root cause analysis, and other quality assurance methods. Responding to customer concern over rising trucking costs, the company developed GraniteXpress, the construction industry's version of an automated teller machine. The service, which operates 24 hours a day, 7 days a week, has reduced the time a truck driver spends at the quarry to 9 minutes instead of the customary 24 minutes. Granite Rock's customer accounts have increased 38% from 1989 through mid-1993, while overall construction spending in its market area declined by more than 40%.[2]

1991—Manufacturing

Solectron Corporation has grown from a small assembly shop when it was founded in 1977 to one that had 2100 employees at the time it submitted the application and which now employs more than 3000 people. By focusing on customer satisfaction, using advanced manufacturing technology, and stressing continuous improvement in operations and services, they have demonstrated that high quality and high efficiency translate into low costs and timely delivery. A team-focused approach to employee involvement is used. This includes training and mentoring for the workforce that is composed of people from more than 20 cultures.

Each customer is supported by two teams that work to ensure quality performance and on-time delivery. A project planning team works with customers in planning, scheduling, and defining material requirements and lead time. A total quality control team meets weekly for monitoring and evaluating production with the aim of preventing potential problems and identifying ways to improve process yield. Statistical process control is used in all departments. The customers have benefited from a defect rate reduction that has fallen to 233 ppm. Solectron, by focusing on customer satisfaction, has seen average yearly revenue growth of 46.8%, and by focusing on process quality has seen average yearly net income growth of 57.3% from 1988 to 1993.[3]

---

[2] *Quality Pays*, NIST, 1993.

[3] *Ibid.*

1991—Manufacturing

Zytec Corporation designs and manufactures electronic power supplies for original equipment manufacturers of computers and electronic equipment. Zytec began operating independently in 1984 and began its quality initiative that year. As the foundation for continuous improvement, Zytec's senior executives chose Dr. W. Edwards Deming's 14 points for management. Design and development of new products are carried out by interdepartmental teams, which are assigned to projects from start to finish. Working closely with the customers, the same cross-functional teams review performance at four key stages: predesign, design, prototype and testing, and preproduction certification.

Zytec is a data-driven company, developing measurable criteria for evaluating performance at all levels. To realize the full potential of its 748 employees, Zytec trains them in analytical techniques, which is a major focus of the 72 hours of quality-related instruction they receive. As a result of these efforts, the mean time between failure of a power supply unit has increased to more than one million hours as measured by field. Further, Zytec's internal manufacturing process yields have improved fivefold from 1988 to 1992, with customer out-of-box quality up from 99% to 99.8%, and on-time delivery improved from 75% to 98%.[4]

1991—Small Business

Marlow Industries started as a five-person operation in 1973 and now employs 160 people with annual sales of $12 million. They manufacture customized thermoelectric coolers for use in electronic equipment. They use a well-integrated, systematic approach for assuring the quality of their products and services. Extensive use of design of experiments techniques has resulted in improved quality and stabilization of its major processes. Supplier partners figure prominently in Marlow's TQM system with "ship to stock" status for its key suppliers. Since 1987, employee productivity has increased at an average annual rate of 10%; the time between new product design and manufactured product has been trimmed; and the cost of scrap, rework, and other nonconformance errors has been cut nearly in half. For customers, these and other gains have translated into products that exceed performance specifications by wider margins, on-time deliveries, extended warranties, and prices that have remained stable or decreased. Marlow Industries' worldwide market share continues to improve, even in Japan where the company has local competition. In 1992, 48% of Marlow Industries' revenues resulted from products introduced since submitting its 1991 Baldrige application.[5]

[4] *Ibid.*
[5] *Ibid.*

1990—Manufacturing

IBM Rochester manufactures intermediate computer systems—currently the AS/400—and computer hard disk drives with an employee base of 8100 people. Rochester formulates its continuous improvement plans based on six critical success factors: an improved product and service requirements definition, an enhanced product strategy, a six sigma defect elimination strategy, cycle time reductions, improved education, and increased employee involvement and ownership. The Rochester quality culture has been transformed from "reliance on technology-driven processes delivering products" to "market-driven processes directly involving suppliers, business partners, and customers delivering solutions."

Since 1986, IBM Rochester has invested more than $300 million in improving their processes, many of them designed to improve the prevention of defects rather than the detection of defects after they have occurred. These investments have paid for themselves. Capital spending on equipment for defect detection declined 75% during the 1980s and write-offs as a proportion of manufacturing output dropped 55%. Customers have benefited from a threefold increase in product reliability and the cost of ownership is among the lowest in the industry. IBM Rochester has had a 25% growth in market share from 1988 to 1992, with 1992 new customer installations at twice the rate of 1988.[6]

1990—Manufacturing

To many car buyers, the **Cadillac** nameplate symbolizes the highest level of automotive quality. To maintain this reputation, improve customer satisfaction, and reverse its decline in market share in the early 1980s, Cadillac implemented simultaneous engineering and integrated the involvement of employees in the running of the business. The effectiveness of simultaneous engineering requires the carefully orchestrated teamwork of employee, dealer, and supplier representatives in defining, engineering, marketing, and continuously improving all Cadillac products. Their coordinated efforts on recent major styling changes trimmed 50 to 85 weeks from what used to be a 175-week process.

Using an open, yet disciplined, planning process, Cadillac pledges to involve all of its 10,000 employees in the development of its annual business plans, which embody short- and long-term quality improvement goals and educational needs. In 1990, for example, skilled hourly personnel received a minimum of 80 hours of formal instruction in support of quality improvement, health, and safety. In step

---

[6] *Ibid.*

with service and product quality, customer satisfaction—as measured by satisfaction with cars, service, and total ownership experience—rose from 70% in 1985 to 96% or better in 1989.

1990—Service

In 1973 **Federal Express** launched the air express industry. By 1990, 90,000 Federal Express employees processed 1.5 million shipments daily at more than 1650 sites. With a target of 100% customer satisfaction, the company has developed a "People—Service—Profit" philosophy to guide management policy and actions. It has a well-developed and thoroughly deployed management evaluation system called SFA (survey/feedback/action). SFA involves a survey of employees, analysis of each work group's results, and a discussion between the manager and the work group to develop written action plans for the manager to improve and become more effective.

Customer satisfaction is measured utilizing a 12 component index that comprehensively describes how the customers view Federal Express' performance. Each item in the service quality index (SQI) is weighted to reflect how significantly it affects overall customer satisfaction. The SQI measurements are directly linked to the corporate planning process and form the basis on which corporate executives are evaluated. Employees are encouraged to be innovative and to make decisions that advance quality goals. The staff quality action teams (QATs) have generated significant savings: $27 million in the Personnel Division from 1986 to 1993; $1.5 million in recovered revenue by a computer automation QAT; and $462,000 in saved overtime payments in six months by a payroll QAT.[7] In an independently conducted survey of air express customers, 53% gave Federal Express a perfect score, as compared to 39% for its next best competitor.

1990—Small Business

Founded in 1942, **Wallace** is a family-owned distribution company serving primarily the chemical and petrochemical industries. In tandem with its move to continuous quality improvement, Wallace shifted its marketing focus from engineering and construction activities to maintenance and repair operations. Although the seeds of Wallace's quality process were sown and cultivated by the company's five top leaders, the company's workforce of 280 associates is responsible for devising and carrying out plans to accomplish objectives under its cooperative, yet centralized, approach to quality improvement.

[7] *Ibid.*

Wallace invested about $2 million in formal training between 1987 and 1991. Top management has received more than 200 hours of intensive training in the methods and philosophy of continuous quality improvement. The team of associates closest to a specific area targeted for improvement are charged with identifying the steps necessary to accomplish a quality objective, standardizing the methods to assure consistent performance, and conducting the necessary training at all departments and offices. The quality initiatives at Wallace have enabled the company to increase its market share from 10.4% in 1987 to 18% in 1990; sales per associate from $180,000 in 1986 to $294,000 in 1991; on-time delivery from 75% to 92%; and operating profits by 7.4 times. *Note*: In 1993, the assets of Wallace were purchased by Wilson Industries, Inc.

1989—Manufacturing

The **Business Products and Systems** (BP&S) Group is one of two Xerox Corporation businesses. It employs more than 50,000 people to produce over 250 types of document producing equipment, predominantly copiers and other duplicating equipment. BP&S's "Leadership Through Quality" thrust has made quality improvement and customer satisfaction the job of every employee. All have received at least 28 hours of training in quality improvement techniques. This is part of the more than four million hours and $125 million invested in the education of employees about quality principles. As part of "Team Xerox," workers are vested with authority over daily work decisions. BP&S estimates that 75% of its workers are members of at least one of more than 7000 quality improvement teams. In 1988, teams were credited with saving $116 million by reducing scrap, tightening schedules, and devising efficiency and quality-enhancing measures.

In its quest to elevate its products and services to world-class status, Xerox BP&S developed a benchmarking system that has become a model for other industries. Returns from the company's strategy for continuous improvement from 1984 to 1990 include a 78% decrease in the number of defects per 100 machines; greatly increased product reliability as measured by a 40% decrease in unscheduled maintenance; and a 27% drop (nearly two hours) in service response time.

1989—Manufacturing

Milliken & Company is a privately owned major textile manufacturer long recognized for quality products and the use of state-of-the-art technology. In its "Pursuit of Excellence," Milliken has achieved a flat management structure in which its 14,000 associates work primarily in self-managed teams, exercising considerable authority and autonomy. The approach has worked so well that Milliken has reduced the number of management positions by 700 between 1981 and 1989, freeing up a large portion of the workforce for assignment in process improvement. There has been a 77% increase in the ratio of production to management associates. Besides participatory management, Milliken's approach to quality includes putting the customer first and a clear focus on quality. As a result of its innovations, Milliken has improved its record for on-time delivery from 75% in 1984 to an industry best of 99% in 1988. The company has been well recognized for achieving success for its customers, receiving 41 major customer quality awards in the five years preceding their Baldrige application.

1988—Manufacturing

Employing 99,000 workers at 53 facilities worldwide, **Motorola** is an integrated company that produces an array of electronic products. Communications systems and semiconductors account for the major portion of its sales. Responding to the rapid rise of the Japanese in the electronic marketplace, Motorola's management began an almost evangelical crusade for quality improvement. But all levels of the company are involved in achieving its quality and total customer satisfaction goals. Motorola concentrates on several key operational initiatives. At the top of the list is its "Six Sigma Quality," a statistical measure of variation from a desired result that Motorola defines to be 3.4 defects per million products. This measure requires that the product designs accommodate expected variation in component parts and that production processes yield consistent acceptable products.

Reduced cycle time—the time from when a Motorola customer places an order until it is delivered—is another vital part of the company's quality initiatives. This requires an examination of the total system, including design, manufacturing, marketing, and administration. Motorola has "benchmarked" the products of some 125 companies against its own standards, verifying that many Motorola products rank as best in their class. Motorola's employee productivity has improved 100% from 1988 to 1993 (an annual compounded rate of 12.2%) through robust design, continuous improvement in defect reduction, and employee education and empowerment.[8]

1988—Manufacturing

Prior to the early 1980s, the **Westinghouse Commercial Nuclear Fuel Division's** (CNFD) quality goals were geared toward satisfying regulatory requirements. By 1988, CNFD had reduced their nonconformance rate to only 50 ppm. Electric

---

[8] *Ibid.*

utilities that are operating nuclear power plants installed with fuel rod assemblies made by CNFD can be confident that each of the thousands of rods supplied will perform flawlessly. To achieve this, CNFD is building a quality culture that has customer satisfaction as its guiding principle, whether it is the ultimate user or the next person in the process. This "Total Quality" approach asks each of the nearly 2000 employees to do "the right things the first time." This philosophy makes every action by every employee a quality initiative. CNFD estimates that quality-related decisions have dictated 75% of its capital allocations during the recent years. Progress is measured by a system called "Pulse Points," which tracks improvements in more than 60 key performance areas identified with statistical techniques and other evaluative tools.

CNFD management deliberately did not include cost concerns in the quality improvement program, believing that gains in quality would spawn cost reductions through increases in efficiency. Results achieved between 1987 and 1994 confirm this belief. For example, first-time yield increased from 50 to 87%, helping CNFD achieve over three years of 100% on-time delivery of high-quality products.

1988—Small Business

In 1985 **Globe Metallurgical** set out to become the lowest cost, highest quality producer of ferroalloys (iron-based metals) and silicon metal in the United States. Three years later, Globe occupied a quality niche above the competition. In customer quality audits the firm's scores have set records. Globe credits these quality successes to its company-wide quality improvement system—termed Quality, Efficiency, and Cost (QEC). Globe has found no scarcity of good ideas for improving product quality and reducing costs, many originating from its 210 employees who participate in quality circles. Through the implementation of these quality improvement ideas, Globe has experienced a 367% increase in productivity, waste reduction of more than $13 million a year, a 77% reduction in inventory, and a 46% reduction in the employee absentee rate. "Despite what seemed to be an enormous initial investment in quality, Globe had gains of a $40 return for every dollar spent on quality."[9]

SUMMARY

Although the focus of this book is on the use of the MBNQA Criteria for self-improvement, an awareness of the Award process was provided to offer a deeper understanding of the Criteria. This chapter examined the Award process and supporting organization. Another reason to understand the process is to use it at the

---

[9] "Baldrige Winner Gains Market Share," *Quality in Manufacturing*, May/June 1993.

state, local, or company level. Many initiatives are using the Baldrige Criteria and process at the state and local level as well as internally at companies. This was done to get the Criteria to organizations that may not have heard about it otherwise.

A summary of the Award recipients was presented that offers insight into their successful strategies, which are varied. There are many approaches and no one right answer. All of the companies have some common threads such as leadership involvement, focus on the customer, developing the full potential of the workforce, and the use of data to guide decisions. Whatever the specific approach, the results show improved business and quality performance. For the ten recipients that analyze productivity enhancement as annual increase in revenue/employee, a median average annual compounded growth rate of 9.4%, and a mean of 9.25% have been achieved.[10]

---

[10] *Quality Pays*, NIST, 1993.

6 | Getting Started: The Assessment Process

INTRODUCTION

Many topics have been discussed so far, dealing with the philosophy, elements, and structure of the Baldrige Criteria and Award process. The emphasis has been on understanding and using the Criteria and Award process for self-assessment and improvement. Ideas were presented for successfully using the Criteria in Chapter 3. This chapter pulls many of the topics together by addressing how to get started. If an organization wants to use the Criteria for self-assessment and improvement, but needs some advice on first steps, this chapter will help.

GETTING STARTED

Although the MBNQA Criteria are not prescriptive, they are built on a set of core values and concepts (see Chapter 1). Before you and your organization begin to use the Criteria for self-improvement, you should make sure that these core values and concepts are consistent with your philosophy and paradigms. Otherwise it is like fitting a square peg in a round hole. And simply deciding to implement a quality initiative is not enough. The organization must be ready for such an initiative. A series of questions is given in Figure 6.1 to assist you in the process of ascertaining your organization's readiness.

Figure 6.1

Questions to Ask Leadership

Questions to Ask Leadership

These questions should be discussed with your organization's leadership to ascertain the readiness to implement a self-improvement (quality) initiative.

1. Is the vision or aim of the organization consistent with the MBNQA core values and concepts?

2. How will the vision or aim of the quality initiative be communicated to all employees? How will their feedback be incorporated?

3. What is your role in quality-related activities such as planning, review of quality plans and progress, teams, and giving and receiving education?

4. What are the strategies to promote cooperation among managers and supervisors at all levels?

5. What strategies will increase the involvement, effectiveness, and productivity of all categories of employees?

6. What is the approach and rationale for deciding what quality education is needed by different categories of employees?

7. How will you assess the effectiveness of the quality initiative?

8. How does the quality initiative address getting and converting customer needs and expectations into product and process requirements (and/or service standards)?

9. How does the quality initiative drive the focus from product/outcome control to process control?

10. How will the suppliers be involved in the quality initiative?

11. What elements of your plan assure the prevention of problems (not just detecting and reacting to problems)?

12. How will you know if observed variation is from the system or is a special cause?

13. What are the main roadblocks to successful implementation of the quality initiative?

14. What can be done to prevent the roadblocks from occurring?

15. What is the process of improving our quality initiative process?

IMPLEMENTATION: KEY POSITIONS

Once the decision has been made to use the Criteria for self-assessment and the *buy-in* has been achieved, a core team should be identified by the organization. The core team will be the facilitators who guide the total process of writing, reviewing, and providing feedback. These individuals should become knowledgeable about the Criteria to help others if they struggle with interpretation, especially during the development of the written response to the Criteria. In small organizations the core "team" is one person. In an organization with 10,000 employees and multiple sites, the core team may consist of 6 to 10 people. This team also ensures that the links between interrelated Items are made.

In addition to the core team, the following should be identified to assist in the development of the written response: category leaders, item leaders, and content experts for areas to address. Also it is helpful to appoint a "chart" champion. That is, someone who can assist with the development of graphs and charts for the results Items as well as general assistance with any computer interface. This person should have good statistical and presentation skills.

The job description for each group needs to be defined. Let's begin with the *category leaders*. Typically there are seven category leaders, one for each category. Their job is to manage the development of the answers to the specific category assigned. They need to communicate pertinent information to item leaders, follow up to assure correct focus is maintained, and integrate various items to compose the Category draft. The category leaders must coordinate among themselves to eliminate redundant activity descriptions (i.e., the activity is described in detail in two Categories) as well as assure consistency in terminology throughout the document. Also, a key job element for them is to identify and remove roadblocks encountered by the item leaders.

The *item leaders* produce a draft response for the Item assigned by analyzing, editing, and integrating the written reports of content experts for areas to address. They identify these content experts, review the Criteria requirements, and help the experts write and describe their activities. It is important for the item leaders to coordinate responses where links exist by integrating various areas to compose the Item draft. Although there are 24 Items in the 1995 Criteria, there are generally fewer than 24 item leaders. Perhaps 10 to 15 people may be involved as item leaders in an organization of 10,000 people.

The *content experts for the areas to address* provide written description of how the system works. They write the responses to the Criteria assigned, providing the reality viewpoint. They should ask questions of the item leader, ask for guidance, and request help in getting over the "speed bump." These experts assure the accuracy of the document, provide examples, and provide the existing data. There may be 50 or more people involved at this level in an organization of 10,000 employees.

A summary of these jobs is provided here.

Category Leaders

☑ Manage development of answers to Category assigned.

☑ Communicate pertinent information to item leaders.

☑ Follow up to assure correct focus is maintained.

☑ Integrate various Items to compose the category draft.

☑ Coordinate among themselves for consistency.

☑ Remove roadblocks that the item leaders encounter.

Item Leaders

☑ Produce draft response for the Item assigned by analyzing, editing, and integrating written reports of content experts for areas to address.

☑ Identify these content experts, review Criteria requirements, and help experts write and describe their activities.

☑ Coordinate responses where links exist by integrating various areas to compose Item draft.

Content Experts for the Areas to Address

☑ Provide written description of how system works.

☑ Write responses to Criteria assigned.

☑ Provide the reality viewpoint.

Continuing with the big picture view of organizing for the self-assessment, the leaders and experts just discussed need to be educated on the Criteria. Although the item leaders and content experts may want to learn only about their particular area, there is a need for them to be exposed to all of the Criteria. Because of the numerous links between Items, one cannot develop a response for one Item independent of others.

DEVELOPMENT PROCESS

The core team meets with all participants to review the process and timetable and to discuss expectations and roles. The core team members should be the ones to answer questions or do the legwork to get answers to questions about the process, timetable, and roles.

All employees should be informed of these activities. If a self-assessment is being done and will be used as input to the business plan, then employees should be informed through an all-employee meeting, company newsletter, company video broadcast, or other means used to communicate within the organization. Updates could be made along the journey by announcing the major findings and plans for improvement.

The category leaders develop and present their plan of attack to their item leaders and content experts. Typically, the initial thrust is to bring together existing information. A working draft, with hand-drawn charts and unpolished descriptions, is pulled together. The category leaders review the consolidated draft and make recommendations for the next steps. For example, if the organization already gets customer feedback as part of the normal process, the category leader might decide to retain the draft response without modification. Without an ongoing customer feedback process, a different category leader might decide to initiate a customer survey or initially try to stretch some limited information and call it customer feedback.

The process continues. The document is updated and completed. All of the employees should be advised of this mile marker and be given an opportunity to read the application. Some companies have placed several copies in central locations for people to read. Other companies encourage the employees to provide comments about the document; of course, this requires that the employees be kept informed of the self-assessment process.

REVIEW PROCESS

At this stage the Examiners, internal or external, review the written response to the Criteria—first individually and then as a team—develop comments and a score, and write the feedback report (see also Chapter 5). This team should write a feedback report that identifies the strengths and areas for improvement. The company then needs to define a plan of action. Why do an assessment unless you are going to use the data? The improvement plan should be linked to the business plan so the organization is not being pulled different ways.

Keep the review process focused on improvement. Don't let it become a numbers game. One way to avoid having units play the numbers game is to minimize the adverse effects of a low score in the early years of using the Criteria. It would seem that a unit that acknowledges its area needs improvement and that has an improvement plan in place would be looked upon more favorably than a unit that may have scored higher but thinks no improvement is needed. The attitude and culture displayed by the first unit would lead to higher levels of achievement in the future than that of the second unit where status quo is thought to be good enough. See Chapter 3 for additional information on using the Criteria successfully.

It is important that the executives and line managers remain involved in the review and improvement process. The more the Criteria are understood, the more likely they will be used to guide the ongoing journey.

SOURCES OF INFORMATION

Data that describe the key measures and indices are an integral part of the Criteria. A description of data being collected and analyzed is the focus of Category 2.0. How the data are used shows up in many places, but specifically in Category 5.0. The actual trends of the data are reported in Categories 6.0 and 7.0. Therefore, a self-assessment requires a review of the flow of current data and information. This should be considered from several viewpoints including who, how, when, why, and what:

"Who" are the customers and suppliers of current data?

"How" are the data gathered, analyzed, and distributed?

"When" does this occur (frequency)?

"Why" are the data collected?

"What" decisions are made based on the data?

There are numerous sources of information in a company. Team results, documentation of corrective or preventive actions, supplier quality data, product or process data, field data, customer surveys, complaints and letters, awards, warranties and guarantees, and cycle time analyses are the more traditional sources. Some other examples include agendas for key meetings, displays of mission and vision, competitor and benchmark studies, process flowcharts, public health and safety reports, regulatory reviews, quality or business plans, personnel statistics, employee opinion surveys, accident reports, product, process or system audit reports, minutes from meetings showing action items—to name just a few.

Determine if you are to describe the data and its usage or are you to report actual data. To do this, review the Award Criteria and determine if approach, deployment, or results are being requested; this information is given at the beginning of each Item. If it is approach and deployment, then consider providing documented processes, systematic and consistent processes, and/or process flow diagrams to demonstrate defined systems. If the words "current levels and trend" are used, results are required so three to five years of data should be shown.

SUMMARY

This chapter provided recommendations and guidelines an organization could follow in developing their written response for self-assessment. Also discussed was the necessity for the organization's philosophy and paradigms to be consistent with the core values and concepts of the Criteria. A series of questions for the leadership of an organization was given to assist them in ascertaining their readiness for a quality initiative based on the Criteria.

7 | Case Study: Collins Technology

INTRODUCTION

This chapter provides an example of a written response to the 1994 MBNQA Criteria. It is a revision and expansion of the case study prepared by NIST for use in the 1992 MBNQA Examiners Preparation Course. All Categories, Items, and Areas to Address have been updated to the 1994 Criteria. Further, several strategies, deployment and implementation activities, and corresponding results have been strengthened and enhanced to provide a high-scoring organization—but one with areas for improvement, which is typical of the Award recipients.

The Collins Technology application and feedback report were developed as if they were being used by the company for self-improvement. If the written response were to be submitted to NIST for consideration in the Award process, an approved eligibility determination form and application form would need to be included as part of the total application package.

This case study can be used either as an example of a completed written response and feedback report or it can be used as an exercise to calibrate the response development team and Examiners to the same yardstick.

AUTHORS' NOTE

Even though the rest of the book focuses on the 1995 Criteria, the Collins Technology Case Study is based on the 1994 Criteria. The primary reason for this was that publication timing constraints did not allow the conversion of the case study to the 1995 Criteria. This was a difficult decision for us because we did not want to complicate this book with two sets of Criteria and yet wanted to have the latest Criteria as the main focus. After much discussion among ourselves and with colleagues, we decided to go with the present configuration. The purpose of the case study is to provide an example of a written response and review documentation as well as a calibration tool for those who wanted to review and score the written response themselves. We feel that this purpose will be well served by the following case study.

Because there are differences between the 1994 and 1995 Criteria, we have included a copy of the 1994 Criteria with the written response and the following section, which compares the 1994 and 1995 Criteria.

DIFFERENCES BETWEEN THE 1994 AND 1995 CRITERIA

With the changes in the 1995 Criteria, the MBNQA process continues its history of continual improvement (see Chapter 14). Figure 7.1 identifies the differences between the 1994 and the 1995 Criteria.[1]

USING THIS CASE STUDY

If your purpose in reading this book is to get an understanding of the MBNQA Criteria and process in order to use it with your organization's self-improvement efforts, then you can simply read the following written response and companion feedback report (in Chapter 10) along with the other chapters of the book. The full written response will show you how all the pieces (the activities of the different Category teams) fit together. The feedback report highlights these strengths and also identifies areas for improvement. Because Collins Technology was developed as a high-scoring organization, you may also get some improvement ideas from this response.

The case study can also be used for training and calibrating Examiners and other persons interested in the evaluation process. The response development team should be included in this training. It is our experience that when the response development team understands the evaluation process, the resulting written response tends to be more concise, detailed, and focused with less redundancy and empty prose. That is, the team has a customer focus and understands that the Examiners are one of their customers.

---

[1] Based on the 1994 and 1995 MBNQA Award Criteria, NIST, Gaithersburg, MD.

Figure 7.1
Differences Between 1994 and 1995 MBNQA Criteria

MBNQA Criteria		Points		Areas to Address		Comments
1994	1995	1994	1995	1994	1995	1995
1.0 Leadership		**95**	**90**	**12**	**7**	Basic purposes are unchanged. Added emphasis on a leadership *system* that promotes performance excellence.
1.1. Senior Executive Leadership		45		4	2	Clarification of senior executives' roles in building, sustaining, and improving an environment for performance excellence. Expanded and clarified Notes.
1.2. Management for Quality	1.2. Leadership System and Organization	25		4	3	Major thrust is toward flexible and responsive organizations and leadership systems—major themes of the 1995 Criteria. Revised Note.
1.3. Public Responsibility and Corporate Citizenship		25	20	4	2	Realignment and clarification of the Areas to Address. Results are no longer requested. Clarified Notes.
2.0 Information & Analysis		**75**		**11**	**6**	Basic purposes are unchanged. Effectiveness of the use of data and information in support of high-performance work units has been added.
2.1. Scope and Management of Quality & Performance Data and Information	2.1. Management of Information and Data	15	20	3	2	The scope is broader than in 1994. The Areas to Address more clearly focus on the process to select and manage data and information and the continual improvement of this process. Revised Notes.

From M. S. Heaphy and G. F. Gruska, *Malcolm Baldrige National Quality Award: A Yardstick for Quality*, Addison-Wesley, 1995

Figure 7.1
Differences Between 1994 and 1995 MBNQA Criteria (*continued*)

MBNQA Criteria		Points		Areas to Address		Comments
1994	1995	1994	1995	1994	1995	1995
	2.2. Competitive Comparisons and Benchmarking	20	15	4	2	More focus on company priorities and effective benchmark practices than on the scope, also addressed by other Items. Expanded Notes.
	2.3. Analysis and Uses of Company-Level Data	40		4	2	Increased emphasis on the process of company-level data analysis in support of decision-making and the improvement of this process. Expanded and revised Notes.
3.0 Strategic Quality Planning	**3.0 Strategic Planning**	60	55	8	5	The title and focus have been changed to reflect the move toward performance excellence as a result of the total system's strategy and deployment.
3.1. Strategic Quality and Company Performance Planning Process	3.1. Strategy Development	35		4	3	Focuses on the operational requirements of strategic initiatives as a key element in performance management. Deployment moved to Item 3.2. Areas to Address simplified. Extensive expansion of the Notes.
3.2. Quality and Performance Plans	3.2. Strategy Deployment	25	20	4	2	Expanded to include deployment of key business drivers. Emphasizes that productivity and cycle time improvement should be included among the key requirements. Areas to Address simplified. Expanded Notes.

Figure 7.1
Differences Between 1994 and 1995 MBNQA Criteria (*continued*)

MBNQA Criteria		Points		Areas to Address		Comments
1994	1995	1994	1995	1994	1995	1995
4.0 Human Resource Development and Management		**150**	**140**	**18**	**9**	Increased emphasis on the alignment between human resource management and the company's performance objectives. Category 4.0 results are now reported in Item 6.2.
4.1. Human Resource Planning and Management	4.1. Human Resource Planning and Evaluation	20		3	2	Evaluation is a crucial element of the alignment process. The title was changed to emphasize that this Item is the key link between company planning and overall human resource management. Expanded and revised Notes.
4.2. Employee Involvement	4.2. High-Performance Work Systems	40	45	4	2	This new Item is a combination of Items 4.2 and 4.4 in the 1994 Criteria. It examines work and job design and the supporting recognition system. It emphasizes that full participation is a design as well as a motivational issue. Results are no longer requested. New Notes.
4.3. Employee Education and Training	4.3. Employee Education, Training, and Development	40	50	4	2	Expanded to include the larger and more continuing issue of development, which has increasing importance in high-performance work designs. Results are no longer requested. Expanded and revised Notes.
4.4. Employee Recognition	(combined with Item 4.2)	25	—	3	—	(combined with Item 4.2)
4.5. Employee Well-Being and Satisfaction	4.4. Employee Well-Being and Satisfaction	25		4	3	Clarification of Areas to Address and Notes. Results are no longer requested. Expanded Notes.

Figure 7.1
Differences Between 1994 and 1995 MBNQA Criteria (continued)

MBNQA Criteria		Points		Areas to Address		Comments
1994	1995	1994	1995	1994	1995	1995
5.0 Management of Process Quality	**5.0 Process Management**	140		14	10	Title of Category and Items changed to minimize confusion due to multiple meanings of "quality." This Category covers the process management and improvement of all operations, processes, and activities of all work units.
5.1. Design and Introduction of Quality Products and Services	5.1. Design and Introduction of Products and Services	40		3		Clarification of Areas to Address and Notes. Expanded Notes.
5.2. Process Managment: Product and Service Production and Delivery	5.2. Process Managment: Product and Service Production	35	40	2		Problem-solving and corrective action activities are not part of the Areas to Address but are tied in via the Notes. Clarification of Areas to Address and Notes. Expanded Notes.
5.3. Process Management: Business and Support Service Processes	5.3. Process Management: Support Services	30		3		Title changed to emphasize that this Item deals with those processes that support the key production and delivery processes, but are usually not designed with them. Clarification of Areas to Address and Notes. Expanded Notes.
5.4. Supplier Quality	5.4. Management of Supplier Performance	20	30	4	2	Title changed to reinforce the focus on overall performance management and improvement. Clarification of Areas to Address and Notes. Expanded Notes.
5.5. Quality Assessment	(included in Item 1.2)	15	—	2	—	This Item has been eliminated due to its overlap with Item 1.2. Responses to this Item in 1994 are appropriate for inclusion in 1.2c.

Figure 7.1
Differences Between 1994 and 1995 MBNQA Criteria (*continued*)

MBNQA Criteria 1994	MBNQA Criteria 1995	Points 1994	Points 1995	Areas to Address 1994	Areas to Address 1995	Comments 1995
6.0 Quality and Operational Results	**6.0 Business Results**	**180**	**250**	**8**	**3**	Title has been changed to reflect the greater emphasis on business-oriented results.
6.1. Product and Service Quality Results		70	75	2	1	No functional changes. Consolidation of Areas to Address. Revised Notes.
6.2. Operational Results	6.2. Company Operational and Financial Results	50	130	2	1	Title changed to reflect that this Item now examines the overall company operational improvements from all processes including human resources processes. Consolidation of Areas to Address. Revised Notes.
6.3. Business and Support Service Results	(combined with Item 6.2)	25	—	2	—	Combined with Item 6.2 to focus on company-level improvements. This is consistent with the more holistic approach to performance management.
6.4. Supplier Quality Results	6.3. Supplier Performance Results	35	45	2	1	No functional changes. Consolidation of Areas to Address. Revised Notes. (See also comments for Item 5.4.)
7.0 Customer Focus & Satisfaction		**300**	**250**	**20**	**14**	Focus includes the systems for "Customer Learning" as well as building and maintaining customer relationships.
7.1. Customer Expectations: Current and Future	7.1. Customer and Market Knowledge	35	30		3	Title has been changed to reflect the Item's broader scope, more analytical approach, and continual nature of activities. Clarification of Areas to Address. Expanded Notes.

Figure 7.1
Differences Between 1994 and 1995 MBNQA Criteria *(continued)*

MBNQA Criteria		Points		Areas to Address		Comments
1994	1995	1994	1995	1994	1995	1995
7.2. Customer Relationship Management		65	30	7	4	Item has been narrowed in scope to distinguish it from Items 5.2 and 5.3. The focus is now on response management. Clarification and reduction of Areas to Address and new and expanded Notes.
7.3. Commitment to Customers	(eliminated)	15	—	2	—	Despite its title, this Item in 1994 dealt primarily with guarantees and warranties, which may be regarded as elements of a company's product, marketing, and/or management strategy. These are addressed or implied in several other Items.
7.4. Customer Satisfaction Determination	7.3. Customer Satisfaction Determination	30	30		3	No functional changes. Clarification of Areas to Address. New and clarified Notes.
7.5. Customer Satisfaction Results	7.4. Customer Satisfaction Results	85	100		2	No functional changes. Clarification of Areas to Address and Notes.
7.6. Customer Satisfaction Comparison	7.5. Customer Satisfaction Comparison	70	60	3	2	No functional changes. Clarification of Areas to Address.
Totals						
7 Categories and 28 Items	7 Categories and 24 Items	1000		91	54	As a result of the reduction in the Items and Areas to Address, the application page limit has been reduced from 85 to 70 pages.

For this later purpose, the reader should follow the guidelines in Chapters 8 and 9. Specifically, the reader should follow these steps (see Chapter 9 for details):

Scoring Steps

- Read entire application

- For each Item

    - Reread the Criteria

    - Read the response

    - Write comments

    - Assign score

    - Review score and comments for consistency

The Examiner teams can then apply the consensus process to develop the final score and feedback report. These results can then be compared to the feedback report in Chapter 10. Although the specific comments may be different between the book and the team, the general flavor of the feedback report should be the same and scores similar.

COLLINS TECHNOLOGY: BACKGROUND

The Collins Technology Case Study describes a fictional company. The company's description is contained in the overview section of the written response. There is no connection between the Collins Technology Case Study and any company, either named Collins Technology or otherwise. Customer and supplier companies cited in the case are also fictitious.

Collins Technology is a small business (one facility with 178 employees) manufacturing multi-layer printed circuit boards (PCBs). It has three business segments: government, industrial, and commercial. Even though their sales are international (16% of shipments are outside North America), Collins Technology does not have any dealers, distributors, or franchises. Instead they deal directly with the original equipment manufacturers (OEMs) and fabricators. Due to the complexity of the circuit boards produced (10 layers or more), consistent quality is of paramount importance. Quality is measured by:

☑ Dimensional precision and stability

☑ Predictable internal capacitance

- ☑ Tolerance for extreme environments
- ☑ General cleanliness and workmanship.

Additional key business factors regardless of business segment include:

- ☑ On-time delivery (design through fabrication)
- ☑ Competitive pricing
- ☑ State-of-the-art performance.

Although Collins Technology is in a specialized niche of the PCB business, it still has several competitors, some of which are captive to large corporations with ready access to significant capital. The key to their success has been customer satisfaction through long-term customer alliances and close direct engineering and support services.

The following is the written response developed by Collins Technology using the 1994 MBNQA Criteria.

1994 Case Study:
Collins Technology Application

The Collins Technology application and feedback report were developed as if they are being used internally by the company for self-improvement. If the application were to be submitted to NIST for consideration in the Award process, an approved eligibility determination form and application form would need to be included.

Because there are differences between the 1994 and 1995 Criteria, a copy of the 1994 Criteria is included within the application. Each Category is presented before the written response to the Category. These are not part of the application.

This is an example of a submission for the 1994 MBNQA. It is a revision and expansion of the case study prepared by NIST for use in the 1992 MBNQA Examiners Preparation Course. All Categories, Items and Areas to Address have been updated to the 1994 Criteria.

The Collins Technology Case Study describes a fictional company. There is no connection between the Collins Technology Case Study and any company, either named Collins Technology or otherwise. Customer and supplier companies cited in the case are also fictitious.

All revisions, expansions, and enhancements © 1993–94 by The Transformation Network, Inc. & The Third Generation, Inc.

OVERVIEW

World-class is the aim of Collins Technology with the successful implementation and operation of a Total Quality Management (TQM) System. Total quality means the full involvement of our customers, employees, management, and suppliers. It requires a stretch of the energy and the imagination of all our employees to continue toward a never-ending destination.

Major Products—Collins Technology incorporated in 1971 and was organized to focus on multi-layer printed circuit boards with concentration on advanced materials for higher frequencies, more demanding environments, and fine lines for high densities of attached components, with very high reliability. These features are accomplished with boards with more than ten layers. Experienced application engineers custom design heat sinks to be used with assembled products to meet customer requirements. The consistency of the high performance of our products is a tribute to the TQM System.

Nature of Major Markets—Our sales are international. We ship to customers in North America, Europe, and the Far East. While North America makes up the majority of our volume, 16% of our shipments are to Europe and the Far East.

Description of Principal Customers—We have three business segments. These are Government, Industrial Products, and Commercial Products. Each of these segments has a Business Segment Manager who is responsible for the customers in that group, coordinating the market information for each segment, and assuring that relevant data flows into the strategic planning process. A Sales Coordinator is responsible for the day-to-day interaction with the customer.

Each customer base has different characteristics for each of our segments.

Government—This group consists of the Department of Defense (DoD), NSA, NASA, and other areas of the government that purchase multi-layer printed circuit boards. The characteristics for this segment are smaller quantities (less than 500), products designed primarily to their specifications with little flexibility, use of assured and proven technology, and contractual agreements.

Industrial Products—This group consists of companies who use printed circuit boards in their own manufacturing processes for process control. They are most interested in product reliability and our control over our manufacturing processes. They are less interested in incorporating new technology until it is fully proven to be reliable.

Commercial Products—These customers use printed circuit boards in products they manufacture to be resold to other businesses. Examples include automotive, electronic consumer products, telecommunications, computer, etc. The key characteristics required by this group of customers include use of new technology that will give them a competitive advantage and ways of differentiating their product. They tend to be very interactive and cooperative in new product design, working together with Collins Technology to improve boards. The Commercial Products customers usually order in larger quantities (over 10,000 boards).

Over the last five years, we have actually reduced our customer base. We concluded that for us to best serve our customers, we have to match their needs with the things we do best. In 1987, we had over 800 customers. Based on our analysis of the market and trends in the industry, we developed a strategy to reduce our customer base. We currently have 384 customers. In this period, however, our sales volume increased by 226%; our profits increased by 311%.

Our customers range from government agencies and Fortune 500 customers, to small specialty manufacturers. In most cases, we select customers where we can supply 100% of their highly complex multi-layer boards. They will typically purchase boards of less than ten layers from other board manufacturers. Many of our customers have the ability to produce printed circuit boards internally.

Key Quality Requirements for Products—Regardless of their business segment, customers are interested in on-time delivery, competitive prices, and state-of-the-art performance. There are also several quality aspects that are significant to all our customers. These additional requirements include dimensional precision and stability, predictable internal capacitance, and tolerance for extreme environments, as well as general cleanliness and workmanship. Quality is of paramount importance for a printed circuit board of more than ten layers to operate with very high frequencies in a hostile environment and yet have the dimensional accuracy and stability to be used with various insertion machines in various manufacturing companies, some of which are located overseas. Very often the value of the piece parts applied to the boards is very high, and the customer depends heavily on the integrity of the raw board.

Our ability to consistently produce complex multi-layer boards is the factor that differentiates Collins Technology from other printed circuit board (PCB) competitors. The achievement of the quality reputation has taken many years to accomplish.

Competitive Environment—
Regardless of being a small business, we aggressively compete in a world-wide market. We maintain a significant market share in the Far East and Europe, even with Eastern and European competitors. There are several competitors in this very specialized niche of the PCB business. Most are small, but some are captive to large corporations and have ready access to significant capital funds, technically trained people, and a ready customer in place. Collins Technology has increased its market share from 25% to 65% of this specialized area of the printed circuit board market over the past five years. By consistently meeting the customers' expectations, we maintain a single source position on many military applications, even with the government's desire to have dual sources and with many customer concerns about the risk of suppliers with complex processes, such as those associated with multi-layer printed circuit boards.

The key to our success has been customer satisfaction through long-term customer alliances and close direct engineering and support service. The majority of the programs begin on a small scale.

Engineering personnel work directly with the customer to establish and implement the specific requirements, as contrasted to just selling a product. In almost all situations a board is unique for a specific application with a given customer. Our approach has developed strong, long-term partnering with our customers. Our reputation of successfully meeting their design, delivery, and quality requirements results in repeat business. Due to the quality of our performance, we have lost only two major customers for competitive reasons since our founding, and none in the last 12 years.

Position in the Industry—In each of the business segments, Collins Technology has achieved a prominent leadership role. Although a small business, we have the major share of our narrow area of specialization worldwide, including the Far East, with a field of several competitors. We have worked with many customers for 18 years. We are proud of our numerous customer awards. We have received special quality recognition from the U.S. Navy and the U.S. Air Force. Our successful market position is a long-term result of supplying our customers with products and services that meet their expectations. The management philosophy results from our dedication to total customer satisfaction. It is significant to note that since 1975 the company has had no new equity capital invested, although this is a very capital-intensive business.

Major Equipment and Facilities—
The equipment we use is essentially the same as that used by other companies in our business. We have developed automatic process control equipment to optimize the many critical processes involved in producing our products. Special test equipment

has been developed to satisfy customer requirements.

Types of Technology Used—The diversified technology base continues to advance the state-of-the art in multi-layer printed circuit boards; particularly in materials and line densities. We have technology specialists in heat transfer, mechanical, electrical, chemical, materials, industrial, and manufacturing engineering.

General Description of Employee Base—Collins Technology is a non-union company. We have grown from 5 to 178 employees since inception, with 70% employed five years or more. Because of the emphasis on support and direct customer relationships, our professional staff represents 26% of our workforce. We provide internal training in the special skills required for our production personnel. Through the tuition reimbursement plan, currently 7% of our employees are upgrading their education level.

We operate with the internal/external customer concept wherein all employees have a customer and spend a minimum of 20 minutes per week communicating with each internal customer. Using the TQM System as a quality guide, action teams demonstrate the positive results of employee empowerment.

Importance of Suppliers and Distributors—The purchasing department updates the critical and active supplier lists annually. Purchasing and quality personnel visit suppliers on an annual schedule. Marketing, engineering, and other personnel visit suppliers during sales trips. The purpose is to develop long-term supplier/customer relationships through early supplier involvement. We have no distributors; all sales are direct to the users of our products.

Occupational Health and Safety—Through our TQM System we have developed and implemented a documented health and safety procedure and related training for our employees. Our extensive use of toxic, controlled materials requires special emphasis on safety awareness to maintain a safe work environment.

Other Factors—We implemented our TQM System because of a desire to improve; it was not the result of competitive pressure, loss of market share, or lack of profits. Collins Technology has a "process" in place which is never-ending, as opposed to a "program" that has a beginning and an end. Culturally we have strived to constantly improve the way we think and act.

Consistent quality is foremost in those applications where a product must be right. For critical applications, customers rely on Collins Technology for their printed circuit boards. Examples are underwater amplification and space applications. Regardless of the application, whether long or short runs, Collins Technology has demonstrated to our customers that they can rely on the quality and performance of our boards.

It is important to note the significance of the MBNQA process relative to our TQM System. Collins Technology was using the MBNQA Criteria as our internal guide even before our decision to submit our application to the review process. It provides a structured program that empowers all of our employees with a method for continual improvement. It is one of the elements in our never-ending quest for customer satisfaction.

Description of Company

- 1 facility

- 178 employees

- manufacturer of multi-layer printed circuit boards

- international sales

- three business segments
  - government
  - industrial
  - commercial

- 75 suppliers

- no dealers, distributors, or franchises

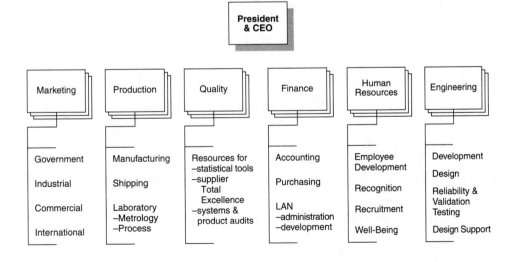

1994 MBNQA Award Examination Criteria

1.0 Leadership *(95 pts.)*

The *Leadership* Category examines senior executives' *personal* leadership and involvement in creating and sustaining a customer focus and clear and visible quality values. Also examined is how the quality values are integrated into the company's management system, including how the company addresses its public responsibilities and corporate citizenship.

1.1 Senior Executive Leadership *(45 pts.)*

Describe the senior executives' leadership, personal involvement, and visibility in developing and maintaining an environment for quality excellence.

☑ Approach[1]
☑ Deployment
☐ Results

Areas to Address

(a) senior executives' leadership, personal involvement, and visibility in quality-related activities of the company. Include: (1) creating and reinforcing a customer focus and quality values; (2) setting expectations and planning; (3) reviewing quality and operational performance; (4) recognizing employee contributions; and (5) communicating quality values outside the company.

(b) brief summary of the company's customer focus and quality values that serve as a basis for consistent understanding and communication within and outside the company.

(c) how senior executives regularly communicate and reinforce the company's customer focus and quality values with managers and supervisors.

(d) how senior executives evaluate and improve the effectiveness of their personal leadership and involvement.

Notes:

1. "Senior executives" means the applicant's highest-ranking official and those reporting directly to that official.

---

[1] Each Item is classified according to the kinds of information and/or data expected. See Chapter 9 for a description of these designations.

2. Activities of senior executives (1.1a) might also include leading and/or receiving training, communicating with employees, benchmarking, customer and supplier interactions, and mentoring other executives, managers, and supervisors.

3. Communication by senior executives outside the company [1.1a(5)] might involve: national, state, and community groups; trade, business, and professional organizations; and education, health care, government, standards, and public service/charitable groups. It might also involve the company's stockholders and board of directors.

1.2 Management for Quality *(25 pts.)*

Describe how the company's customer focus and quality values are integrated into day-to-day leadership, management, and supervision of all company units.

☑ Approach
☑ Deployment
☐ Results

Areas to Address

(a) how the company's customer focus and quality values are translated into requirements for all managers and supervisors. Describe: (1) their principal roles and responsibilities within their units; and (2) their roles and responsibilities in fostering cooperation with other units.

(b) how the company's customer focus and quality values (1.1b) are communicated and reinforced throughout the entire workforce.

(c) how overall company and work unit quality and operational performance are reviewed. Describe: (1) types, frequency, content, and use of reviews and who conducts them; and (2) how the company assists units that are not performing according to plans.

(d) how the company evaluates and improves managers' and supervisors' effectiveness in reinforcing the company's customer focus and quality values.

Notes:

1. Communications throughout the entire workforce (1.2b) should emphasize the overall company approach and deployment. Some of this communication may be done by senior executives, as addressed in Item 1.1.

2. The evaluation (1.2d) might utilize employee input or feedback on managers' and supervisors' leadership skills in reinforcing a customer focus and quality values.

1.3 Public Responsibility and Corporate Citizenship *(25 pts.)*

Describe how the company includes its responsibilities to the public in its quality policies and improvement practices. Describe also how the company leads as a corporate citizen in its key communities.

☑ Approach
☑ Deployment
☑ Results

Areas to Address

(a) how the company integrates its public responsibilities into its quality values and practices. Include: (1) how the company considers risks, regulatory and other legal requirements in setting operational requirements and targets; (2) a summary of the company's principal public responsibilities, key operational requirements and associated targets, and how these requirements and/or targets are communicated and deployed throughout the company; and (3) how and how often progress in meeting operational requirements and/or targets is reviewed.

(b) how the company looks ahead to anticipate public concerns and to assess possible impacts on society of its products, services, and operations. Describe briefly how this assessment is used in planning.

(c) how the company leads as a corporate citizen in its key communities. Include: (1) a brief summary of the types and extent of leadership and involvement in key communities; (2) how the company promotes quality awareness and sharing of quality-related information; (3) how the company seeks opportunities to enhance its leadership; and (4) how the company promotes legal and ethical conduct in all that it does.

(d) trends in key measures and/or indicators of improvement in addressing public responsibilities and corporate citizenship. Include responses to any sanctions received under law, regulation, or contract.

Notes:

1. The public responsibility issues addressed in 1.3a and 1.3b relate to the company's impacts and possible impacts on society associated with its products,

services, and company operations. They include business ethics, environment, and safety as they relate to any aspect of risk or adverse effect, whether or not these are covered under law or regulation.

2. The term "targets" as used in Item 1.3 and elsewhere in the Criteria refer to specific performance levels, based upon appropriate measures or indicators.

3. Major public responsibility or impact areas should be addressed in planning (Item 3.1) and in the appropriate process management Items of Category 5.0.

4. Health and safety of employees are not included in Item 1.3. They are covered in Item 4.5.

5. The corporate citizenship issues appropriate for inclusion in 1.3c relate to actions by the company to strengthen community services, education, health care, environment, and practices of trade or business associations. Such involvement would be expected to be limited by the company's available human and financial resources.

6. If the company has received sanctions under law, regulation, or contract during the past three years, include the current status in responding to 1.3d. If no sanctions have been received, so indicate. If settlements have been negotiated in lieu of potential sanctions, give explanations.

1.0 LEADERSHIP

1.1 Senior Executive Leadership

Our President/Chief Executive Officer (CEO) personally initiated our TQM System and has a very active daily role in its implementation. All senior executives are permanent members of our TQM Forum. Senior executives have been involved in the creation and implementation of the prevention-based TQM System. Our commitment is to achieve continual improvements through leadership by example.

1.1a *Leadership, Involvement, and Visibility*

1. *Creating and reinforcing a customer focus and quality values*—The items in Figure 1.1 are tools or methods used to build and reinforce quality values, including a customer focus, for all employees. These quality values are systematically integrated company-wide through the TQM Forum.

The TQM Forum consists of all six senior executives as permanent members, seven rotating members, and invited members. The rotating members serve one-year terms. The invited employees may change weekly.

All members of the TQM Forum have equal status. The purpose of having invited employees is twofold: (1) employees may be participating in a team or activity being reviewed by the Forum and (2) it allows all employees over a period of time to observe and have input into how the Forum functions. Rotating membership encourages leadership through personal involvement in quality goals and values.

Figure 1.1
Building Quality Values

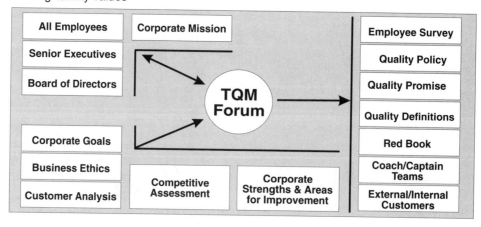

The TQM Forum, chaired by our President/CEO, is the day-to-day operational structure of our TQM System. Senior executives build quality values into the organization beginning with the corporate mission and the corporate goals. The corporate mission defines the relationship with our customers, employees, shareholders, technology, business ethics, and community. The corporate goals include the achievement of world-class status on a worldwide basis and continual improvement in technology and customer satisfaction to increase our market share. The TQM Forum was responsible for developing the quality policy and the quality promise. Figure 1.1 illustrates how the Board of Directors (BoD), senior executives, and our employees all provide leadership in building the quality values.

These values reinforce and support our customer focus. Collins Technology operates with an internal/external customer concept wherein all employees have a customer and regularly spend time communicating with customers. Each senior executive is responsible for specific business customers. It is the executive's responsibility to regularly contact and visit these customers at least twice each month and to share the customer information at the monthly all-employee meetings and in the newsletter. Other employees often accompany the executive. It is important to employees to hear customer feedback because each employee's performance review includes customer satisfaction as measured by this customer feedback to senior executives and through the customer survey.

2. *Planning and setting expectations*—Senior executives within the TQM Forum initiate the planning required to meet quality goals. This effort is accomplished by working with the employees. Goals are defined in the four-year Strategic Business Plan (SBP). Annually, off-site strategic planning meetings result in definitive quality goals and expectations for each senior executive's department.

3. *Reviewing progress*—The TQM Forum, meeting weekly, continually reviews the goals that are assembled in the *"Red Book."* Overall, company performance is reviewed monthly in the all-employee meetings.

   The TQM Forum has the authority and responsibility to review and assess quality and operational performance and is empowered to implement actions and redirect resources.

4. *Recognizing employee contributions*— Understanding that employees play a major role in customer satisfaction, the TQM Forum fosters our employee recognition program. Figure 4.7 shows,

details of this program, which currently consists of over 27 methods of recognition. As with any program of this type, there is always a danger that the program will degrade into a series of competitive contests and/or a substitute for normal compensation. This not only destroys the employees' intrinsic motivation to do well, but also undermines our quality values. A cross-functional team continually reviews and upgrades our employee recognition program.

Most of these recognitions are presented by senior executives, typically at the monthly all-employee meeting. In addition to formal recognition, senior executives are directly involved with employees through participation in various teams and training classes. Through these informal personal interactions, senior executives recognize and reinforce outstanding contributions.

5. *Communications outside the company*—Our senior executives, on a worldwide basis, communicate quality excellence to groups outside the company. The following activities illustrate the depth and breadth of these presentations.

National—Our President/CEO, who serves on the National Board of Directors of the Institute of Printed Circuits (IPC), promotes quality as the method to keep America competitive. He has made many presentations at the annual IPC meetings, describing the importance of quality and customer focus and how they make stronger, competitive companies.

State—We are a cofounder of the annual Quality Expo. The expo is held in conjunction with the American Society for Quality Control's (ASQC) National Quality Month presentations. Our senior executives and employees participate in the organization and sometimes serve as speakers at the Quality Expo. The Quality Expo consists of speakers on various aspects of quality systems. We provide financial support and personal involvement of all of our employees.

Community—We cofounded the Local Quality Consortium in 1987. The consortium is a structured program on how to start and implement a TQM System. We are one of the ten companies in the first group. We participate by teaching and learning. The consortium was the cornerstone of our TQM System. We promote quality excellence through local publications.

Trade/Business—Senior executives provide leadership to assist local companies with their quality education programs. Typical subjects are strategic quality plans, quality goals, employee recognition, and customer and employee surveys. Presentations have been to our customers, suppliers, and local businesses. Similar presentations have been made internationally to Tyler Chemical (England), Anton (France), and our 1992 International Sales Meeting (Germany).

1.1b *Company Quality Values*

Communication of the company quality values to outside communities is described in Item 1.1a. Communications within the company are as follows.

Quality Values—In November 1986, the TQM Forum defined two documents that are the basis of our company's quality values. These documents are the *Quality Policy*, which defines our business philosophy, and the *Quality Promise*, which is each employee's individual commitment.

Quality Policy—We will meet or exceed the customers' expectations in every product or service we provide, without exception. Our policy is:

1. Quality is to consistently meet or exceed all of our customers' expectations.

2. Measurements are via the customers' perspective.

3. The objective is delighted customers.

4. Quality is attained through prevention.

5. Senior executive commitment and involvement leads the quality process.

Quality Promise—I promise my customers I will strive to meet their expectations and will continually improve in everything I do.

Basis for Consistent Communication—During employee orientation, senior executives review our *Quality Policy* and *Promise* and show our video "Total Quality Management at Collins Technology," the story of how we implemented TQM. Thus, from the very beginning employees understand our philosophy. Signing of the quality promise is a voluntary and visual expression of the employee's acceptance and commitment to this philosophy. This understanding, acceptance, and commitment is reinforced through recitation, TQM Forum meetings, company meeting activities, Quality Commitment Days, senior executive coaching, and performance appraisal focus on quality to ensure consistent, continued behavior.

1.1c *Communicate & Reinforce Values—Managers*

The senior executives are all permanent members of the TQM Forum, which meets weekly. This meeting agenda always includes a review of customer satisfaction indicators. Minutes of the TQM Forum are posted. The company uses a coach/captain concept, which is described in Area 4.2a. A senior executive is assigned to each team as coach and in this function continually supports team members with required assets and removal of any roadblocks that may be preventing an individual team from accomplishing its goals.

Our senior executives have all received quality training in classes with other employees. Classes are mixed with managers, supervisors, and workers. As new employees receive training and refresher courses are given, a senior executive is in attendance at *all* training classes. Each senior executive is expected to teach a portion of a class each week.

But the communication and reinforcement of quality values is not only restricted to meetings and classes— see Area 1.2b. The reinforcement of our cultural change is also reflected in the agendas of our operational meetings and our review process. Quality, which used to come last on the agenda, now comes first. Further, each employee's performance review includes quality and customer satisfaction performance measurements.

1.1d *Senior Executive Self-Assessment*

The most effective indicator of senior executive leadership is the achievement of company goals. Progress toward these goals is regularly reviewed in the TQM Forum as described in Area 1.2c.

In addition, the annual employee surveys that are given to all employees each year are structured to provide information feedback to the executives. Questions include those dealing with the quality values, such as: "walking the talk," cooperation from others, importance of all customers, the value of employee opinions. The review process is accomplished with the CEO and all the senior executives together in an off-site meeting. After reviewing the output, each executive develops a "corrective action" personal plan for improvement.

1.2 Management for Quality

Quality is integral to our systems. The basic management system for quality is our normal operating system with direction and oversight by the TQM Forum. The TQM Forum's task is to:

1. oversee and observe the TQM System;

2. establish, review and assess strategic quality goals and attainment plans; and

3. identify opportunities for continual improvement of the total system.

1.2a *Values Translated into Requirements*

1. *Roles within units*—The Strategic Business Plan (SBP) for the entire company is updated annually, consistent with the basic company aims of total customer satisfaction by means of total company quality

management. Each department develops annual attainment plans to meet the corporate goals listed in the SBP. The plans are a joint effort of the department management and the employees within the department. The department executive assures that teams are in place to adequately cover ongoing efforts to meet the various goals. Often teams are formed by the employees themselves to address a specific business problem.

The executive serves as a coach to these teams to be sure that adequate resources are available and no roadblocks are in the way.

All managers and supervisors assure that the employees within their departments are adequately recognized and will devise new methods of recognition when needed. Both the department executives and the employees in the department are accountable to the TQM Forum for goal accomplishment.

The annual employee survey serves as feedback that the leader is adequately performing his/her position and the proper level of leadership is being accomplished.

2. *Cooperation with other units*—The TQM Forum includes representatives from all departments. The SBP is developed from a total company perspective with input from each department.

Cross-functional teams are established to observe and resolve corporate-wide problems. The meetings of these teams serve to demonstrate to the members the broad, interactive cooperation between departments, promoted by all levels of company management.

Monthly company meetings are held to keep all employees informed on the overall company performance and the reinforcement of the commitment to our TQM System. The meeting is interactive.

Each department's goals are set based on the department (internal and external) customers' expectations. These expectations are determined and documented in a series of meetings between representatives from the customer and supplier departments.

Each employee spends approximately 20 minutes each week with their internal customer, reviewing progress made toward existing goals and resolving new concerns or questions.

1.2b *Communicate & Reinforce Values—Employees*

The company actively supports an open-door policy by all senior executives. The senior management follow the practice of "Management by Walking Around" (MBWA). They regularly "wander around," through all areas of the company at least twice-a-day and stop and talk with workers and supervisors to hear concerns, answering questions from employees and asking questions which reinforce the quality values. Questions that cannot be answered immediately are answered to the concerned employee within 24 hours. Any significant issues are discussed at the monthly all-employee meetings.

The CEO leads these monthly meetings which all employees attend. Company performance indicators are reviewed, customer survey results are discussed, and employee questions are answered. Following these meetings, there is always a social period intended to encourage dialogue with employees who may feel apprehensive about speaking out in a formal meeting.

Figure 1.2 shows some of the other methods of communicating with employees.

Figure 1.2
Internal Communications

• Employee Survey	• Newsletter
• Team Meetings	• Luncheon Awards
• Bulletin Boards	• Key Indices
• Employee Orientation	• Annual Reviews
• Open-Door Policy	• Quality Training
• Collins Quality Video	• Departmental Meetings
• Suggestion Program	• Graphs
	• All-Employee Meetings
	• Charts

1.2c Types, Frequency, and Content of Reviews

The TQM Forum reviews the performance of three teams and each department at each meeting. The assessment process is continuous. The purpose of the review is to determine if the unit is progressing toward the goals on schedule, or, if not, what the forum can do to remove the roadblocks. The TQM Forum monitors results and provides regular course corrections to the teams. Senior executives review the operational performance (which includes quality objectives) at the monthly meeting. Each department reviews the key product and process quality data charts weekly.

If the trend data show the goal will be achieved, there is no further action. Adverse trends, noted during TQM Forum or departmental reviews, may warrant a change in schedule, direction, scope, manpower, or financial support. The TQM Forum will then assist in defining areas for improvement that will allow achievement of the goal. These areas for improvement could be in the form of additional employee training, budget increases, or manpower.

1.2d Indicators for Improvement Effectiveness

Indicators —The key indicators the company uses to evaluate the effectiveness of the approach to integrating quality values into our business operations are Customer Satisfaction (Customer Surveys), Awareness and Empowerment (Employee Surveys), and Performance (*Red Book*).

Evaluation to Improve Effectiveness — The trends of each key indicator are how we determine the effectiveness of this approach to improving quality values. Key indicator trends assist all management levels in defining areas for improvement. Positive trends of a key indicator confirm the approach is effective.

1.3 Public Responsibility and Corporate Citizenship

1.3a *Public Responsibilities in Quality Policies*

Collins Technology recognizes that it is part of a larger system—in a cultural, societal, and economical sense. To assist in overall system optimization, a team is assigned to each of the following activities. The team identifies areas for improvement and compliance to the TQM Forum.

Business Ethics

1. Risks are identified through discussions with other companies and our accounting firm. The consortium mentioned in Item 1.1a(5) has been very useful.

2. Our Business Ethics Policy, derived from the corporate mission, defines the goals set by the senior executives and approved by the BoD.

3. Improvements are made through monitoring similar policies of 12 other businesses.

The best indicators relate to the long-term relationships with our customers, suppliers, bankers, accounting firms, and law firms.

The TQM Forum reviews the policy annually. We have had no ethics violations.

Environmental

1. The risks are identified by the EPA, OSHA, and comparisons with other companies through our benchmarking activities.

2. Initiated in 1988, our goal is to maintain waste water and gaseous/particulate emissions one order of magnitude less than the Environmental Protection Agency (EPA) and the state air control board requirements.

3. The methods include process improvements, reduction of discharge, and recycling.

Key indicators such as heavy metals and toxic organic compounds are monitored through independent laboratories and the state Water Utilities Board.

Reports are submitted to EPA and the TQM Forum for review. We are in compliance with all key indicators.

Public Safety

1. Risks are identified through the state regulatory boards and our insurance carriers.

2. Quality improvement goals are set by the Safety Team in cooperation with the local fire department and an emergency planning committee.

3. Goals are improved through cooperation with the local fire department and the emergency planning committee.

A principal indicator to monitor the effectiveness is the involvement of our employees. Based on our internal training, three of our employees saved lives outside the company during 1992.

The public safety goals are reviewed annually.

1.3b *Anticipate Concerns*

By setting our goals a magnitude better than required, we are addressing potential concerns in advance. One example is waste management. Initiated in 1988 when disposal was neither difficult nor costly, our goal was the elimination of solid waste disposal. The methods used are process changes, in-plant reclamation, and recycling. As a result of our recovery and recycling program, our actual disposal needs have been reduced by 70% since 1990 (Figure 1.3).

Figure 1.3

Total Waste Stream Reduction

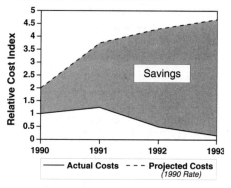

We monitor key indicators such as the amount of particulate emissions and solid waste disposal set by the state water commission and the EPA. Annual Waste Summary and Waste Minimization Reports are submitted to the state, EPA, and the TQM Forum for progress monitoring. We are in compliance with all key indicators by a margin of 1.5 or more. Recognized by the state as a "model system," it recommends other companies to visit our facility.

1.3c *External Promotion of Quality Awareness*

1.3 *National*—As a corporate member of the ASQC and the IPC, Collins Technology promotes quality through leadership, participation, and quality presentations.

State/Community—Our President/CEO has made multiple presentations to state and local governments concerning how they can apply TQM principles to accomplish more efficient and effective public organizations.

Trade/Business—Senior executives encourage employees to participate in public expression of our TQM System. We hosted a breakfast meeting for 20 local companies. We presented our TQM System and a facility tour demonstrating our commitment to quality.

Professional—Our quality video is available through the Institute of Industrial Engineers (IIE) catalog. We distributed the video to over 50 companies throughout the world. We participate in the Consortium Quality interchange. Quality presentations are made at IPC, ASQC, and IIE meetings.

Education—We have an outreach program with our local schools and communities to communicate our quality values and foster diversity. We have made quality presentations to students of three state universities. Our emphasis illustrates that TQM Systems will dictate the requirement for local manufacturing on a worldwide basis. We are members of the Koalaty Kids program, a national program that teaches the quality concepts to elementary students.

Government—Our President/CEO has lobbied in Washington, D.C., with a panel of congress-men for the continuation of National Quality Month.

ASQC—We have an employee on the National Quality Month Committee which defines the local program. The company supports the local section financially.

National Quality Expo—We have an employee on the Advisory Committee which defines the local program. Our company also provides financial support.

State Quality Consortium—Our senior executives participate on the Advisory Board. The purpose is to advise and critique the quality curriculum.

IPC—Our President/CEO is on the National Board of Directors and the Quality Commitment Task Force.

2. Through allowances for financial support and personal involvement, our BoD and TQM Forum encourage our company to participate in public responsibilities that promote quality awareness.

The TQM Forum is the focal point for employee participation in quality activities outside of the company. We encourage employee involvement through payment of society membership dues, time off to attend meetings, time for writing technical and quality papers, recognition in *Collins News*, etc.

4. The philosophy that we promote with our employees is

> *"Do not do anything you would not be proud to tell your family or read about in the newspaper."*

By following this philosophy, we are confident our decisions reflect our high standards.

Business ethics was the topic at our community consortium (1.1a5) for two months last year.

1.3d *Trends*

Figure 1.4 shows that the number of presentations regarding quality continues to increase. The growth in presentations demonstrates the leadership role that our senior executives and managers have taken in sharing our quality philosophy. Another measure is eighteen of our employees are ASQC Certified Quality Engineers.

Collins Technology has not received any sanctions under law, regulation, or contract since its incorporation in 1971.

Figure 1.4

External Quality Presentations

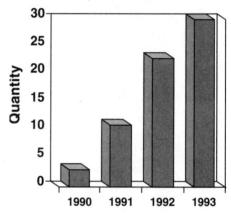

2.0 Information and Analysis *(75 pts.)*

The *Information and Analysis* Category examines the scope, management, and use of data and information to maintain a customer focus, to drive quality excellence, and to improve operational and competitive performance.

2.1 Scope and Management of Quality and Performance Data and Information *(15 pts.)*

Describe the company's selection and management of data and information used for planning, day-to-day management, and evaluation of quality and operational performance.

☑ Approach

☑ Deployment

☐ Results

Areas to Address

(a) criteria for selecting data and information for use in quality and operational performance improvement. List key types of data and information used and briefly outline the principal roles of each type in improving quality and operational performance. Include: (1) customer-related; (2) product and service performance; (3) internal operations and performance, including business and support services, and employee-related; (4) supplier performance; and (5) cost and financial.

(b) how reliability, consistency, and rapid access to data are assured throughout the company. If applicable, describe how software accuracy and reliability are assured.

(c) how the company evaluates and improves the scope and management of data and information. Include: (1) review and update; (2) shortening the cycle from data gathering to access; (3) broadening access to all those requiring data for day-to-day management and improvement; and (4) aligning data and information with process improvement plans and needs.

Notes:

1. Item 2.1 permits the applicant to demonstrate the *breadth and depth* of its data. Applicants should give brief descriptions of the data under major headings such as "internal operations and performance" and subheadings such as "support services." Note that information on the scope and management of competitive and benchmark data is requested in Item 2.2.

2. Actual data should not be reported in Item 2.1. These data are requested in other Items. Accordingly, all data reported in other Items should be part of the base of data and information to be described in Item 2.1.

3. Improving the management of data and information (2.1c) might also include company mechanisms and/or incentives for units to contribute and to share data and information.

2.2 Competitive Comparisons and Benchmarking *(20 pts.)*

Describe the company's processes, current sources and scope, and uses of competitive comparisons and benchmarking information and data to support improvement of quality and operational performance.

☑ Approach
☑ Deployment
☐ Results

Areas to Address

(a) how competitive comparisons and benchmarking information and data are used to help drive improvement of quality and operational performance. Describe: (1) how needs are determined; and (2) criteria for seeking appropriate competitive comparisons and benchmarking information—from within and outside the company's industry.

(b) brief summary of current scope, sources, and principal uses of each type of competitive comparisons and benchmarking information and data. Include: (1) customer-related; (2) product and service quality; (3) internal operations and performance, including business and support services and employee-related; and (4) supplier performance.

(c) how competitive comparisons and benchmarking information and data are used to improve understanding of processes, to encourage breakthrough approaches, and to set "stretch" targets.

(d) how the company evaluates and improves its overall processes for selecting and using competitive comparisons and benchmarking information and data to improve planning and operational performance.

Notes:

1. Benchmarking information and data refer to processes and results that represent superior practices and performance and set "stretch" targets for comparison.

2. Sources of competitive comparisons and benchmarking information might include: (1) information obtained from other organizations through sharing; (2) information obtained from the open literature; (3) testing and evaluation by the company itself; and (4) testing and evaluation by independent organizations.

2.3 Analysis and Uses of Company-Level Data *(40 pts.)*

Describe how data related to quality, customers and operational performance, together with relevant financial data, are analyzed to support company-level review, action, and planning.

> ☑ Approach
> ☑ Deployment
> ☐ Results

Areas to Address

(a) how customer-related data and results (from Category 7.0) are aggregated with other key data and analyses and translated via analysis into actionable information to support: (1) developing priorities for prompt solutions to customer-related problems; and (2) determining key customer-related trends and correlations to support reviews, decision-making, and longer-term planning.

(b) how quality and operational performance data and results (from Category 6.0) are aggregated with other key data and analyses and translated via analysis into actionable information to support: (1) developing priorities for improvements in products/services and company operations, including cycle time, productivity and waste reduction; and (2) determining key operations-related trends and correlations to support reviews, decision-making, and longer-term planning.

(c) how the company relates overall improvements in product/service quality and operational performance to changes in overall financial performance to support reviews, decision-making, and longer-term planning.

(d) how the company evaluates and improves its analysis for use as a key management tool. Include: (1) how analysis supports improved data selection and use; (2) how analysis strengthens the integration of overall data use for improved decision-making and planning; and (3) how the analysis-access cycle is shortened.

Notes:

1. Item 2.3 focuses primarily on analysis for company-level purposes, such as reviews (1.2c) and strategic planning (Item 3.1). Data for such analysis come

from all parts of the company. Other Items call for analyses of specific sets of data for special purposes. For example, the Items of Category 4.0 require analyses to demonstrate effectiveness of training and other human resource practices. Such special-purpose analyses are assumed to be part of the overall information base of Category 2.0, available for use in Item 2.3. These specific sets of data and special-purpose analyses are referred to in 2.3a and 2.3b as "other key data and analyses."

2. "Actionable" means that the analysis provides information that can be used for priorities and decisions leading to allocation of resources.

3. Solutions to customer-related problems [2.3a(1)] at the company level involve a process for: (1) aggregation of formal and informal complaints from different parts of the company; (2) analysis leading to priorities for action; and (3) use of the resulting information throughout the company. Note the connections to Item 7.2e, which focuses on day-to-day complaint management, including prompt resolution.

4. The focus in 2.3a is on analysis to improve customer-related decision-making and planning. This analysis is intended to provide additional information to support such decision-making and planning that result from day-to-day customer information, feedback, and complaints.

   Examples of analysis appropriate for inclusion in 2.3a are:
   - how the company's product and service quality improvement correlates with key customer indicators such as customer satisfaction, customer retention, and market share;
   - cross-comparisons of data from complaints, post-transaction follow-up, and won/lost analyses to identify improvement priorities;
   - relationship between employee satisfaction and customer satisfaction;
   - cost/revenue implications of customer-related problems and problem resolution effectiveness;
   - rates of improvements of customer indicators;
   - customer loyalty (or positive referral) versus level of satisfaction; and
   - customer loyalty versus level of satisfaction with the effectiveness of problem resolution.

5. The focus in 2.3b is on analysis to improve operations-related decision-making and planning. This analysis is intended to *support* such decision-making and planning that result from day-to-day observations of process performance.

Examples of analysis appropriate for inclusion in 2.3b are:

- how product/service improvement priorities are determined;

- evaluation of the productivity and cost impacts of improvement initiatives;

- rates of improvement in key operational indicators;

- evaluation of trends in key operational efficiency measures such as productivity;

- comparison with competitive and benchmark data to identify improvement opportunities and to set improvement priorities and targets.

6. The focus in 2.3c is on the links between improvements in product/service quality and operational performance and overall financial performance for company goal and priority setting. Analyses in 2.3c might incorporate the results of analyses described in 2.3a and 2.3b, and draw upon other key data and analyses.

Examples of analysis appropriate for inclusion in 2.3c are:

- relationships between product/service quality and operational performance indicators and overall company financial performance trends as reflected in indicators such as operating costs, revenues, asset utilization, and value added per employee;

- comparisons of company financial performance versus competitors based on quality and operational performance indicators;

- allocation of resources among alternative improvement projects based on cost/revenue implications and improvement potential;

- net earnings derived from quality/operational performance improvements;

- comparisons among business units showing how quality and operational performance improvement have improved financial performance; and

- contributions of improvement activities to cash flow and/or shareholder value.

2.0 INFORMATION AND ANALYSIS

2.1 Scope and Management of Quality & Performance Data and Information

Our TQM System relies on extensive data and information analysis as an essential element of both the planning and improvement of our performance. Information alone does not make quality happen, but it does allow us to set and communicate tangible goals, steer our plans, and measure and track our progress. It is a basic tool for attaining continuous improvement and management by fact.

2.1a *Criteria, Scope, and Types of Data*

Criteria for Selection—For each plan or program within our company, we define measurements and goals. This includes externally oriented programs (customer satisfaction or regulatory compliance), and internally oriented efforts throughout the company by senior executives, supervisors, employees, and across the breadth of operations (marketing, quality assurance, manufacturing, engineering, finance, personnel). Data selection criteria include the following:

- Data must measure progress toward a goal, department plan, or TQM actions, or support day-to-day operations.

- It must be meaningful, objective, verifiable, linked to a prevention

orientation, and customer (internal and external) oriented.

We developed a Many-Data Checklist for each data or information point. The checklist identifies the "owner/user," specifies the data/information point, goals and/or benchmarks, method of tracking (run charts, control charts), location of database, frequency of review, validation, basis of calculation, etc.

Types of Data—Over 600 data/information points are in active use in all areas of our company. The types of data are illustrated by the following general data categories and examples:

1. Customer Related

   Performance—Customer satisfaction survey data (Figures 7.3, 7.4), field quality (Figure 7.5), customer-reported delivery and quality complaints (Figure 7.6), customer requirements, visits/calls, sales/forecast by customer, returned material, historical files, on-time delivery, contracts/agreements, etc.

   Business segment related—Market forecast and trend data, alternative technologies, potential new customer background.

2. Product and Service Performance

   Internal audits, production test records, quality systems procedures, reliability test data, process capability information, yields, failure analysis data/ information, customer audit results, awards given by customers, first article inspection results, returned material response time, Pareto data on internal and external failures, productivity and cycle time, labor effectiveness, customer contact to order time, and order to delivery time.

3. Internal Operations

Manufacturing operations— Statistical process control data (Figure 5.4), material yields (Figure 6.1), material cost (Figure 6.2), team involvement (Figure 6.3), operational productivity charts (Figures 6.4–6.6), first pass yields (Figure 6.7), process performance indices (Figures 6.11, 6.17), first article acceptance rates (Figure 6.12), product return rates (Figure 6.13), as well as cycle times, Pareto data on failures/rework, throughput, process documentation, training levels, housekeeping indices, process flowcharts, and procedures.

TQM Forum—Minutes, team results, *Red Book* (financial and SBP data).

Engineering operations—On-time delivery (Figure 6.8), design cycle times (Figures 6.9, 6.10, 6.21), artwork costs (Figure 6.22), as well as schedule/cost performance, internal customer satisfaction surveys, training, team involvement, new designs released, ECNs.

Accounting—Overdue receivables (Figure 6.20), billing accuracy (Figure 6.23a), accounts payable late (Figure 6.23b), days to close books end of period (Figure 6.23d).

Education and training—Team involvement (Figure 4.2), quality training (Figures 4.3, 4.6), classes and seminars attended (Figures 4.4–4.6), operator certifications, safety, professional certifications, educational levels achieved.

Teams—Employee involvement (Figure 4.2), active teams and members (Figure 4.2), employee team training (Figure 4.6).

Recognition—Employee of the Month (Figure 4.7), special presentations at company meetings (Figure 4.8), "Clean Aide," housekeeping.

Company-wide—Cost of Quality (Figure 6.17), productivity (Figure 6.18), well-being and satisfaction indicators (Figure 4.9), interview/hire ratios, staffing levels, affirmative action plans/goals.

4. Suppliers

Number of suppliers (Figure 6.26), Supplier Results Indicator (SRI) (Figure 6.27), which combines quality and delivery performance, number of suppliers in the "dock to stock" program.

5. Cost and Financial

Government information (IRS), business data (sales, receivables, inventory, cash flow, etc.), sales per employee, Cost of Quality tracking, financial audit data, product and operations cost/budget/performance, SBP.

2.1b *Reliability, Rapid Access, and Software Quality*

We use multiple cross-checks of data selected and information to ensure reliability, consistency, and standardization. Examples include our Many-Point Data Checklist, comparisons to "downstream" customer data, quality engineering assistance when adding new data points, and verification of data to source documents through periodic audits/reviews. During our annual SBP cycle, monthly operation reviews, and weekly TQM Forum meetings, we evaluate each operation to identify new data measurements, review existing data points for elimination or enhancement, and update our data sources.

A data checklist described in Area 2.1a is completed for all data/information used. Checklists are centrally located, easily accessible, and reviewed annually by senior management and the "owner" under the guidelines of our document management program. Data are added, deleted, or improved as warranted.

We maintain an extensive PC-based local area network (LAN) with over 80 connected stations. The LAN features timely update and rapid access of information through local data collection and on-line displays. When updates are made, all data files are updated and information is available immediately.

Internally we utilize primarily purchased application software in combination with internally developed software. New software is selected based on its ability to satisfy our current/projected needs, its compatibility with our existing hardware/software, prior experiences of our employees, and the history and reputation of the vendor for quality, technical assistance, and customer service. Before new or internally developed or modified software is made available to all users, it is tested and evaluated by a control group to ensure data quality and reliability. We are sensitive to maintaining the quality of our overall system through software testing, multi-level access control, and accurate database management. We have a LAN Steering Team, consisting of employees from each functional area, which oversees training, system development, system quality and validity, and provides "expert user" guidance on the use of our system. The training of employees is very important so all employees understand how to use the LAN and how to analyze and utilize the data.

2.1c *Key Indicators the Company Uses for Evaluation of Data Scope and Quality*

Our annual Strategic Business Plan (SBP) cycle is our primary means of evaluating and improving the use, scope, and quality of data. This evaluation includes, for each data point, questioning why we maintain the data/information, the need to improve or discontinue a data point, move it "up stream," tie multiple points together, etc. This assures the necessary data is being collected to further guide us on our journey defined in our SBP.

We use Collins' Seven Quality Tools taught to all employees (see Area 4.3a), which include Pareto charts, cause-and-effect diagrams, and histograms, to analyze the need, scope, and quality of the data being collected. If the data is needed but not being used, we examine the type of analysis employed, the skill level—required and actual—as well as the availability and timeliness of the results.

Similar evaluations during the year result from management, TQM Forum, employee team or operation reviews. Shortened cycle times stem from the use of the LAN because operators update the files directly, as opposed to generating a document for a data entry clerk. Approximately 90% of the data points are maintained on the LAN with a station-to-employee ratio of 1:2. Placing control charts near the end user further reduces cycle time.

2.2 Competitive Comparisons and Benchmarking

2.2a *Criteria for Benchmarks*

As described in Area 2.1a, we require goals for all key strategic (market share, equity growth), product (performance, reliability, cost), and customer satisfaction (sales, on-time delivery, product quality) indices. We use external benchmarks to validate our goals.

1. *Needs determined*—Benchmarks and competitive comparisons are sought to support both our SBP and quality plans. To be selected, they must be meaningful, customer satisfaction oriented, related to our quality goals, and represent best in class. Benchmark data are obtained from customers, suppliers, other companies, and industry publications. If the benchmark is different from our goals, we determine why, develop an action plan and schedule to close the gap, and assign a team, if required. (See also 2.2c.)

2. *Within and outside the industry*—We compare ourselves to our major competitors. We maintain comparisons in terms of product performance, delivery, reliability, and visual characteristics. We use outside industry data to satisfy our criteria for world-class or best-in-class comparisons in company-wide manufacturing, and support services. Through the cooperation of our customers,

we benchmark against our customers' best suppliers. These include KTFL, Ace Circuits, Ridgeford Tech, Arnold-Harrison, TECON Instruments, GCAM, and Worldwide Corp. We also compare ourselves to America's best, prior Malcolm Baldrige National Quality Award recipients: Globe Metallurgical, Marlow Industries, Xerox Business Products Division, Westinghouse Nuclear Fuel Division, IBM Rochester Division, Zytec, Solectron, and Motorola. We use industry benchmarks developed from various sources, such as IPC surveys, by selecting the upper quartile figures for our Standard Industrial Classification (SIC) code.

2.2b *Scope of Competitive and Benchmark Data*

1. *Customer satisfaction*—Benchmark sources are the Delta Company (customer service data), Gamma Industries (total customer satisfaction), our customers' best suppliers (delivery, quality), and our customers' requirements for supplier awards (Wrightman Certified Supplier, Kenan Supplier). These data assist us in increasing employee empowerment, improving on-time delivery requirements (Figure 6.10) and identification of our customers' expectations.

2. *Product and service quality*— Our competitors are the primary sources of comparisons and benchmark data. Scope of competitive data includes product manufacturing costs and techniques (via analysis and product testing), product and process capabilities (via performance testing and inspection, process studies, and long-term reliability testing), and materials technology (in-house test/analysis of current and potential competitor's material performance). These data, along with the SBP, help us to allocate process engineering, R&D, and manufacturing resources to areas where gaps exist between our goals and the benchmark. We have taken actions which have improved our material manufacturability and performance, and automated portions of our production process, which resulted in lower cost.

3. *Internal operations and support services*—Sources of benchmarks include Peterson, KTFL (yields), Gamma Industries, M-TECH, GCAM (customer response time, training, and involvement), and industry publications. We use these benchmarks to establish yield goals, minimum annual training requirements, cycle time goals, payroll processing goals, financial ratio goals, and customer response time criteria.

Employee related—Sources of benchmarks include IPC and Department of Commerce surveys on turnover, absenteeism, safety, and compensation. We use these benchmarks to establish compensation levels, employee benefits, and safety programs. We also perform our own surveys of IPC and consortium companies when making specific policy decisions.

4. *Supplier performance*—We use our customers' best suppliers' benchmarks (delivery, quality), M-TECH and Worldwide Corp. (supplier base reduction, strategic partnerships, and involvement). We use these data to communicate requirements to our suppliers on quality, technical support, service, and price. We have established our supplier indicator to rate our suppliers based on these benchmarks. We have set goals for supplier reduction and track progress toward them.

2.2c Process Focus

The TQM Forum reviews all benchmark comparisons. We recognize benchmarking is more than getting a measurement from a good company and making that measurement our new goal. Understanding how our process works and how the process which generated the good results works is essential. Typically when evaluating benchmark information of other companies, we find our processes must be modified to meet the benchmark results. A team is formed to develop a list of improvement ideas and action plans to close the gap. The TQM Forum reviews the benchmark comparison, the recommended list of improvement plans,

and sets priorities and resource allocations. Teams are assigned improvement activities and are encouraged to be innovative in their approach— meeting offsite, including customers, suppliers, other outside resources on their team, etc. The President's "Patent Idea" award is used to recognize innovation and is presented at a monthly all-employee meeting.

2.2d Evaluation and Improvement of Benchmark Data

As part of our planning process, we tie each data point to an action or desired goal, thus using this cycle to update our goals and benchmarks. We identify "holes" in our benchmarking process and review any new ideas for benchmarking. When significant gaps or lack of closure is evident, we use Pareto charts and cause-and-effect diagrams to develop enhanced or alternative action plans and resource allocations.

To improve the effectiveness of our benchmarking analysis, senior executives participate in benchmarking training, personally visit companies we benchmark to understand their benchmarking process, and utilize consulting expertise in this area.

2.3 Analysis and Uses of Company-Level Data

2.3a How Data Support Actions

1. *Developing priorities for solutions*—We have a common form that is used by everyone to

document customer complaints. Our documented procedure is described in Area 7.2e. With all customer problems, our first objective is to establish containment activities to assure the customer will not receive an additional unacceptable products and sort out any unacceptable products in the hands of the customer. Teams are formed as required to determine the root cause of the problem and to develop and implement solutions. When needed, Pareto analyses of aggregated losses due to the problems (including the loss to the customer as well as the costs of comtainment) are used to prioritize efforts.

2. *Determining trends and correlations*—Customer trend information is collected, using the LAN system, as standard data within the marketing department attainment plans. The results are monitored in department reviews, TQM Forum reviews, and the monthly all-employee meeting. Data points tracked are analyzed for autocorrelation and seasonal effects, for correlation between related indicators, and for multiple correlation among groups of related measures. The customer survey is used to validate these correlations with customer satisfaction. This information is used to develop goals for product and service quality improvements.

2.3b *Operational Performance*

1. *Priorities for short-term improvements*—Area 6.2a discusses how cost of quality, productivity, and waste stream data are collected and analyzed. The TQM Forum uses these data to set short-term goals at the SBP review. Goals and priorities are set by comparing past and current performance to benchmarks and resource allocation requirements.

2. *Trends in operational performance*—We hold monthly operation reviews and weekly TQM Forum reviews of our business processes (productivity charts, financial results), programs (safety, training, supplier reduction), and *Red Book* indices. Process control, output, yield, and delivery charts are placed throughout the site so employees can monitor process stability, take corrective action as needed, and track progress toward goals and benchmarks.

We review our cost of quality trend chart monthly to determine our overall progress toward total quality. We prepare a Pareto chart of the components and review the actions in place which address them. At company meetings, we review prior month and year-to-date achievement of the plan, on-time shipping performance, quality indices, and cycle times.

These trend data are used to make changes as required and to set future goals.

2.3c *Quality and Financial Performance*

The TQM Forum initially sets goals in the annual planning process. Because the TQM Forum is cross-functional, all data (customer, process, and financial) are reviewed from a company-wide perspective. The respective departments are responsible for tracking against the goals they have set which, in turn, directly correlate with the corporate goals listed in the SBP. Cost of Quality, inventory, and productivity are tracked and compared to their respective goals.

2.3d *Evaluation and Improvement—Analysis Process*

1. *Improve data selection*—As our quality improved, the analysis identified the need to collect different data. For example, percent nonconforming to parts per thousand to parts per million (ppm).

   One study showed that achieving customer satisfaction was not sufficient; you had to delight the customer! To make this distinction, yes/no data could not be used on customer surveys. A five-point

scale from very dissatisfied to very satisfied was needed.

2. *Shortening cycle of analysis*—We strive to improve our planning and goal-setting process at the initial session of our annual business planning cycle. Using uniform data point definitions and a common database for data collection and analysis has significantly improved our analytical capabilities and speed of planning and decision-making. We have expanded our computer network throughout the company to decrease the cycle time of data input to availability of results. We place charts of results in close proximity to the areas they affect to speed access and allow for faster action. We teach our employees how to read and analyze the different types of charts we use (Collins' Quality Tools). Our TQM Forum and operational managers review and make recommendations on the scope of our analysis, results, and comparisons.

3. *How analysis strengthens decision-making*—By having the common database in the LAN network available to all employees, and having terminals close at hand for all employees, the employees are well aware of the status of the various performance goals. These issues are discussed at the monthly all-employee meetings.

3.0 Strategic Quality Planning *(60 pts.)*

The *Strategic Quality Planning* Category examines the company's planning process and how all key quality and operational performance requirements are integrated into overall business planning. Also examined are the company's short- and longer-term plans and how plan requirements are deployed to all work units.

3.1 Strategic Quality and Company Performance Planning Process *(35 pts.)*

Describe the company's business planning process for the short term (1–3 years) and longer term (3 years or more) for customer satisfaction leadership and overall operational performance improvement. Include how this process integrates quality and operational performance requirements and how plans are deployed.

☑ Approach
☑ Deployment
☐ Results

Areas to Address

(a) how the company develops strategies and business plans to address quality and customer satisfaction leadership for the short term and longer term. Describe how plans consider: (1) customer requirements and the expected evolution of these requirements; (2) projections of the competitive environment; (3) risks: financial, market, technological, and societal; (4) company capabilities, such as human resource and research and development to address key new requirements or market leadership opportunities; and (5) supplier capabilities.

(b) how strategies and plans address operational performance improvement. Describe how the following are considered: (1) realigning work processes ("reengineering") to improve customer focus and operational performance; and (2) productivity and cycle time improvement and reduction in waste.

(c) how plans are deployed. Describe: (1) how the company deploys plan requirements to work units and to suppliers, and how it ensures alignment of work unit plans and activities; and (2) how resources are committed to meet plan requirements.

(d) how the company evaluates and improves: (1) its planning process; (2) deploying plan requirements to work units; and (3) receiving planning input from work units.

Notes:

1. Item 3.1 addresses overall company strategies and business plans, not specific product and service designs. Strategies and business plans that might be addressed as part of Item 3.1 include operational aspects such as manufacturing and/or service delivery strategies, as well as new product/service lines, new markets, outsourcing, and strategic alliances.

2. Societal risks and impacts are addressed in Item 1.3.

3. Productivity and cycle time improvement and waste reduction (3.1b) might address factors such as inventories, work in process, inspection, downtime, changeover time, set-up time, and other examples of utilization of resources such as materials, equipment, energy, capital, and labor.

4. How the company reviews quality and operational performance relative to plans is addressed in 1.2c.

3.2 Quality and Performance Plans *(25 pts.)*

Summarize the company's specific quality and operational performance plans for the short term (1–3 years) and the longer term (3 years or more).

☑ Approach
☑ Deployment
☐ Results

Areas to Address

(a) for planned products, services, and customer markets, summarize: (1) key quality requirements to achieve or retain leadership; and (2) key company operational performance requirements.

(b) outline of the company's deployment of principal short-term quality and operational performance plans. Include: (1) a summary of key requirements and associated operational performance measures or indicators deployed to work units and suppliers; and (2) a brief description of resources committed for key needs such as capital equipment, facilities, education and training, and new hires.

(c) outline of how principal longer-term (3 years or more) quality and operational performance requirements (from 3.2a) will be addressed.

(d) two-to-five-year projection of key measures and/or indicators of the company's quality and operational performance. Describe how quality and operational performance might be expected to compare with key competitors and key benchmarks over this time period. Briefly explain the comparisons, including any estimates or assumptions made regarding the projected quality and operational performance of competitors or changes in benchmarks.

Notes:

1. The focus in Item 3.2 is on the translation of the company's business plans, resulting from the planning process described in Item 3.1, to requirements for work units and suppliers. The main intent of Item 3.2 is alignment of short- and long-term operations with business directions. Although the deployment of these plans will affect products and services, design of products and services is not the focus of Item 3.2. Such design is addressed in Item 5.1.

2. Area 3.2d addresses projected progress in improving performance and in gaining advantage relative to competitors. This projection may draw upon analysis (Item 2.3) and data reported in results Items (Category 6.0 and Items 7.5 and 7.6). Such projections are intended to support reviews (1.2c), evaluation of plans (3.1d), and other Items.

3.0 STRATEGIC QUALITY PLANNING

3.1 Strategic Quality and Company Performance Planning Process

We greatly simplified and clarified our TQM vision and goals by developing the annual four-year Strategic Business Plan, which effectively combines all quality, operational, financial, human resources, training, and capital goals and attainment plans.

3.1a *Strategies and Plans Development*

When the company started in business in 1971, it recognized that a very competitive market existed and a specific niche market was required. Collins selected the high-tech end, concentrating on advanced materials, high densities of lines, and designs for demanding environments. The initial goals were for products and services of the highest quality and reliability. Further, the boards would consist of ten or more layers. These basic strategies were documented in "The Corporate Strategic Goals" and the "Collins Corporate Mission." The analysis process is portrayed in Figure 3.1.

Performance requirements are developed during the SBP process wherein the company's position within the

Figure 3.1
Strategic Business Planning Process

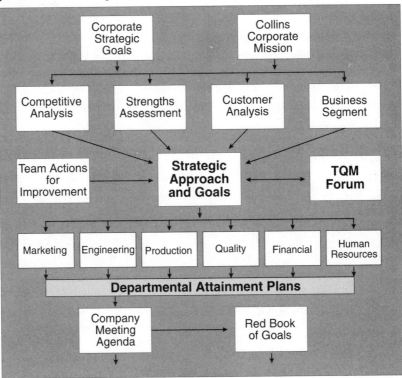

marketplace is assessed with respect to competitors and customer expectations.

1. *Customer requirements*—Our long-term close relationship with the majority of our customers has resulted in earlier and earlier engagement in their planning and development programs. These working relationships extend beyond just marketing and purchasing, including direct contact by engineering, quality, and manufacturing. We let our customers guide us with not only their requirements, but also by developing a sense of the expectations of *their* end customers. Other inputs include our customer surveys, supplier ratings, and on-site audits. Our Business Segment Managers are responsible for collecting and analyzing customer and industry data for each business segment. These short-term and longer-term analyses include trending, gap identification, prioritizing needs, and recommending actions and milestones.

2. *Projections of competitive environment*—We continually test and analyze our competitors' product and materials performance, including statistical analysis and long-term reliability testing. Our analysis also includes emerging competitors' materials and product capabilities, such as Japanese, German, and British companies. We analyze these data

for comparison and emerging technology trends. We annually commission an independent study to review state-of-the-art technology in our industry and evaluate competitive threats. We benchmark our best-in-class products and service quality, not only to our competitors, but relative to our customers' "best suppliers." We use these analyses to identify competitive gaps and closure rates to set our short-term and longer-term product, technology, and quality goals.

3. *Risks: financial, market, technological, societal*—Risk analysis is done through utility theory and risk simulation techniques using the analysis from Failure Mode and Effects Analysis (FMEA), warranty analysis, lab testing, and field testing. (See 5.1a2 for more information on FMEAs). Each year, the SBP is updated and the risks are reviewed. Benchmark data are used extensively to evaluate risk in setting goals and priorities.

4. *Company capabilities*—The overall goals and resource commitments of the company (operational, capital, training, financial, quality, human resources, technology, suppliers, etc.) are contained within each departmental attainment plan. Each department develops their individual technical and human resource requirements. As the SBP is

updated, each of the six departmental attainment plans is structured to meet the short-term and longer-term goals resulting from the SBP. The approval of the attainment plans commits the annual financial budget to support the execution. Independent study and benchmark data are used to determine key requirements and indicators for the SBP.

As a small company we rely heavily on the R&D capabilities of our suppliers.

5. *Supplier capabilities*—Similar to our close-working relationship with our customers, we have concurrent working relationships with our key suppliers. They provide us with valuable in-depth expertise and resources, which we, as a small business, could not afford in-house. Our major material suppliers help us during our planning and goal-setting with design trade-offs, new technologies, etc. We also rely on the capabilities of our service suppliers, such as our insurance companies, government agencies (OSHA, EPA, EEOC, etc.), and our accounting firm. They provide help in identifying needs, developing short-term and longer-term goals, and making improvements in their specialty areas.

3.1b *Reengineering and Elimination of Waste*

For key processes throughout the company, cross-functional teams are established to address three broad questions:

1. What should the process be doing?

2. What can go wrong?

3. What is the process doing?

Tools such as process flow diagrams, operational definitions, FMEAs, and capability studies are used. Gaps between the "what should ..." and the "what is ..." determines the answer to "what is needed to be done?" This leads to streamlining the process—the elimination of waste. Answers to all the questions lead to prevention activities to eliminate or minimize the occurrence of potential problems. Comparisons of "what should ..." to benchmark results and customer responses enables us to focus on innovation and quantum improvements.

3.1c *How Plans Are Deployed*

1. *Methods used for deployment*— An output of the SBP process is the setting of goals, both short term and longer term. The deployment of requirements in all departments is through the departmental attainment plans (DAPs), which directly address the SBP documents. Each department produces an attainment plan that addresses the company goals of the SBP. The departmental

attainment plans are developed and written by the respective departments. Each department involves its employees in this process, which is handled by assigned teams and reviewed for concurrence and knowledge in the monthly departmental meetings.

The DAP goals are recorded in the *Red Book,* which is regularly reviewed by the TQM Forum to assure that all the departmental goals are aligned with the corporate goals. Departmental goals are reviewed at departmental meetings and during individual performance reviews. Suppliers are covered in Item 5.4.

2. *Resources committed*—As the departmental attainment plans are developed and the goals addressed, the necessary resources are identified and recorded. Figure 3.2 is a

Figure 3.2

1994 "Top Ten" Quality Issues

External Focus	Resource* Investment
Improve on-time delivery to > 98%	2.5%
Reduce customer response time 30%	2.0%
Reduce return rate < 0.9%	2.0%
Double number of supplier partners	0.5%
Reduce hazardous waste stream 60%	2.0%
Internal Focus	
Cpk > 2.00 for all key processes	10.0%
Achieve design to cost goals	9.0%
Improve cycle time	5.0%
Expand involvement in teams 100%	3.0%
Expand Station Qualification System certification to 85%	3.0%

* Resource Investment as % of Total Operating Expenses

listing of 1994 short-term goals and the necessary allocation of resources. During the year, if changes are required, the TQM Forum approves necessary modifications.

3.1d *Improvement of the Planning Process*

The TQM Forum reviews the planning process for improvements each year. We benchmark our process as described in Item 2.2 and analyze the process as described in Item 3.1b.

The progress toward our goals is an indicator of the effectiveness of our process. We use feedback from individual departments and employees (see Item 4.2). Finally, we have formed a team to oversee improvements to this process.

3.2 Quality and Performance Plans

3.2a *Chosen Directions*

1. *Key quality factors*—Our corporate strategic goal is clear leadership in all key quality areas: product performance, customer support, technology, and cost. We combine our short-term and longer-term customer goals in each target business segment with our corporate purpose and strategic goals, our own strengths and improvement analysis, analysis of competitors, and our current plans and actions. These form an integrated set of current and

future business and quality leadership goals for all operations in the company. We address our current and projected position relative to customer quality requirements and our competitors' quality and trends, and set aggressive goals to improve our current dominant position in our specific market niches.

2. *Key performance requirements—* To maintain and improve our position in the market, it is necessary for Collins Technology to continually improve in all phases of our business. It is also necessary to maintain our technical leadership. These factors must be combined with cost performance to maintain our level of customer expectations.

3.2b Short-Term Goals and Plans

1. *Key requirements and indicators—* Our TQM Forum's 1994 "Top Ten" Quality Issues (Figure 3.2) best describe our short-term quality plans and resource commitments. These quality issues summarize our highest priority short-term plans and goals for 1994. The departmental attainment plans contain the execution details.

   Each department addresses these goals with applicability to that department. The purchasing department reviews these requirements with our suppliers.

2. *Resources committed—*Figure 3.2 also lists the overall company resources required. Each departmental attainment plan lists

specifically what resources are required in the categories of space, equipment, and people, with the people category grouped by skill levels and necessary improvements in skills.

3.2c Longer-Term Goals and Plans

Our principal longer-term quality plans are best described by the key quality goals of our current four-year SBP, as follows.

1. Continue to expand our market share and customer base through improved levels of service via:

   - 100% on-time delivery by achieving high stability process and reduced cycle times

   - Improving our customer satisfaction index to 100%

   - Achieving a product return rate of < 100 ppm by continuous design improvement

   - 100% "first article" acceptance rate

   - Reducing cycle times by 80% by prioritizing cycle time reduction goals in all areas of the company and our supplier base.

2. Continue to improve our internal operational capabilities by:

   - Improving processes with non-robust product/process characteristics to a minimum Cpk of 2.0 using SPC, Design of Experiments (DOE), and process engineering improvements

- Eliminating rework and reducing all scrap and waste streams by first-pass yield improvement and recycling

- Expanding skills, responsibility of our employees by improved training methods and empowerment policies

- Ensuring that we maintain our worldwide quality leadership position and are also the lowest cost producer in our industry by 1995. This will be achieved through execution of our materials, design, and manufacturing plans.

- Maintaining our technological lead through continued advanced materials and process development.

Key to achieving these longer-term plans is the further deployment of our TQM System. Resource commitments are outlined in the current four-year SBP.

3.2d *Two-to-Five-Year Projections*

We fully expect that our competitors are developing similar objectives for improvement. Our customers indicate that we lead our competition in all their important quality areas, including

Figure 3.3

Performance Relative to Competitors

		Projected	
	Current	1995	1998
On-time delivery	+12%	+15%	+18%
Performance	+ 5%	+ 8%	+10%
Unit Price	− 5%	+ 1%	+ 5%

response time, delivery, performance, and reliability. A sole exception is in unit pricing, where competitors have aggressively tried to penetrate our markets. Our superior service and performance have allowed us to defend our position and maintain premium pricing. However, our TQM actions are having a major positive impact on reducing our costs and will allow us to lead on price as well. Our plans are to "raise the benchmarks" by continuously increasing the standards of performance our competitors must reach.

Our principal plan for success is long-term sustained continuous improvement.

Figure 3.3 shows where we stand in comparison with our competitors and where we expect to be in 1998. This projection will require continued improvements in all functions of Collins Technology.

4.0 Human Resource Development and Management *(150 pts.)*

The *Human Resource Development and Management* Category examines the key elements of how the work-force is enabled to develop its full potential to pursue the company's quality and operational performance objectives. Also examined are the company's efforts to build and maintain an environment for quality excellence conducive to full participation and personal and organizational growth.

4.1 Human Resource Planning and Management *(20 pts.)*

Describe how the company's overall human resource management, plans, and processes are integrated with its overall quality and operational performance plans and how human resource planning and management address fully the needs and development of the entire workforce.

☑ Approach

☑ Deployment

☐ Results

Areas to Address

(a) brief description of the most important human resource plans (derived from Category 3.0). Include: (1) development, including education, training and empowerment; (2) mobility, flexibility, and changes in work organization, work processes, or work schedules; (3) reward, recognition, benefits, and compensation; and (4) recruitment, including possible changes in diversity of the workforce. Distinguish between the short term (1–3 years) and the longer term (three years or more), as appropriate.

(b) how the company improves key personnel processes. Describe key improvement methods for processes such as recruitment, hiring, personnel actions, and services to employees, including support services to managers and supervisors. Include a description of key performance measures or indicators, including cycle time, and how they are used in improvement.

(c) how the company evaluates and improves its human resource planning and management using all employee-related data. Include: (1) how selection, performance, recognition, job analysis, and training are integrated to support improved performance and development of all categories and types of employees; and (2) how human resource planning and management are aligned with company strategy and plans.

Notes:

1. Human resource plans (4.1a) might include the following:

   - mechanisms for promoting cooperation such as internal customer/supplier techniques or other internal partnerships;

   - initiatives to promote labor-management cooperation, such as partnerships with unions;

   - creation and/or modification of recognition systems;

   - creation or modification of compensation systems based on building shareholder value;

   - mechanisms for increasing or broadening employee responsibilities;

   - creating opportunities for employees to learn and use skills that go beyond current job assignments through redesign of processes;

   - creation of high-performance work teams;

   - education and training initiatives; and

   - forming partnerships with educational institutions to develop employees or to help ensure the future supply of well-prepared employees.

2. The personnel processes referred to in 4.1b are those commonly carried out by personnel departments or by personnel specialists. Improvement processes might include needs assessments and satisfaction surveys. Improvement results associated with the measures or indicators used in 4.1b should be reported in Item 6.3.

3. "Categories of employees" (4.1c) refers to the company's classification system used in its personnel practices and/or work assignments. It also includes factors such as union or bargaining unit membership. "Types of employees" takes into account other factors, such as workforce diversity or demographic makeup. This includes gender, age, minorities, and the disabled.

4. "All employee-related data" refers to data contained in personnel records as well as data described in Items 4.2, 4.3, 4.4, and 4.5. This might include employee satisfaction data and data on turnover, absenteeism, safety, grievances, involvement, recognition, training, and information from exit interviews.

5. The evaluation in 4.1c might be supported by employee-related data such as satisfaction factors (Item 4.5), absenteeism, turnover, and accidents. It might also be supported by employee feedback.

4.2 Employee Involvement *(40 pts.)*

Describe how all employees are enabled to contribute effectively to meeting the company's quality and operational performance plans; summarize trends in effectiveness and extent of involvement.

☑ Approach
☑ Deployment
☑ Results

Areas to Address

(a) how the company promotes ongoing employee contributions, individually and in groups, to improvement in quality and operational performance. Include how and how quickly the company gives feedback to contributors.

(b) how the company increases employee empowerment, responsibility, and innovation. Include a brief summary of principal plans for all categories of employees, based upon the most important requirements for each category.

(c) how the company evaluates and improves the effectiveness, extent, and type of involvement of all categories and all types of employees. Include how effectiveness, extent, and types of involvement are linked to key quality and operational performance improvement results.

(d) trends in key measures and/or indicators of the *effectiveness* and *extent* of employee involvement.

Notes:

1. The company might use different involvement methods and measures or indicators for different categories of employees or for different parts of the company, depending on needs and responsibilities of each employee category or part of the company. Examples include problem-solving teams (within work units or cross-functional); fully-integrated, self-managed work groups; and process improvement teams.

2. Trend results (4.2d) should be segmented by category of employee, as appropriate. Major types of involvement should be noted.

4.3 Employee Education and Training *(40 pts.)*

Describe how the company determines quality and related education and training needs for all employees. Show how this determination addresses company plans and supports employee growth. Outline how such education and training are evaluated, and summarize key trends in the effectiveness and extent of education and training.

☑ Approach
☑ Deployment
☑ Results

Areas to Address

(a) how the company determines needs for the types and amounts of quality and related education and training for all employees, taking into account their differing needs. Include: (1) linkage to short- and long-term plans, including company-wide access to skills in problem solving, waste reduction, and process simplification; (2) growth and career opportunities for employees; and (3) how employees' input is sought and used in the needs determination.

(b) how quality and related education and training are delivered and reinforced. Include: (1) description of education and training delivery for all categories of employees; (2) on-the-job application of knowledge and skills; and (3) quality-related orientation for new employees.

(c) how the company evaluates and improves its quality and related education and training. Include how the evaluation supports improved needs determination, taking into account: (1) relating on-the-job performance improvement to key quality and operational performance improvement targets and results; and (2) growth and progression of all categories and types of employees.

(d) trends in key measures and/or indicators of the *effectiveness* and *extent* of quality and related education and training.

Notes:

1. Quality and related education and training address the knowledge and skills employees need to meet their objectives as part of the company's quality and operational performance improvement. This might include quality awareness, leadership, project management, communications, teamwork, problem solving, interpreting and using data, meeting customer requirements, process analysis, process simplification, waste reduction, cycle time reduction, error-proofing, and other training that affects employee effectiveness, efficiency, and safety. In many cases, this might include job enrichment skills and job rotation that enhance employees' career opportunities. It might also include basic skills such as reading, writing, language, arithmetic, and

basic mathematics that are needed for quality and operational performance improvement.

2. Education and training delivery might occur inside or outside the company and involve on-the-job or classroom delivery.

3. The overall evaluation (4.3c) might compare the relative effectiveness of structured on-the-job training with classroom methods. It might also address how to best balance on-the-job training and classroom methods.

4. Trend results (4.3d) should be segmented by category of employee (including new employees), as appropriate. Major types of training and education should be noted.

4.4 Employee Performance and Recognition *(25 pts.)*

Describe how the company's employee performance, recognition, promotion, compensation, reward, and feedback approaches support the improvement of quality and operational performance.

☑ Approach

☑ Deployment

☑ Results

Areas to Address

(a) how the company's employee performance, recognition, promotion, compensation, reward, and feedback approaches for individuals and groups, including managers, support improvement of quality and operational performance. Include: (1) how the approaches ensure that quality is reinforced relative to short-term financial considerations; and (2) how employees contribute to the company's employee performance and recognition approaches.

(b) how the company evaluates and improves its employee performance and recognition approaches. Include how the evaluation takes into account: (1) effective participation by all categories and types of employees; (2) employee satisfaction information (Item 4.5); and (3) key measures or indicators of improved quality and operational performance results.

(c) trends in key measures and/or indicators of the *effectiveness* and *extent* of employee reward and recognition.

Notes:

1. The company might use a variety of reward and recognition approaches—monetary and non-monetary, formal and informal, and individual and group.

2. Employee satisfaction (4.4b) might take into account employee dissatisfaction indicators such as turnover and absenteeism.

3. Trend results (4.4c) should be segmented by employee category, as appropriate. Major types of recognition, compensation, etc., should be noted.

4.5 Employee Well-Being and Satisfaction *(25 pts.)*

Describe how the company maintains a work environment conducive to the well-being and growth of all employees; summarize trends in key indicators of well-being and satisfaction.

☑ Approach
☑ Deployment
☑ Results

Areas to Address

(a) how employee well-being factors such as health, safety, and ergonomics are included in quality improvement activities. Include principal improvement methods, measures or indicators and targets for each factor relevant and important to the company's employee work environment. For accidents and work-related health problems, describe how root causes are determined and how adverse conditions are prevented.

(b) what special services, facilities, activities, and opportunities the company makes available to employees to enhance their work experience and/or to support their overall well-being.

(c) how the company determines employee satisfaction. Include a brief description of methods, frequency, and the specific factors for which satisfaction is determined. Describe how these factors relate to employee motivation and productivity. Segment by employee category or type, as appropriate.

(d) trends in key measures and/or indicators of well-being and satisfaction. Explain important adverse results, if any. For such adverse results, describe how root causes were determined and corrected, and/or give current status. Compare results on the most significant measures or indicators with appropriately selected companies and/or benchmarks.

Notes:

1. Special services, facilities, activities, and opportunities (4.5b) might include: counseling; recreational or cultural activities; non-work-related education; day care; special leave; safety off the job; flexible work hours; and outplacement.

2. Examples of specific factors for which satisfaction might be determined (4.5c) are: safety; employee views of leadership and management; employee development and career opportunities; employee preparation for changes in technology or work organization; work environment; teamwork; recognition; benefits; communications; job security; and compensation.

3. Measures or indicators of well-being and satisfaction (4.5d) include safety, absenteeism, turnover, turnover rate for customer-contact employees, grievances, strikes, worker compensation, and results of satisfaction determinations.

4. Comparisons (4.5d) might include industry averages, industry leaders, local/regional leaders, and key benchmarks.

4.0 HUMAN RESOURCE DEVELOPMENT & MANAGEMENT

Collins Technology is a high-technology company that requires special materials and process technology. The key element to our success is the dedication and involvement of our people. A company can be no better than the people who compose it. With dedicated, enthusiastic, trained, and empowered people, the accomplishments are limitless. Collins Technology is this type of company. We are a small, dynamic company that is highly sensitive to our employees' needs and focused on realizing the maximum benefit of an employee's talents for both the advancement of the individual and the company. Our corporate goal of "Total Customer Satisfaction" requires "Total Employee Satisfaction."

4.1 Human Resource Planning and Management

4.1a *Human Resource Plans*

The Collins Technology annual four-year Strategic Business Plan (SBP) cycle is the beginning source for all plans, including human resource (HR) planning (see Figure 3.1). The SBP receives inputs/data from multiple sources, then develops short-term and longer-term goals. The annual HR plan/goals are developed by the Human Resource Department from three key sources: annual employee surveys, direct employee communications, and these strategic initiatives.

Short-term goals are in the areas described below. Longer-term goals incorporate short-term goals plus areas critical to our future growth as identified in the SBP. An annual HR plan is developed to include specific programs and resources to meet these needs.

1. *Development*—Using the TQM System as a quality guide, action teams demonstrate the positive results of employee empowerment (see Item 4.2). Development plans, including education and training, are developed jointly by the employee and his/her supervisor. Our Training Action Team works with each department to define a training plan as described in Item 4.3. Our goal is 100% training by 1996, as compared to current levels given in Figure 4.6. Collins Technology also provides tuition assistance to all employees to increase their competence in present jobs and to prepare for advancement in the future.

2 *Mobility, flexibility*—In the interest of employee development, we emphasize cross-training to increase existing skills. Our facility is configured with many different

operations housed within a cell. This configuration allows for easy in-house cross-training and gives Collins Technology the flexibility to utilize employees in various tasks as work schedules require. In-house training on new equipment and procedures, along with outside training and certification, is provided to all levels of employees. Open hourly positions are posted internally to provide mobility and advancement wherever possible.

3. *Recognition*—There is an emphasis on recognition at Collins Technology and it is an important element in our HR program. Recognition is more fully described in Item 4.4. Benefits and compensation are recognized as part of the basic requirements of any organization. As such, they are handled as any other system requirement in the SBP process.

4. *Recruitment*—Requirements and plans for new employee needs are developed jointly with the affected departments, as each department develops its own departmental attainment plan. We understand the positive impact a diversity of backgrounds and knowledge can have on the long-term growth of a company such as ours. We foster diversity with an outreach program to the local community and schools, communicating our quality values and recruiting new employees.

Because of our quality reputation, stable employment, and above-average wage rate, we are able to attract high-quality recruits. Our intent is to continue to be in this position by monitoring our competition.

4.1b *Key Performance Indicators*

Key quality goals in the HR plan are training, empowerment, and involvement. Short-term goals are 100% involvement in 1994 and 100% certified operators in the station qualification system (SQS), described in Area 4.3b. Hiring methods have been improved by more thorough interviewing techniques, interviewing more candidates, and by networking through quality and electronics associations. Career development occurs primarily through our training and education program, educational assistance, and by added responsibility and authority available in our structure/philosophy.

Cycle time reduction is one of the Top Ten Quality actions, is included in each departmental attainment plan, and is regularly reviewed in the TQM Forum as a *Red Book* item.

The key indicator used to improve practices is the employee survey. Survey trends are shown in Table 4.1.

4.1c *Evaluate and Improve*

An annual employee satisfaction survey is given to all employees. To ensure we are progressing toward

Table 4.1

Selected Employee Survey Results

Selected Employee Survey Results

Criterion	1989	1990	1991	1992	1993
			(In Percent)		
Overall Satisfaction					
Collins is a good place to work	96	96	97	98	98
Job Satisfaction					
I like the work I do	70	75	79	86	89
Teamwork					
People cooperate to get the job done	80	87	93	96	97
Career Development					
Opportunity for advancement	82	87	93	96	98
Opportunity to improve skills	50	62	73	89	93
Education and Training					
I am trained to do my job	40	55	70	87	95
I have adequate input into training	35	65	80	88	95
Recognition & Rewards					
Satisfaction with pay	48	73	82	86	89
Efforts are recognized	50	61	76	89	95

"Total Employee Satisfaction," we monitor and analyze several key measurements (Item 4.5). We use quality analysis tools, particularly Pareto and cause-and-effect techniques, for this analysis. These measurements include training, turnover, absenteeism, accidents, and recognition. It is through these data that improvements are recognized and implemented. The data are reviewed by the TQM Forum. If the need for improvement is determined, a manager and/or team is assigned the responsibility for its solution and implementation. When we reviewed our turnover rate in 1990, concentrated turnover was shown among newly hired employees. This trend and its root causes were reviewed by the TQM Forum. As a result, we improved interviewing techniques, revised hiring policies, and developed orientation and "pal" programs for new employees. These changes reduced the turnover rate among newly hired employees by 180%.

We utilize measurements for all types of employees, as our high-tech capabilities are such an important factor in our corporate mission—to be the best with total customer satisfaction.

4.2 Employee Involvement

4.2a *Promote Contributions*

Collins Technology's approach to continual quality improvement requires the involvement of all employees. Employees contribute to the quality objectives through both individual and team participation. Our primary strategy for employee involvement/participation is by using teams. There are three types of formal team participation and involvement:

- Cross-Functional Team (CFT)

- Department Functional Team (DFT)

- Collins Employee Teams (CET)

The TQM Forum approves/establishes goals and priorities, assigns resources, and oversees all the action teams.

Each CFT has a TQM Forum member assigned as a coach. The coach provides support and guidance, eliminates barriers, and provides feedback to the TQM Forum. The captain runs the activities and reports on the progress of the team. Cross-functional teams have the highest priority in regard to quality improvement and meeting quality goals.

CFTs draw their members from all of the departments of the company. An example of a recently completed CFT was one whose mission was to stabilize a key front-end process. This team worked for one year on this project. The results were further process stabilization, more than 200% increase in quality, and a 250% increase in productivity.

DFTs attack priority issues within each department, with outside support as required. CETs are voluntary/self-managed teams composed primarily of hourly employees. Their mission is to solve the "useful many" problems of their work groups. Meetings are held weekly and are documented and reviewed by the CET. The CET program provides feedback, support, and helps with final team presentation to the TQM Forum. Special action teams are designated for tasks relating to external customer requests, programs, and issues. In addition, teams work with suppliers for qualification, and solving supplier problems. Figure 4.1 shows some of the current teams. Teams at Collins Technology have been effective for many activities.

This overall participatory team approach serves as a highly effective employee involvement and problem-solving vehicle. In addition to team involvement, our employees have many ways to contribute and be recognized on an individual basis. One of these mechanisms is the Error Cause Identification (ECI) system, which allows employees to notify supervisors, management, and support personnel of areas of improvement in performing their jobs or any problems they see within the company. Employees must receive a response to the ECIs within 24 hours. In 1993, 534 ECIs were written compared to 410 in 1992, 205 in 1991, and 93 in 1990.

Another formal method for individual contribution to the quality objectives is by the employee quality survey. This year's survey produced over 200 suggestions/comments. The results of the survey are compiled and reviewed by the TQM Forum. The top quality issues are investigated and solutions implemented. Feedback and communication are essential for employee involvement and improvements. This is accomplished through the many action teams, TQM Forum, coaches, supervisors, and through close personal interaction that is enhanced by our small business atmosphere.

All action teams routinely report to our TQM Forum to provide an update on the current status of problems and recommendations, to review future plans, and to receive feedback and suggestions. The coach/captain concept has proven to be most effective in providing timely, detailed, and accurate feedback. We have a monthly company meeting, and each department also has meetings on a monthly basis. The purpose of these meetings is to communicate our business environment, and organizational issues, provide recognition, give and receive feedback, and, most importantly, reinforce our quality values. Special methods to facilitate communication have also been developed, such as the Chief Operating Officer's monthly supervisor council, where all supervisors and managers can ask any questions, discuss issues, etc.

Figure 4.1

Teams and Involvement Programs

Involvement Programs	Teams *(continued)*	Teams *(continued)*
• TQM Forum	• Recycle	• LAN
• Safety	• Vendor Survey	• CET Program
• Quality Survey	• Customer Contact	• Customer Survey
• Groundbreakers	• Strategic Quality Planning	• Emergency Response
• Error Cause Identification	• Transition to Production	• Hole Plating
• Station Qualification System (SQS)	• Cycle Time Reduction	• Board Layout
Teams	• Design Guide	• Drilling
• Housekeeping	• Material	• Contouring
• Recognition	• Test Equip. Improvement	• Lamination
• Training	• Automation	• Electro-less Copper
• Chemical Handling	• Supplier Quality	• Test Equipment Fixturing
• Process	• Board Survivability	• Mask Fabrication
	• Design to Cost (DTC)	

Our open-door policy and small company size allow easy access for employees to express issues to senior management. Open-door issues are a management priority, and response to these issues is immediate. MBWA is a key tool for all managers, including upper-level management, who make frequent walks throughout all areas of the company to review the quality/delivery of our products and employee well-being. Individuals are encouraged and supported to take the initiative to solve problems. We try to create a "stop me in the hall" environment so opportunities/problems are recognized and addressed quickly.

4.2b *Approaches to Empowerment*

We train and encourage our employees to take individual action immediately when the quality of products or services is involved. The company has an established history of empowering teams. A space to meet is provided, $300 can be spent without any higher approval, and a coach is assigned from senior management. An example is the ability of manufacturing operators to shut down a manufacturing line. Process control charts are used in assembly operations, and if control limits are exceeded, the operator stops production until the problem is corrected. Assembly lines are equipped with green, yellow, and red status lights which show the condition of the line. SQS certification, described in Area 4.3b, gives operators additional responsibility for

quality issues. In 1990 we reduced the number of layers in our organization structure and gave assembly line supervisors full authority/responsibility. This responsibility includes all manufacturing functions, such as assembly, test, inspection, and pack. The business segment team, with personnel from quality, marketing, engineering, and manufacturing, is now responsible for product build, quality, and customer satisfaction by business segment. Employee involvement is also the focal point of other aspects of our business. For example, engineers exercise design and budget authority on their projects, and marketing personnel make on-the-spot decisions with respect to customer satisfaction.

These decisions previously required the joint action of the various department managers. The result is better quality, performance, and faster delivery of products to our customers. Innovation is encouraged by the openness with which our company accepts new ideas and suggestions. This innovation is crucial to remaining the industry leader. Our market position and the number of patents received are evidence of our innovation in multi-layer board technology.

Requirements are determined using multiple inputs: employee survey, ECIs, TQM Forum goal progress review, internal customer/supplier 20-minute weekly meetings, performance reviews, customer surveys, and customer complaint analysis.

Survey indicators are presented in Area 4.1b and involvement goals are presented in Figure 4.2.

4.2c Employee Involvement Indicators

The key indicators we use for employee involvement include the overall percentage of employees on teams, especially hourly and senior management personal involvement and percent of suggestions implemented. The goal is 100% involvement in all categories of the workforce by 1994, as shown in Area 4.2d. The TQM Forum evaluates these indicators, monitors trends, and gives direction, as necessary, to improve employee involvement. Team reviews with the TQM Forum and analysis of the cost of quality (COQ) help determine the effectiveness of employee involvement. The results of employee involvement are faster response time to customer demands as well as overall improved customer satisfaction. When customer satisfaction is an issue, everyone knows that this always gets top priority.

Figure 4.2

Employee Team Involvement

Indicators	1991	1992	1993	Goal 1994
Number of Teams	25	41	50	75
Employees Involved (%)	50	88	95	100
Hourly Members (%)	40	82	90	100
Salaried Members (%)	70	100	100	100
Senior Exec. on Teams	100	100	100	100

4.2d Trends in Employee Involvement

The trends (Figure 4.2) in employee involvement have been dramatic over the past several years. The number of teams increased by 100% over the past three years, even with the successful completion of many of these teams. Employee involvement has steadily risen and remains high, with over 95% involvement. More important is the participation of the hourly employees. Their involvement has risen dramatically by over 120% in the past three years. Our goal is 100% involvement. Increasing the number of employee teams and the emphasis on the team concept, coupled with individual involvement and the empowerment philosophy, will help us easily accomplish this goal. In 1993 over 80% of the suggestions were implemented.

4.3 Employee Education and Training

4.3a Needs

1. One of Collins Technology's philosophies is that to be successful, quality training and education are of paramount importance. We began our TQM System with an extensive quality education series focusing on employee quality awareness and statistical quality control. One of the first teams formed by our TQM Forum was the "Training Action Team." This team reviews the specific training and education needs from each department, as identified during performance

appraisals and department meetings, and then defines a training plan. This training and education process is comprehensive and encompasses both individual and group training ranging from quality awareness/orientation to job specific training. Figure 4.3 identifies the elements in this program. The company recognized at the beginning that to accomplish our mission in the marketplace with our high-tech product, we would need a highly trained workforce at all levels.

2. Core courses have been provided to all employees with specialized courses, such as designed experiments, being provided to employees with special needs. Initially all employees received training in quality awareness, basic statistical process control, problem solving, process simplification, and

Figure 4.3

Training and Education Courses and Support Programs

Quality
- Awareness
- SPC
- Recognition
- Problem Solving
- COQ
- Supplier Quality
- Statistical Tools
- Design of Experiments
- Sensitivity Analysis
- Teachings of Deming
- Teachings of Juran

Training Reinforcement
- Company Newsletter
- Performance Reviews
- Pal System

Management Seminars
- Effective Delegation
- Communications
- MBWA
- Time Management
- Employee Motivation

Management Development
- Quality Training
- Supervisor Development
- Strategic Planning
- Supervisor Council
- Tuition Assistance
- Business Segment Alignment

Special Training
- Customer Service
- Supplier Education
- MRP
- Tuition Assistance
- Proposal Writing
- New Employee Orientation

Job Skills
- SQS
- Desktop Publishing
- Editors
- Spreadsheets

Safety
- Awareness
- MSDS Reading
- CPR
- Emergency Response
- Container Labeling
- Hazard Detection
- Emergency Evacuation
- Respiratory Protection
- Hearing Conservation
- Back Safety
- Medical Testing

Team Training
- Coach Program
- Collins Employee Teams (CET)
- Cross-Functional Teams (CFT)
- Department Functional Teams (DFT)

group dynamics. Employees can attend classes to prepare themselves for a new job.

3. Education and training plans are developed jointly between the employee and supervisor.

4.3b *Delivery*

Our method for quality education and training began with outside training. We have since shifted to self-developed training taught by our own management. Teaching by management serves as another means to reinforce our commitment to quality and training. By 1995, our goal is to have 100% of quality, safety, skills, and team training taught in-house.

We also build on-the-job reinforcement into our training processes and practices to ensure retention and effective use of this knowledge. Reinforcement of knowledge and skills acquired by the operators is accomplished through daily reinforcement and through annual recertification in job-specific tasks. In our

SQS training program, operators take a written test and demonstrate their skills for certification and annual recertification. These test the understanding and use of documented procedures for their process. A unique feature of our operator skills training is that the supervisors must be tested and certified in the job-specific tasks of their work group. Among the benefits have been added attention to the details and improvements to individual processes.

4.3c *Evaluate and Improve*

Education and training effectiveness are determined through the training team, which is a cross-functional team. This team analyzes training for all categories and types of employees. The training team presents their plans, recommendations, and reasons to the TQM Forum for review, approval, and additional recommendations. The key indicators used by this committee are the measurements shown in Figures 4.4 and 4.5, the training matrix shown in Figure 4.3, the number of CET teams,

Figure 4.4

Training Hours/Employee

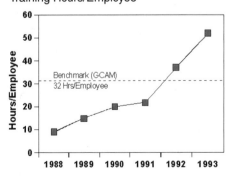

Figure 4.5

Training Expenditures/Employee

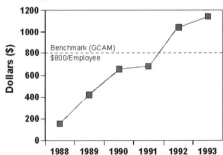

and percent of certified operators. Key additional inputs and feedback on training are derived from the employee quality survey, which has a complete section devoted to training, and individual inputs from MBWA and performance reviews.

1. *On-the-job improvements*—As our workforce becomes more proficient in their jobs, one of the best indicators is the number of certified operators. The SQS has grown to include 45% hourly, with supervisors at 60%. Other function improvements are shown by increased performance and reduced cycle times as reviewed by the TQM Forum.

2. *Employee growth*—The company encourages all employees to grow in their jobs and personal life. The SQS program permits operators to cross-train on various stations in their areas. Office workers and engineers take courses at local colleges. In the last four years we have had twelve people attain bachelor's degrees and three receive master's degrees through our tuition refund program.

4.3d *Trends in Education/Training*

1. *Quality orientation*—All new employees receive initial quality orientation. It consists of a senior executive providing an overview of the company background and the total quality philosophy, viewing of our President/CEO's videotape "Total Quality Management at Collins Technology," a personal review, and initial safety training. A "pal" is assigned to each new employee on the first day. A unique feature of our orientation program is that all employees (even temporary employees) go through this training. Thus, employees are introduced to the Collins Technology's quality culture from the very beginning, and they feel more comfortable and knowledgeable about the company. They understand not only how, but why we do what we do. Evidence of quality awareness is the personal commitment our employees have made by signing the quality promise. The signing of the quality promise is voluntary. Since the initial signing ceremony in February 1990, 100% of the employees have made this important commitment. This is a good example of the significant difference that going from no orientation training to 100% training today has made.

2. *Percent of employees (by category)*—The percent of workers receiving quality training and education is shown in Figure 4.6. Especially significant is the total training in the areas of quality awareness and safety, and initial training in SPC. Management commitment to quality is evidenced by the participation in all areas.

3. *Average hours*—devoted to quality education/training per

Figure 4.6

Quality Training and Education

Quality Training	% of Population Trained			
	Senior Management	Middle Management	Supervisor	Hourly Employees
Quality Awareness	100	100	100	100
Supervisory Skills	100	100	80	N/A
Safety	100	100	100	100
Problem Solving	100	100	90	85
Prevention—Total Systems Approach	100	100	90	80
SQS Certification	N/A	N/A	60	45
Statistical Tools	100	100	100	85

employee has increased by over 550% and expenditures by 725% in the last six years. Figures 4.4 and 4.5 show these trends.

4. *Percent of employees*—100% of our employees have received quality education/training. Figure 4.6 gives specific details by course and category of employee.

5. *Statistical/problem-solving training*—100% of our employees have received initial training in statistical and quantitative problem-solving methods. This training includes the Collins' Seven Quality Tools: Pareto charts, cause-and-effect diagrams, control charts, process maintenance charts, yield charts, scatter plots, and flowcharts. Comprehensive statistical training has been given to over 85% (goal = 100%) of our employees. Quality training by

Figure 4.7

Recognition and Reward Methods

- Profit Sharing—401(k)
- Individual/Team Bonuses
- Good Housekeeping Award
- Clean Aide Award
- Service Award
- Customer Awards
- SQS Certification

- Quality Commitment Days
- Newsletter
- Team Awards
- Promotion
- Performance Reviews
- CET Recognition
- Company Meetings
- Milestone Celebrations
- Christmas/New Year's Party

- Verbal Thanks
- Perfect Attendance
- Company Picnics
- Birthday Cakes
- Sirloin & Brew Tickets
- Free Breakfast Snacks
- Thanksgiving Turkey
- Baseball Game Tickets

senior executives has more than doubled in the past two years, which further emphasizes our management's commitment to quality.

4.4 Employee Performance and Recognition

4.4a *Strategies for Recognition*

1. *Quality emphasis relative to other business considerations —* Employee recognition is an important aspect of Collins Technology's quality program. We feel recognition is a key to creating/maintaining quality consciousness, motivation, and communicating values. We emphasize many different employee recognition programs, which are shown in Figure 4.7. Recognition is focused on achieving the SBP objectives.

   Our recognition programs seek to reward all key quality contributions whether by an individual or a team. A balance between team and individual recognition is sought. Our monthly all-employee meeting and the *Collins News* are the primary vehicles for our public recognition program. Quality relative to the financial results is reinforced by the TQM Forum monthly review of the cost of quality. Company meetings review our performance to shipments forecast with emphasis on the importance of on-time

delivery in relation to future business. At Collins Technology, quality and financial results are important, but quality considerations come first. This assertion is supported by the performance appraisal which gives most weight to quality indicators.

2. *Employee involvement in performance measurements —* The recognition program is the responsibility of the TQM Forum. It approves and administers all company recognition to ensure equity and a high visibility emphasis. There is a cross-functional team to help assist and provide input to the recognition program. The Recognition Team is composed of hourly employees and coached by the Chief Operating Officer. This committee reviews the existing recognition programs, provides feedback, reviews and sets goals, and presents suggestions/plans to the TQM Forum. Senior management and department managers develop performance measurements and goals. Each individual or work group will develop goals/plans to support the performance measurement. Individuals also develop and modify the measurements based on the results of their process and process limits.

4.4b *Indicators for Improvement*

The key indicators and methods used to evaluate and improve the recognition program are feedback received from our Recognition Team, our annual employee quality survey, performance appraisal feedback, benchmark data, and quantifying the number of individual/team recognition (Figure 4.8). The Recognition Team is responsible for using these data as part of the annual planning process.

1. Recognition is tracked by Managers, Supervisors, and Hourly to assure all categories of employees are included.

2. Correlation analysis is done on the answers from the employee survey which asks for feedback on "Efforts are recognized" and "Satisfaction with pay."

3. Our total quality effort requires total employee involvement. Leaders remove the roadblocks that keep people from taking pride in their job. There is a positive link between employee satisfaction and quality levels achieved.

Figure 4.8
Special Recognitions per Year

	1989	1990	1991	1992
Individual:				
Manager	10	11	11	15
Supervisor	20	22	30	35
Hourly	18	26	44	60
Team	10	25	40	51
Total	58	84	125	161

4.4c *Trends in Recognition*

Our recognition programs are growing and continually being refined. In addition to the well-established group and individual recognition programs, we recently began the certification program (SQS). This program recognizes those individuals (both operators and supervisors) who successfully complete process certification courses in their specialty.

Figure 4.8 shows trends in employee recognition. We categorize our recognition in terms of individuals and teams. Recognition of employees shows a very positive trend and has more than doubled over the past four years. In addition to increased individual recognition, team recognition has increased fivefold and will continue to be emphasized to increase the synergistic benefit of teamwork.

4.5 Employee Well-Being and Satisfaction

4.5a *Incorporation of Well-Being and Satisfaction*

Collins Technology provides a safe and productive work place for all employees. The health and safety of our employees are of paramount concern, and special safety teams specifically address these needs. Our health and safety goal is to have no employee injuries. The method used to accomplish this is through extensive safety training and a constant emphasis on the health of all employees by upgrading equipment, even when the health risk is minimal (e.g., fume hoods, special gloves). Accident measurements together with tools such as Pareto analysis, cause-and-effect diagrams, FMEA, and/or fault tree analysis, are used to identify the underlying causes and solutions.

Employee satisfaction is monitored and improved from the results of the employee quality survey. The goals for employee satisfaction are to continuously improve upon the high scores from the quality survey and to use this survey as a primary vehicle for feedback and plans.

Both individuals and teams make suggestions on ergonomic improvements. Ergonomic goals and methods will be attained by total involvement of employees in designing/selecting their equipment and facilities. Monthly safety audits focus on ergonomics.

Safety and housekeeping quality goals are established to ensure that employees have a safe and clean environment in which to work. During the last several years we have made significant strides in the improvement of the work environment. Our housekeeping program has been used as a model for industry in our area.

Accidents and work-related health problems are prevented by using several methods. The first is our Safety Team, which is responsible for reviewing the health and safety

factors at Collins Technology. This team reviews accident reports each month and recommends preventive measures immediately. All accidents must be reported and investigated and solutions implemented. Additionally, safety audits are performed quarterly with periodic voluntary safety audits by our insurance carrier. Collins Technology seeks to protect the safety and health of all its employees. Our safety coordinator and personnel coordinator both serve as resources to ensure a safe work environment.

Collins Technology takes a proactive approach toward overall employee health. The personnel coordinator is in contact with any ill employees on a constant basis and reports to the management staff each week. Prevention is the emphasis in the safety program. One example is that the Safety Team noticed an increased trend in accidents from our chairs (25% of all accidents). The employee team selected newly designed chairs that not only eliminated the accidents, but also provided increased comfort and reduced fatigue.

Another resource is our insurance company. Collins Technology and the insurance company have worked cooperatively resulting in a perfect score on a safety audit. Also, through this cooperative effort, air quality testing has been conducted in all areas, emergency procedures and safety equipment reviewed, standards set for hearing/medical testing, and suggestions obtained to minimize physical lifting and repetitive motion disorders.

4.5b *Special Services*

Our EAP offers referrals to various agencies to address employees' personal needs such as counseling and financial assistance. Monthly sport outings are organized by the Human Resource Department. Child care assistance is provided in cooperation with the local Day Care Center. We have a tuition reimbursement program for employees and a merit scholarship program for children of employees.

4.5c *Determination of Employee Satisfaction*

Employee satisfaction is determined through the annual employee quality survey and by close involvement and communication with all employees. The survey has consistently reported high marks (95%) for employee satisfaction, identities top concerns by employees, and is thoroughly reviewed and analyzed by the TQM Forum. The data are reviewed using Pareto techniques, and the top issues are then solved. Quality improvement plans are developed and implemented from this employee survey. Collins Technology also has an open-door policy and a grievance procedure, which provide opportunities to ascertain existing problem areas and to find solutions.

4.5d *Trends in Key Indicators*

We monitor several key indicators for employee well-being and satisfaction.

Figure 4.9 shows the trends in these indicators for the past four years. During 1991, our accident rate declined 45% from the previous year. Our absenteeism rate (1.7%) is well below the industry average (7.7%), and we are working toward the benchmark of 1%. Our turnover rate shows a steady decrease. Area 7.2e addresses attrition of customer-contact personnel. As explained in Area 4.5c, employee satisfaction, as measured by our quality survey, is very high. Collins Technology's workers compensation experience modifier decreased each year through 1993. We reduced our unemployment compensation tax rate by over 75%, having a direct financial impact. As can be seen in Figure 4.9, we have not experienced any adverse trends. This is attributed to our TQM program and our HR operations. Our employees work together and all employees are concerned with Collins Technology continuously improving.

Figure 4.9

Well-Being and Satisfaction Indicators

	1990	1991	1992	1993
Safety (# accidents)	40	25	15	9
Absenteeism (%)	2.7	2.3	1.9	1.7
Turnover (%)	22	20	15	12
C.C. Attrition	—See Category 7.0—			
Satisfaction (%)	96	97	98	98
Recruitment success (%)	25	30	40	60
Degrees attained	2	4	3	6
Grievances (%)	—None—			
Strikes (#)	—None—			
Worker Comp. Modifier	.90	.88	.74	.68
Unemployment Comp. rate	6.32	5.45	2.50	1.50

5.0 Management of Process Quality *(140 pts.)*

The *Management of Process Quality* Category examines the key elements of process management, including design, management of day-to-day production and delivery, improvement of quality and operational performance, and quality assessment. The Category also examines how all work units, including research and development units and suppliers, contribute to overall quality and operational performance requirements.

5.1 Design and Introduction of Quality Products and Services *(40 pts.)*

Describe how new and/or modified products and services are designed and introduced and how key production/delivery processes are designed to meet both key product and service quality requirements and company operational performance requirements.

☑ Approach
☑ Deployment
☐ Results

Areas to Address

(a) how products, services, and production/delivery processes are designed. Describe: (1) how customer requirements are translated into product and service design requirements; (2) how product and service design requirements, together with the company's operational performance requirements, are translated into production/ delivery processes, including an appropriate measurement plan; (3) how all product and service quality requirements are addressed early in the overall design process by appropriate company units; and (4) how designs are coordinated and integrated to include all phases of production and delivery.

(b) how product, service, and production/delivery process designs are reviewed and validated, taking into account key factors: (1) overall product and service performance; (2) process capability and future requirements; and (3) supplier capability and future requirements.

(c) how designs and design processes are evaluated and improved so that new product and service introductions and product and service modifications progressively improve in quality and cycle time.

Notes:

1. Design and introduction might address modifications and variants of existing products and services and/or new products and services emerging from

research and development or other product/service concept development. Design also might address key new/modified facilities to meet operational performance and/or product and service quality requirements.

2. Applicants' responses should reflect the key requirements for their products and services. Factors that might need to be considered in design include: health; safety; long-term performance; environment; measurement capability; process capability; manufacturability; maintainability; supplier capability; and documentation.

3. Service and manufacturing businesses should interpret product and service design requirements to include all product- and service-related requirements at all stages of production, delivery, and use.

4. In 5.1a(2), company operational performance requirements relate to operational efficiency and effectiveness—waste reduction and cycle time improvement, for example. A measurement plan should spell out what is to be measured, how measurements are to be made, and performance levels or standards to ensure that the results of measurements will show whether or not production/delivery processes are in control.

5. Results of improvements in design should be reported in 6.1a. Results of improvements in design process quality should be reported in 6.2a.

5.2 Process Management: Product and Service Production and Delivery Processes *(35 pts.)*

Describe how the company's key product and service production/delivery processes are managed to ensure that design requirements are met and that both quality and operational performance are continuously improved.

☑ Approach
☑ Deployment
☐ Results

Areas to Address

(a) how the company maintains the quality and operational performance of the production/ delivery processes described in Item 5.1. Describe: (1) the key processes, their requirements, and how quality and operational performance are tracked and maintained. Include types and frequencies of in-process and end-of-process measurements used; (2) for significant (out-of-control) variations in processes or outputs, how root causes are determined; and (3) how corrections of variation [from 5.2a(2)] are made, verified, and integrated into process management.

(b) how processes are improved to achieve better quality, cycle time, and operational performance. Describe how each of the following is used or considered: (1) process analysis/simplification; (2) benchmarking information; (3) process research and testing; (4) use of alternative technology; (5) information from customers of the processes—within and outside the company; and (6) stretch targets.

Notes:

1. Manufacturing and service companies with specialized measurement requirements should describe how they assure measurement quality. For physical, chemical, and engineering measurements, describe briefly how measurements are made traceable to national standards.

2 Variations [5.2a(2)] might be observed by those working in the process or by customers of the process output. The latter situation might result in formal or informal feedback or complaints. Also, a company might use observers or "mystery shoppers" to provide information on process performance.

3. Results of improvements in product and service production and delivery processes should be reported in 6.2a.

5.3 Process Management: Business and Support Services Processes (30 pts.)

Describe how the company's key business and support service processes are designed and managed so that current requirements are met and that quality and operational performance are continuously improved.

☑ Approach
☑ Deployment
☐ Results

Areas to Address

(a) how key business and support service processes are designed. Include: (1) how key quality and operational performance requirements for business and support services are determined or set; (2) how these quality and operational performance requirements [from 5.3a(1)] are translated into delivery processes, including an appropriate measurement plan.

(b) how the company maintains the quality and operational performance of business and support service delivery processes. Describe: (1) the key processes, their requirements, and how quality and operational performance are tracked and maintained. Include types and frequencies of in-process and end-of-process measurements used; (2) for significant (out-of-control) variations in processes or outputs, how root causes are determined; and (3) how corrections of variation [from 5.3b(2)] are made, verified, and integrated into process management.

(c) how processes are improved to achieve better quality, cycle time, and overall operational performance. Describe how each of the following are used or considered: (1) process analysis/simplification; (2) benchmarking information; (3) process research and testing; (4) use of alternative technology; (5) information from customers of the business processes and support services—within and outside the company; and (6) stretch targets.

Notes:

1. Business and support services processes might include units and operations involving finance and accounting, software services, sales, marketing, public relations, information services, purchasing, personnel, legal services, plant and facilities management, basic research and development, and secretarial and other administrative services.

2. The purpose of Item 5.3 is to permit applicants to highlight separately the improvement activities for functions that support the product and service design, production, and delivery processes the applicant addressed in Items 5.1 and 5.2. The support services and business processes included in Item 5.3 depend on the applicant's type of business and other factors. Thus, this selection should be made by the applicant. Together, Items 5.1, 5.2, 5.3, and 5.4 should cover all operations, processes, and activities of all work units.

3. Variations [5.3b(2)] might be observed by those working in the process or by customers of the process output. The latter situation might result in formal or informal feedback or complaints.

4. Results of improvements in business processes and support services should be reported in 6.3a.

5.4 Supplier Quality *(20 pts.)*

Describe how the company assures the quality of materials, components, and services furnished by other businesses. Describe also the company's actions and plans to improve supplier quality.

☑ Approach
☑ Deployment
☐ Results

Areas to Address

(a) how the company's quality requirements are defined and communicated to suppliers. Include a brief summary of the principal quality requirements for key suppliers. Also give the measures and/or indicators and expected performance levels for the principal requirements.

(b) how the company determines whether or not its quality requirements are met by suppliers. Describe how performance information is fed back to suppliers.

(c) how the company evaluates and improves its own procurement processes. Include what feedback is sought from suppliers and how it is used in improvement.

(d) current actions and plans to improve suppliers' abilities to meet key quality, response time, or other requirements. Include actions and/or plans to minimize inspection, test, audit, or other approaches that might incur unnecessary costs.

Notes:

1. The term "supplier" refers to providers of goods and services. The use of these goods and services may occur at any stage in the production, delivery, and use of the company's products and services. Thus, suppliers include businesses such as distributors, dealers, warranty repair services, contractors, and franchises as well as those that provide materials and components.

2. Generally, suppliers are other-company providers of goods and services. However, if the applicant is a subsidiary or division of a company, and other units of that company supply goods/services, this relationship should be described as a supplier relationship.

3. Determining how quality requirements are met (5.4b) might include audits, process reviews, receiving inspection, certification, testing, and rating systems.

4. Actions and plans (5.4d) might include one or more of the following: joint planning, partnerships, training, long-term agreements, incentives, and recognition. They might also include supplier selection. "Other requirements" might include suppliers' price levels. If this is the case, suppliers' abilities might address factors such as productivity and waste reduction.

5.5 Quality Assessment *(15 pts.)*

Describe how the company assesses the quality and performance of its systems and processes and the quality of its products and services.

> ☑ Approach
> ☑ Deployment
> ☐ Results

Areas to Address

(a) how the company assesses: (1) systems and processes; and (2) products and services. For (1) and (2), describe: (a) what is assessed; (b) how often assessments are made and by whom; and (c) how measurement quality and adequacy of documentation of processes are assured.

(b) how assessment findings are used to improve: products and services; systems; processes; supplier requirements; and the assessment processes. Include how the company verifies that assessment findings are acted upon and that the actions are effective.

Notes:

1. The systems, processes, products, and services addressed in this Item pertain to all company unit activities covered in Items 5.1, 5.2, 5.3, and 5.4. If the assessment approaches differ appreciably for different company processes or units, this should be described in this Item.

2. Adequacy of documentation should take into account legal, regulatory, and contractual requirements as well as knowledge preservation and knowledge transfer to help support all improvement efforts. Adequacy should take into account completeness, timely update, usability, and other appropriate factors.

5.0 MANAGEMENT OF PROCESS QUALITY

Collins Technology provides its customers with highly engineered products and services for unique applications. The thrust of our TQM System is to achieve customer satisfaction through constant feedback and control of all business operations. Each business operation continuously improves its processes by understanding customer needs and reducing variation.

5.1 Design and Introduction of Quality Products and Services

5.1a *Meeting Customer Design Needs and Expectations*

We follow a thorough product development plan. The plan starts by defining customer requirements/expectations and ends with full production. Both new and modified products are treated the same. The Sales Coordinator initiates a design team at the onset. The design team is a cross-functional group consisting of members from design, manufacturing, sales, and quality. In some cases customer representatives are members of the team. Where purchased material is critical to the product, suppliers are included on the team. The team follows the product all the way through production.

1. *Translating customer requirements—* How we translate customers' requirements and expectations varies depending on which business segment they are

in. For the Government and Industrial Products segments, translation of requirements is relatively easy since detailed specifications are supplied by the customer. These are usually designs they have developed internally and often transmit the detailed drawings directly to our engineering computers. There is typically little room for modification from these designs.

With customers in our Commercial Products the process is different. Here the designs and requirements are much more flexible and fluid based on our input. In approximately 50% of these projects, a design engineer and/or marketing person from the customer is included on the design team. Where a customer's representative is not on the team, the Sales Coordinator is responsible for refining and communicating expectations between the customer and the design team.

2. *Process control plan*—The cross-functional design team (see 5.1a4) is responsible for developing a process control plan. This plan includes a production flowchart (PFC) with related sources of variation for the product identifying all processes that will be used in its manufacture. The PFC includes identification of the deliverables and sources of

variation at each step of the process. The deliverables include productivity, reliability, safety, ergonomic, and housekeeping as well as product requirements.

There is a strong emphasis on robust designs, i.e., designs that are not sensitive to variation. The loss function is used to determine if a design is robust and to determine its key characteristics. When sensitivity to variation is identified, key control characteristics are determined. These are process characteristics that are needed to be kept on target with reduced variation in order to maintain the key product characteristic on target. For the key product and control characteristics, an interrelationship matrix is developed as well as measurement systems analysis and capability studies performed (see 5.2a1). The appropriate control method(s) for the dominant source(s) of variation (i.e., setup, material, maintenance, etc.) of the process are established. At each stage of the design process, the process control plan is reviewed and updated.

Key processes—For each manufacturing process, we determine its capability by using Statistical Process Control (SPC) techniques. These capabilities are inputs for the *Multi-layer Board Design Guide.* The employees using these processes are trained in the use of SPC techniques, which enable them to control the process within the control limits and identify and correct out-of-control conditions.

Key indicators—Performance information in the LAN and the *Red Book* are reviewed at all TQM Forum meetings and at the monthly departmental meetings. Customer feedback and testing results are evaluated at these meetings. When action is required, usually a CET, DFT, or CFT is established. The teams use the statistical tools in which they have been trained. Often DOE is utilized. Through the team process, root causes are determined and process improvements established. The new and modified processes are reviewed and evaluated in the process laboratory (described in Area 5.2c), prior to incorporation. Our company's productivity per employee is monitored quarterly by management to verify that the sum of all the individual quality efforts is truly having a substantial bottom-line impact.

Cost of Quality (COQ)—In 1991, our TQM Forum perceived the need for a "bottom-line" trend indicator of the results of our TQM System. A team of operations and finance personnel was appointed to devise a cost of quality (COQ) measurement. The quality cost of nonconformance measure we developed includes both internal and external failure costs measured as a percentage of our sales revenue. Internal failure cost encompasses such metrics as scrap, yield loss, rework, receivables, and inventory. External failures include any returns and warranty costs. The COQ data are reviewed monthly by both operations management and our TQM Forum. The COQ drivers are identified and prioritized for allocation of resources, such as engineers and action teams.

Waste Stream Reduction Program— We have a comprehensive program to recover and recycle our industrial waste. When we began the program, disposal was neither difficult nor costly. However, we proactively anticipated the escalating costs and environmental impact associated with disposal. We addressed solvents, plating system effluent, and filter media (see Figure 1.3).

3. *Addressing requirements early*—In 1989, a cross-functional team consisting of all disciplines of the company established a *"Multi-layer Board Design Guide."* This guide, which contains all design criteria including design-for-manufacturing considerations, is under document control. As the primary source of design information, it contains detailed information about each of our key processes including all established capabilities of process parameters. It also contains all checklists used during the design

and it documents any design limitations encountered by prior design teams.

The product development process requires a Preliminary Design Review (PDR) very early in the program. Using the *Multilayer Board Design Guide*, the team reviews a checklist set of topics (customer requirements, deviation from our design standards, testing, preliminary manufacturing plans, cost estimates, risk management), along with the actual preliminary product/process design. This review forces discussion of key issues before finalizing designs. The majority of topics discussed at the reviews are quality and design related. One of the tools used is failure mode & effects analysis (FMEA), which is a systematic way of asking what can go wrong. This is done so that problems can be prevented.

Product cost is another key requirement. We use Design to Cost (DTC) techniques to ensure that we meet or exceed the customer's requirements.

At this review and all of the subsequent reviews, there is full representation from both customers and company units (marketing, manufacturing, engineering, quality, purchasing, drafting, etc.). All representatives must approve the completion of the

PDR before the program can proceed.

4. *Integrating designs*—The design team controlling product design is cross-functional, representing all areas of the company including manufacturing and assembly. This assures that we coordinate and integrate all product designs and modifications into all phases of production and delivery.

5.1b *Design Validation*

The design of new and modified products undergoes a series of design reviews and transition checks. This diligent development process ensures meeting the customer's requirements and expectations the first time.

1. *Product and service performance*— A mathematical model is developed to determine expected capacitance and tolerances for extreme environmental conditions. The mathematical model is validated by the use of designed experiments. A computer simulation is done to consider various conditions and customer requirements. Results from the mathematical model are compared to the initial run inspection and capability studies. The model is then updated, if necessary, for future design and continual improvement activities.

2. *Process capability*—There are two design reviews (preliminary and critical) and three formal releases

before production. At each review, the team evaluates the current product manufacturing process capabilities and future goals. Appropriate statistical measures define the capability of all related processes. This information is incorporated into the process control plan. Process improvement plans for key characteristics and marginal processes are also reviewed.

3. *Supplier quality*—The design team reviews supplier capability during the two design reviews. These reviews identify long lead times or design constraints for components required in our products. Suppliers attend design reviews or participate on design teams to align capabilities and capacities with product designs. The close relationships we maintain with our suppliers allows them to be a part of understanding and defining future requirements.

5.1c *Improving Design Process Effectiveness*

The *Multi-layer Board Design Guide* is the primary source of information related to designs and how they relate to our process capability. At the end of each new design, the design team reviews and updates the design guide to continually improve the effectiveness of designs. The design guide prevents future problems by documenting any design limitations. We also incorporate these standards into our computer-based product modeling software. We then compare actual test results to the model's prediction and use the results to update our model. The Design Cycle Review Team monitors the design cycle and recommends improvements.

Since 1990 we have achieved significant reductions in our design cycle time (see also Area 6.3a). Our short-term goal (two years) is for another 80% improvement.

5.2 Process Management: Product and Service Production and Delivery Processes

5.2a *Controlling Processes*

1. *Performance determined and maintained*—Process control, rather than product inspection, is our major focus. The early detection of process problems reduces our quality costs. Processes that are in control provide products that consistently meet our customers' expectations. Figure 5.1 shows typical key process and product measurements. These characteristics detail the critical points for board fabrication.

   The control plan (described in 5.1a2) is updated as the measurement system quality, process capability, or process flow changes.

   Collins Technology has developed several systems that control the processes by computer. Plating baths are analyzed and chemicals added, when required, automatically. Periodic measurements are taken and recorded on control charts to assure automatic systems are performing correctly.

   Measurement system analysis is performed to determine the bias, repeatability, and reproducibility of the measurement system. Collins Technology became aware of these procedures and their importance during a visit with the Ford Motor Company. The procedures used are defined in the Ford, GM, and Chrysler *Measurement Systems Analysis* manual, which was cosponsored by the Automotive Division of ASQC and is administered by AIAG.

2. *Out-of-control occurrences*—When an out-of control condition occurs, we use a six-step problem-solving process (Figure 5.2). This process helps us to systematically determine the problem's root cause.

Figure 5.1

Key Indices

Characteristic	Freq	Type	
		Product	Process
Inner layer thickness	1%		✔
Thickness after lamination	2%		✔
Plating Baths	hourly		✔
• pH	hourly		✔
• Nickel concentration	hourly		✔
• Copper concentration	hourly		✔
• Additive A concentration	hourly		✔
• Additive B concentration	hourly		✔
Plating thickness	4×/shift		✔
Plating adhesion	3×/shift		✔
Hole cross-section	Sample	✔	
Completed board — final test	100%	✔	

Figure 5.2

Six-Step Problem-Solving Process

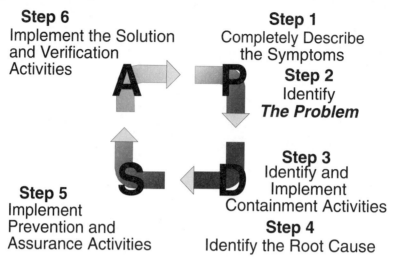

Step 6
Implement the Solution and Verification Activities

Step 1
Completely Describe the Symptoms

Step 2
Identify
The Problem

Step 5
Implement Prevention and Assurance Activities

Step 3
Identify and Implement Containment Activities

Step 4
Identify the Root Cause

Test runs verify predicted results. A problem-solving team then implements solutions which prevent future occurrences of the same problem. The key features of this problem-solving process are involvement of all key employees and use of Collins' Seven Quality Tools [see Area 4.3d(5)]. Analysis of this formatted data drives preventive-based actions to resolve the root causes.

3. *Correction Verification and integration*—The SPC chart documents corrections made to all out-of-control conditions. The applicable drawing and process maintenance chart document product and/or process or equipment changes.

The existing system of SPC charts verifies that the solution did indeed correct the problem. All out-of-control conditions are recorded in the LAN. This allows the TQM Forum to review process results and any corrective action taken. It also allows team members to monitor the process on a longer term basis to assure that root causes of problems are resolved.

5.2b *How Processes Are Improved*

All processes are reviewed on a department level by the department head through analysis of the data collected in the LAN and review of process flowcharts. From this, processes are identified for special improvement by the use of Pareto charts. Either teams are created or individuals are assigned to work on making these improvements. We also encourage individuals to continuously review their processes for potential improvements.

Figure 5.3

Simplified Multi-layer Board Process Flow

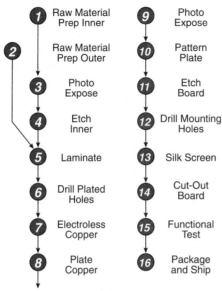

① Raw Material Prep Inner
② Raw Material Prep Outer
③ Photo Expose
④ Etch Inner
⑤ Laminate
⑥ Drill Plated Holes
⑦ Electroless Copper
⑧ Plate Copper
⑨ Photo Expose
⑩ Pattern Plate
⑪ Etch Board
⑫ Drill Mounting Holes
⑬ Silk Screen
⑭ Cut-Out Board
⑮ Functional Test
⑯ Package and Ship

Figure 5.3 shows a simplified flow diagram of our manufacturing process flow. The initial part of the process flow (steps 1–6) is the preparation of the boards for plating, the center process flow (steps 7–11) is the plating, and the final part of the

process flow (steps 12–16) is final operations. Because of our many process steps with close tolerances (4 mil lines and 4 mil spaces), excellence in process control and stability is critical in ensuring a reliable, high-performance product for our customers. In this section we will use this flow diagram to illustrate the TQM methods and techniques we use throughout our product manufacturing operations. Statistical, effectiveness, yield, and performance results given in this section are representative of similar results in all areas of our business.

As an example, the etch process (step 4) is critical for producing a high-quality board. Using SPC analysis, we measure the stability and Cpk of the process for all operators. Figure 5.4 shows actual data for two different operators over a sequence of unit production. Although both operators were stable (i.e., in statistical control), Operator 1's technique resulted in an

Figure 5.4

Etching Process Control Charts

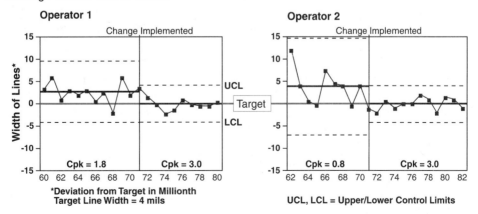

Operator 1

Change Implemented

Width of Lines*

Cpk = 1.8   Cpk = 3.0

60 62 64 66 68 70 72 74 76 78 80

*Deviation from Target in Millionth
Target Line Width = 4 mils

Operator 2

Change Implemented

Cpk = 0.8   Cpk = 3.0

62 64 66 68 70 72 74 76 78 80 82

UCL, LCL = Upper/Lower Control Limits

initial Cpk of 1.8 while Operator 2's Cpk was 0.8 and both operators were off target.

The two operators and their supervisors shared and analyzed the results and refined the procedures. The process is now operating in statistical control, on target, and with a Cpk of 3.0.

This continuing learning cycle and sharing of information between operators is illustrative of our continuous improvement principles. Heavy use of design of experiments (DOE), such as Taguchi and Response Surface Methods, and SPC, has enabled our etching team to define their procedures and continually improve their quality, yield, and cost.

During this time our etching team improved step 11, so the combination of steps 4 and 11 was improved by a factor of five.

One of the measures reviewed by the TQM Forum on a quarterly basis is the level of process improvements made by department. These include both quantity of improvements made, impact of improvements, and frequency of out-of-control conditions.

1. *Process simplification*—By reviewing information on cycle time and number of out-of-control conditions for each process, we identify which processes need to be simplified. Individuals and teams are also constantly looking for ways to simplify processes.

2. *Benchmark information*—The most significant yield has been the review of processes used by the company being benchmarked (see Item 2.2b). If their processes accomplish the same results in a better, less costly manner with less variability, we evaluate the process for incorporation.

3. *Process research and testing*—We maintain a process laboratory where processes are researched, modified, and tested prior to incorporation into the production process line.

4. *Alternative technologies*—Our engineering department is constantly identifying alternative technologies. These technologies can either be improvements to existing processes or the incorporation of a new process to meet a change in design requirements. When alternative technologies are identified, we use our process laboratory to research, modify, and test the new process before incorporation into production.

5. *Customer feedback*—Formal and informal feedback from any customer is captured and considered to be an opportunity to improve. The control plans are reviewed to see if the problem had been anticipated. If not, our control plan process is modified to provide more accurate results in the future. Either way, the customer input is captured and acted upon.

6. *Stretch targets*—Through our SBP process, stretch targets are established for cycle time, product quality, and cost.

5.3 Process Management: Business and Support Service Processes

5.3a *Designing Quality of Business and Support Processes*

1. *Key process definitions*—Key processes are identified for external customers through their input and feedback. These are established by the Business Segment Managers. For internal customers, key processes are identified by the internal customer. The other source for identifying key processes is through the TQM Forum. When a process is identified as needing improvement, it is recommended to the department for inclusion in the key measures. Many of the measures developed have been a direct result of identifying our cost of quality measures (see Item 6.2a). When a process is defined, key measures are established based on the needs of the process customer (either internal or external).

2. *Translation of processes to measures*— We use many measures of performance throughout the company. Principal indicators are defined for each process by means of team participation and individual involvement. Some

examples of indicators include cycle times, accuracy, timeliness, or response times. The data on these measures are maintained in either the *Red Book* or the LAN.

5.3b *Maintaining Quality of Business and Support Processes*

Figure 5.5 shows some of the current performance measures maintained for support areas. (Also see Item 6.3a.) When out-of control conditions are identified (usually by the person performing the function), the root cause of the problem is determined by using the Collins' Seven Quality Tools. Many times a CET will be formed by an individual, the department, or the TQM Forum. Once the root cause of the problem is identified, a corrective action plan is developed which includes implementation and subsequent review to assure that the problem has been corrected.

5.3c *Process Improvements*

The collection of a wide variety of performance measures in each of the key areas of the business provides for a base of information upon which to evaluate our overall performance and identify areas for improvement.

1. *Process simplification*—The development of process flowcharts by internal supplier/customer teams identifies opportunities for improvement when the "what should the process be doing?" is compared to the "what is the

Figure 5.5

Measures of Performance for Business and Support Processes

Area	Measure	Department Review	TQM Forum Review
Billing Accuracy	Customer Complaints	Monthly	Quarterly
Design Cycle Time	Days to Complete	Project	Monthly
A/P Checks	Discounts Missed	Weekly	Quarterly
	Late Pays	Weekly	Quarterly
P/R Checks	Errors on Check	Weekly	Quarterly
Machine Maintenance	Fixed Fast Time	Weekly	Monthly
	Response Time	Weekly	Monthly
Purchase Orders	Errors on P/O	Monthly	Quarterly
	Issued within Lead Time	Monthly	Quarterly
Process Lab	Cycle Time	Weekly	Monthly
Financial Statements	Days after Period	Monthly	Annually
Cash Flow Accuracy	Variance to Actual	Monthly	Quarterly
Customer Support	Time to Return Calls	Weekly	Monthly
	Time to Resolve Complaints	Monthly	Quarterly
	Timely Response to RFQs	Monthly	Quarterly
	Rings to Answer Phone	Monthly	Annually

process doing?" Expectations for improvements are established by each department and included in their plans (DAPs).

2. *Benchmark information*—The company utilizes benchmarks from several companies to define and better manage the business processes. We have examined companies in the local area, as well as suppliers to past Baldrige winners, to better understand their processes (see Area 2.2b). Many changes have been incorporated.

3. *Process research and testing*—To achieve several magnitudes of improvement in quality, cost, and cycle time, Collins Technology recognizes the need to make major changes in our processes. New ideas are encouraged and

initially tried out on a small scale. For example, eight forms, one to two pages in length, which were used in accounting were consolidated, updated, and improved resulting in four one-page forms.

4. *Use of alternate technologies* — Collins utilizes a wide LAN network of computers with a terminal readily available to all employees. Business process employees have at least two terminals for every three employees. Software is continually acquired or developed and applied to the network to allow each and all service functions to perform better, with shorter cycle times, and at lower costs. We are continuously reviewing new technology that might improve our performance.

5. *Information from customers*—Most customers of business processes are internal customers. Our policy to spend 20 minutes per week with internal customer/suppliers has improved our performance and resulted in the elimination of many internal reports. We regularly ask our external customers for feedback relating to our services.

6. *Stretch targets*—The DAPs contain both short- and long-term goals that address the company SBP goals. In addition, each departmental plan contains specific goals for the particular department. These goals are usually "stretch" goals and are reviewed in the monthly departmental and TQM Forum meetings.

5.4 Supplier Quality

5.4a *Defining Quality Requirements to Suppliers*

Our supplier base includes suppliers of production material, general use material, and services (de-ionized water, telephone, insurance, banking, office supplies, legal services, etc.). We have established special relationships with our top 30 suppliers (representing 74% of our production material). These suppliers represent the most critical inputs affecting either purchased dollar amount or production consistency. As a small company we rely heavily on these supplier partners who provide us with quality products on time. Our supplier quality

programs are documented in our supplier *Total Excellence* procedures.

1. *Principal supplier quality requirements*—We provide all of our suppliers with a purchase order containing our goals for quality, services, and price. These orders also contain information relating to Collins SPC procedures, our purchasing policy, explanation of our supplier rating system, and other pertinent information. We communicate requirements to all of our suppliers via Specification Control Drawings (SCD) containing critical control items, incoming inspection, and first article inspections. The suppliers who participate on product design teams become a part of the product specification development process.

2. *Key indicators*—We modeled our supplier assurance and rating system after one of our major customer's system. This system provides procedures for supplier surveys and audits as well as our Supplier Results Indicator (SRI) rating system. Each supplier's delivery and incoming quality data are extracted from our computer database and a monthly supplier quality rating (SQR) is calculated. The key indicator of delivery performance is the percentage of products received on time to our delivery window. The key indicator of quality performance is the parts per million (ppm) of acceptable units received.

5.4b *Assuring Supplier Quality*

Our supplier quality assurance system, *Total Excellence*, is documented and in full use. In addition to the SRI rating system, we conduct supplier surveys and audits on all new and/or critical suppliers. Purchasing, design engineering, and quality assurance perform these audits and surveys for new products. Our emphasis is on the supplier's quality system rather than product characteristics. We conduct critical characteristic incoming inspection (MIL-STD-105 AQL) when required by the government. For some items we conduct additional testing, such as burn-in or test run samples, through our processes for verification. Each purchased part has an Incoming Inspection Requirement (IIR) defined at the design stage. The IIR is available through the LAN and defines what inspections are required when the material is received.

5.4c *Procurement Improvements*

A Supplier Council was formed in 1991 when we asked representatives from six commodity groups to meet with our Purchasing Director and other Collins Technology employees. The reason for forming the council was to get feedback on system issues. The council has met three times in the two years and plans to invite four new members to attend the September 1994 meeting. Feedback from the council has been helpful in increasing communications with our supplier base and identifying proce-dures that have caused confusion in the supplier community.

5.4d *Strategy to Improve Supplier Quality*

Our product quality is built on our supplier's quality. Our goal is to have ship-to-stock, ship-to-line suppliers who require no inspection. We currently certify 30% of the top suppliers as ship-to-stock. Our goal for 1994 is 60%. Our success with utilizing the skills and knowledge of our top 30 supplier partners has led to plans of expansion of our supplier program. We are planning a two-tiered supplier program with (1) formal recognition for quality and technical support and (2) certification for ship-to-stock/line.

5.5 Quality Assessment

5.5a *Principal Approach of Assessments*

1. *Systems and processes*—We use a multilevel approach to assessing the ability of our systems, processes, practices, products, and services. The four levels are

   - TQM Forum oversight,

   - customer audits,

   - internal quality assurance/ operational audits, and

   - local operational checks by our operating units, teams, and individuals.

   A cross-functional team (CFT) has been established to review our total business systems. That team makes quarterly reports to the TQM Forum and remains active for continuing improvements. Each department has at least two CETs that continually review its processes. The company performs internal audits to assure that each service function operates in compliance with written procedures. Each department has completely documented operating procedures. A cross-functional team reviews all departmental procedures and reviews the output of individual departmental teams.

2. *Products and services*—In addition to internal audits and departmental and cross-functional team activities, we have extensive customer audits. We have had 40 customer audits within the past four years with 100% first pass success. The audit criteria include ISO-9002, U.S. Navy Class 1, MIL-1-45208A, and various customer-specific criteria.

5.5b *Using Assessment Findings*

The TQM Forum reviews all audit results and recommendations. The responsible operational unit develops an action plan, and the TQM Forum approves the plan and monitors implementation and effectiveness through results based evidence. We review and establish new goals and standards quarterly through the TQM Forum to include our entire quality assurance system, training requirements, and supplier performance. In addition, each year during our annual business planning process, we review all programs and actions. We assess our closure to our goals and benchmarks, and track repeat audit findings.

6.0 Quality and Operational Results *(180 pts.)*

The *Quality and Operational Results* Category examines the company's achievement levels and improvement trends in quality, company operational performance, and supplier quality. Also examined are current quality and operational performance levels relative to those of competitors.

6.1 Product and Service Quality Results *(70 pts.)*

Summarize trends and current quality levels for key product and service features; compare current levels with those of competitors and/or appropriate benchmarks.

        ☐ Approach
        ☐ Deployment
        ☑ Results

Areas to Address

(a) trends and current levels for the key measures and/or indicators of product and service quality.

(b) comparisons of current quality level with that of principal competitors in the company's key markets, industry averages, industry leaders, and appropriate benchmarks.

Notes:

1. Key product and service measures are measures relative to the set of all important features of the company's products and services. These measures, taken together, best represent the *most important factors that predict customer satisfaction and quality in customer use*. Examples include measures of accuracy, reliability, timeliness, performance, behavior, delivery, after-sales services, documentation, appearance, and effective complaint management.

2. Results reported in Item 6.1 should reflect all key product and service features described in the Business Overview and addressed in Items 7.1 and 5.1.

3. Data reported in Item 6.1 are intended to be objective measures of product and service quality, not the customers' satisfaction or reaction to the products and/or services. Such data might be of several types, including: (a) internal (company) measurements; (b) field performance (when applicable); (c) proactive checks by the company of specific product and service features (7.2d); and (d) data routinely collected by other organizations or on behalf of the company. Data reported in Item 6.1 should provide information on the company's

performance relative to the specific product and service features that best predict customer satisfaction. These data, collected regularly, are then part of a process for monitoring and improving quality.

4. Bases for comparison (6.1b) might include: independent surveys, studies, or laboratory testing; benchmarks; and company evaluations and testing.

6.2 Company Operational Results *(50 pts.)*

Summarize trends and levels in overall company operational performance; provide a comparison with competitors and/or appropriate benchmarks.

☐ Approach
☐ Deployment
☑ Results

Areas to Address

(a) trends and current levels for key measures and/or indicators of company operational performance.

(b) comparison of performance with that of competitors, industry averages, industry leaders, and key benchmarks.

Notes:

1. Key measures of company operational performance include those that address productivity, efficiency, and effectiveness. Examples should include generic indicators such as use of manpower, materials, energy, capital, and assets. Trends and levels could address productivity indices, waste reduction, energy efficiency, cycle time reduction, environmental improvement, and other measures of improved *overall company performance*. Also include company-specific indicators the company uses to track its progress in improving operational performance. Such company-specific indicators should be defined in tables or charts where trends are presented.

2. Trends in financial indicators, properly labeled, might be included in Item 6.2. If such financial indicators are used, there should be a clear connection to the quality and operational performance improvement activities of the company.

3. Include improvements in product and service design and production/delivery processes in this Item.

6.3 Business and Support Service Results *(25 pts.)*

Summarize trends and current levels in quality and operational performance improvement for business processes and support services; compare results with competitors and/or appropriate benchmarks.

☐ Approach
☐ Deployment
☑ Results

Areas to Address

(a) trends and current levels for key measures and/or indicators of quality and operational performance of business and support services.

(b) comparison of performance with appropriately selected companies and benchmarks.

Notes:

1. Business and support services are those covered in Item 5.3. Key measures of performance should reflect the principal quality, productivity, cycle time, cost, and other effectiveness requirements for business and support services. Responses should reflect relevance to the company's principal quality and operational performance objectives addressed in company plans, contributing to the results reported in Items 6.1 and 6.2. Responses should demonstrate broad coverage of company business and support services, and work units. Results should reflect the most important objectives for each service or work unit.

2. Comparisons and benchmarks for business and support services (6.3b) should emphasize best practice performance, regardless of industry.

6.4 Supplier Quality Results *(35 pts.)*

Summarize trends in quality and current quality levels of suppliers; compare the company's supplier quality with that of competitors and/or with appropriate benchmarks.

    ☐ Approach
    ☐ Deployment
    ☑ Results

Areas to Address

(a) trends and current levels for the key measures and/or indicators of supplier quality performance.

(b) comparison of the company's supplier quality levels with those of appropriately selected companies and/or benchmarks.

Notes:

1. The results reported in Item 6.4 derive from quality improvement activities described in Item 5.4. Results should be broken down by major groupings of suppliers and reported using the principal quality measures described in Item 5.4.

2. Comparisons (6.4b) might be with industry averages, industry leaders, principal competitors in the company's key markets, and appropriate benchmarks.

6.0 QUALITY AND OPERATIONAL RESULTS

Achieving totally satisfied customers has been the basis for our success. In order to meet and exceed ever-increasing requirements and expectations of world-class customers, we have built a dedication to quality excellence in every facet of our operation.

Customer satisfaction requires a dependable product, delivered on time, and at a competitive price. Stable processes, dedicated team improvement actions, and broad use of quality tools and methods throughout our operations have enabled us to build our customer satisfaction results on a firm product, process, and performance foundation. This section presents our quality-based achievements in manufacturing and supporting services and describes how these results directly tie to our key customer satisfaction drivers.

6.1 Product and Service Quality Results

6.1a *Quality Trends and Levels*

Figure 6.1 illustrates the 75% in yield improvement achieved since 1991 (see discussion in Area 5.2b). By using DOE, the MLB team also developed procedures to reprocess out-of-specification material and other improvements, which has resulted in a 60% overall material cost reduction since 1991 (Figure 6.2). We also achieved major reductions in

Figure 6.1

Material Yield

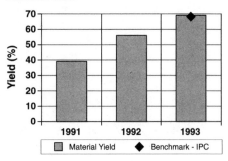

our waste stream environmental disposal (see Figure 1.3).

These results are illustrative of similar results achieved in all areas of the company. They are a direct result of our TQM training, teamwork, and application of our quality tools on a continuous basis.

These improvements have almost tripled our production capacity (Figure 6.3).

Figure 6.2

Material Cost per Panel

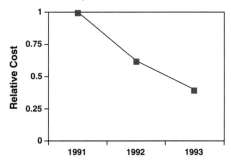

Figure 6.3

Production Capacity Gains

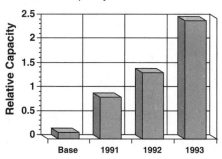

The resulting 35% improvement in cycle time also reduced the overtime hours for expediting from 6% of labor hours to below 3% for the second half of 1993.

We organized each manufacturing area department as a cellular facility with full supervisor ownership and authority for all actions. They are responsible for operations, measurements, and improvements within their facility. To "scale" this concept, each of our facilities consists of about 15 to 20 employees plus attached engineering support. We monitor each factory's output through the use of key internal indices, plus an overall productivity chart which we developed as an integrated index. This chart typically trends per-unit output cost, thus aggregating raw material, full labor, facility expenses, yields, quality, and cycle time. An example productivity chart for our plating line (steps 7–11 of Figure 5.3) is shown in Figure 6.4. Figure 6.4a has cost per

Figure 6.4a

Plating Process Effectiveness—Cost per Unit

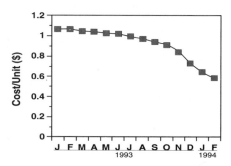

Figure 6.4b

Plating Process Effectiveness—Yield

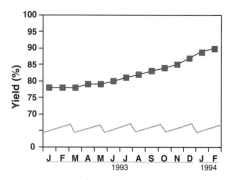

unit indices, and Figure 6.4b contains yield data. At our monthly operations review, each facility's productivity chart is reviewed and included in our *Red Book* for TQM Forum review.

Combining the productivity charts (Figures 6.4a and b), we can show the overall cost improvement for all processes. The results of this Composite Cost Improvement Index are shown in Figure 6.5.

Figure 6.5

Composite Cost Reduction

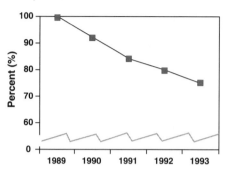

The composite quality improvements in materials and process stability have had major contributions to "downstream" improvements in our board production process.

Our assembly process lines use DOE, design for manufacturability, employee problem-solving teams, training, cellular production, on-line engineering resources, SPC, self-inspection techniques, and business segment teams to accomplish continuous improvements in their areas. Figure 6.6 shows the dramatic 6×

Figure 6.6

Final Operations Performance

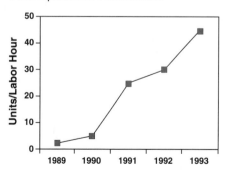

Figure 6.7

Test Yield—First Pass

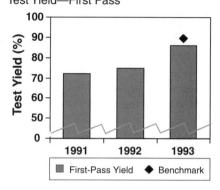

improvement in net assembly units per labor hour for one of our assembly lines. Their goal is 8.5× improvement by the end of 1994.

Figure 6.7 shows the improvement in our composite first-pass test yield at step 15 of Figure 5.3. Although we have internally benefited through these cost savings and efficiencies, the ultimate beneficiary has been our customers. The resultant improvements in on-time delivery, quality, stability, board performance, reliability, cycle time, and cost described here have allowed us to satisfy our long-term customers and acquire new ones.

The following output measures reflect our achievements as seen through our customers' eyes.

On-time delivery—Figure 6.8 shows our records for our overall on-time delivery performance to our key customers. The improvements since 1990 have been a direct result of our internal improvement efforts, especially in

Figure 6.8

On-Time Delivery

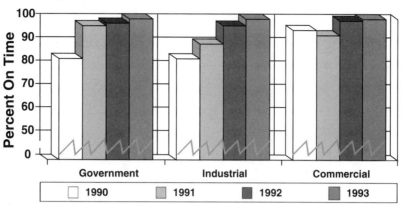

the manufacturing production area and from closer partnerships with suppliers. We have benchmarked on-time delivery to be 92% from a survey of our customers. Our overall customer on-time rate exceeds 95%. Our goal is 100% on-time on every order.

A critical requirement for our customers is the length of time it takes to design or redesign a board. We started tracking these measures in 1990. The adoption of the CAD equipment and use of the *Multi-layer Board Design*

Guide has led to a dramatic improvement in our design lead times (see Figures 6.9 and 6.10).

Quality—Our key output quality indicator is our customers' quality acceptance results. Our outgoing quality level is at a world-class 100%, as measured by our major customers (see Figure 7.4 in the following section). Internally, we measure our outgoing process control stability through statistical analysis of our product test results.

Figure 6.9

Total Design Cycle Time—New Products

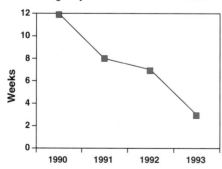

Figure 6.10

Total Design Cycle Time—
Redesigned Products

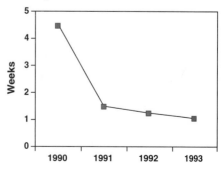

Figure 6.11

Process Performance—Cpk

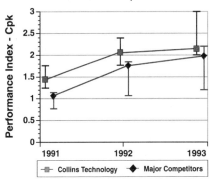

Figure 6.12

First Article Acceptance Rate

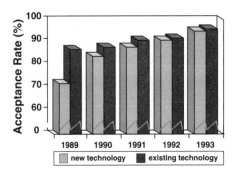

Figure 6.13

Product Return Rate

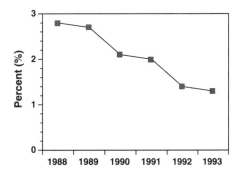

We have steadily improved the board quality and test distribution from a Cpk of 1.45 in 1991, to a Cpk of 2.05 in 1992, and to a Cpk of 2.15 at the end of 1993. Figure 6.11 shows a comparison of our Cpk results to our major competition (based on internal testing).

Figure 6.12 shows our performance on first article acceptance rates for new products. The products designed for our Government and Industrial segments typically do not use any types of new technology. The Commercial segment, however, uses new technology

in a significant number of their designs. To reflect this difference, we track the acceptance rates for products using new technology and the ones using existing technology.

With the adoption of the *Multi-layer Board Design Guide* and increasing control over our manufacturing processes, we have had a corresponding improvement in product return rates (see Figure 6.13).

Our improvements in process control and in our ability to match designs with processes have led to our being able to reduce the lead time needed to produce product. While the impact in the Government Business Sector is not that great, there have been tangible benefits to both the Industrial and Commercial sectors (see Figure 6.14).

Performance—We continually strive to advance the state-of-the-art in frequency performance through material and manufacturing process

Figure 6.14

Manufacturing Lead Times
(Quoted to Customers)

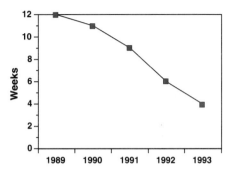

improvements. Our frequency performance is a critical value-added feature our customers depend on to achieve systems performance. Figure 6.15 illustrates our progress in our final test performance margins by improving materials and manufacturing processes. The advances in manufacturing, coupled with our research and development, provides us with even better processes for new products.

Price—Our manufacturing cost reductions have allowed us to hold our price increase over the past five years to a total of 5% for our customers in spite of inflationary pressures and increased product quality requirements.

6.1b *Quality Level Comparisons*

We constantly test and evaluate competitive products and benchmark comparative processes in other industries. Figure 6.16 outlines some of our key comparisons/benchmarks, types, sources, and cross-references to preceding figures and sections.

Objectivity and validity of our benchmarking are maintained by actual test data, customer data, and outside sources in that order. Efforts made to continually upgrade the confidence level of our information by improving the quality of our data have been fruitful.

Figure 6.15

Frequency Performance Margin

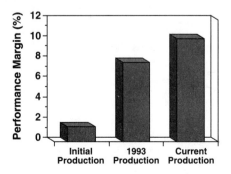

Figure 6.16

Operational Comparisons
and Benchmarking

Comparison/ Benchmark	Type	Source	Reference
Process Cpk	3	Worldwide Corp.	Fig. 6.11
Material Yield	1	IPC	Fig. 6.1
Test Yield	2	Wrightman Technology	Fig. 6.7
Inspect Yield	2	Kenan Corp.	Fig. 6.7
Board Performance	3	Competitive Testing	Fig. 6.15
On-Time Delivery	1	Customer Survey	Fig. 6.8, Narrative

(1) = Independent Sources   (3) = Internal Testing
(2) = Benchmarking

6.2 Company Operational Results

6.2a *Trends in Operational Results*

Area 5.1a2 discusses our COQ activity and Figure 6.17 shows the over 50% reduction in cost of nonconformance we achieved in the past two years. Many of the results of manufacturing process improvement were discussed in Item 6.1. However, numerous team and individual efforts have reduced errors and made improvements in every area of our business. Our company productivity per employee is monitored quarterly by management to verify that the sum of all the individual quality efforts is truly having a substantial bottom-line impact. The graph shown in Figure 6.18 illustrates our 45% improvement in total company productivity per employee achieved since 1989. As a comparison,

Figure 6.18
Productivity per Employee

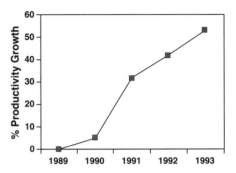

the industry's annual growth rate over this period was 7.8% while the rate for Collins Technology was 10.6%.

Waste Stream Reduction Program—In 1990, we disposed of our solid waste in approved landfills. We computed that by 1994, the amount of solid waste generated would have grown 450%. As a result of our recovery and recycling program, our actual disposal needs have been reduced by 70%, as a result of our aggressive reuse and recycling programs (see Figure 1.3).

6.2b *Operational Comparisons*

In addition to competitive product evaluations and manufacturing process benchmarking, we also benchmark our operational functions. Where applicable, we have shown the benchmark on the figures. Figure 6.19 outlines key comparisons, source, and cross references.

Figure 6.17
Cost of Quality Trend

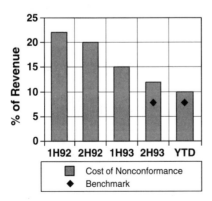

Figure 6.19

Typical Comparisons

Comparison/ Benchmark	Source	Reference
Productivity	U.S. Labor Dept.	Fig. 6.18
Environmental Standards	U.S. EPA	Standards
COQ	Wrightman Tech.	Fig. 6.17

We strive to continuously upgrade our benchmarks/comparisons against Malcolm Baldrige National Quality Award winners, government, trade, and industry publications, customers and suppliers, and other members of the consortium.

6.3 Business and Support Service Results

6.3a *Trends of Results*

A few examples are listed of our accomplishments in the business process and support service areas.

Accounts receivable—Overdue receivables represent assets which could be profitably employed to fund growth. The Accounting Department developed an action plan to reduce outstanding receivables. There was an assignment of responsibility for reduction of receivables and maintaining personal contact between companies for continuity and consistency. Another contributor to reduction has been our quality supplier awards from our customers, some of which offer preferred payment terms to their top suppliers. We have cut

our percentage of receivables over thirty days by 70% since 1991, resulting in improved cash flow, as shown in Figure 6.20. The benchmark is from an IPC survey and represents the top quartile companies.

Expense Reports—Cycle time on paying employee expense reports was reduced from six weeks in 1990 to 24 hours.

Design cycle time—We introduce an average of six new board designs per month, most of which are custom developed to customers' special requirements. Rapid response is vital. In 1989 we developed computer-aided modeling and CAD systems to speed up our design cycle time. Through use of cross-functional design teams, the *Multi-layer Board Design Guide*, standardized in-house formats, design from manufacturing rules using macro instructions, and computer systems integration, we have significantly improved our response time for design and layout

Figure 6.20

Receivables over 30 Days

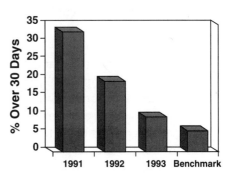

Figure 6.21

Design/Layout Hours—New Designs

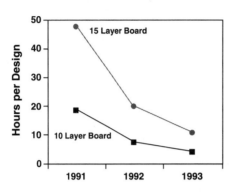

Figure 6.22

Artwork Generation Cost Reduction

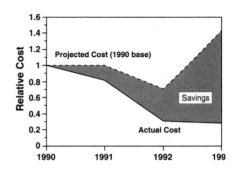

of a new board. We reduced the 15-layer board design time by 80% and 10-layer board design time by 75% since 1991 (Figure 6.21) as well as improving first-pass quality. Our total cycle time has also been reduced by 80% in the last three years (see Figures 6.9 and 6.10).

By working with our photo artwork supplier, we also linked our CAD systems for electronic data interchange (EDI) of digital information. Previously the patterns had been plotted and shipped by air. This accelerated the entire artwork cycle, resulted in better first-pass yield, and reduced our cost 75%, as shown in Figure 6.22.

Failure analysis cycle time—When a failure occurs, it is important to determine the cause quickly so that corrective and preventive steps can be taken. In the cases where a customer has experienced the failure, the need for timeliness of response is doubly critical. We use in-house electron

microscopy, metallurgical analysis, and other techniques to isolate root causes of failure. In 1991, we determined that our response time of 12 working days was unacceptable. The TQM Forum established a CFT team, whose mission was to reduce the cycle time to two days. The team consisted of representatives from customer service, quality assurance, production control, and manufacturing. After applying our problem-solving techniques, the team established new procedures for reducing the number of steps, defined a tracking system, and created a visual control system so that failure analysis items would not get "lost." Management reviews the cycle time monthly. During 1993, the team succeeded in reducing the analysis cycle time from 12 days to 4 days with a 1994 goal of 2 days.

Figure 6.23(a–e) presents results related to the tracking of performance measures for support areas.

Figure 6.23a

Billing Accuracy

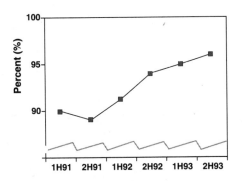

Figure 6.23b

A/P Checks—Paid Late

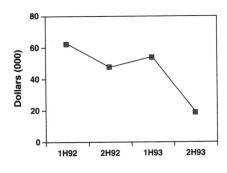

Figure 6.23c

PM—Successful Repairs

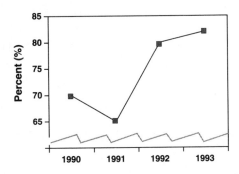

Figure 6.23d

Month-End Statement—Closure

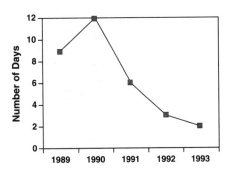

Figure 6.23e

P/Os with Errors (Average per Day)

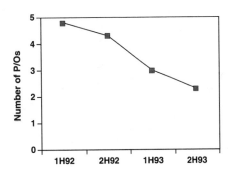

Figure 6.24 (a–c) provides examples from the customer support area.

6.3b Business Process Quality Comparisons

We benchmark our business processes in the same manner that we benchmark our production products and operational processes. Figure 6.25 outlines key comparisons, sources, and cross-references.

Figure 6.24a

Average Time to Return Calls

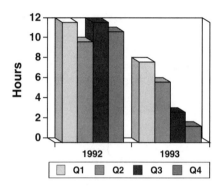

Figure 6.24b

Days to Resolve Complaint

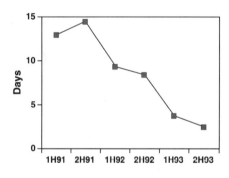

Figure 6.24c

Technical Aid on First Contact

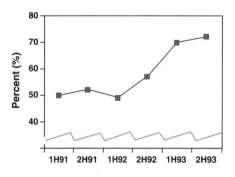

Figure 6.25

Typical Comparisons—
Business Processes

Comparison/ Benchmark	Source	Reference
Receivables	IPC Operating Ratios Study	Fig. 6.20
Financials	IPC Growth Resources Surveys	Various
Design/Layout	CAD Industry Publications	Area 6.3a

6.4 Supplier Quality Results

6.4a *Supplier Quality Levels and Trends*

We began our formal supplier program in early 1990. Our top 10 critical suppliers (comprising 40% of our production material purchases) were the first selected to participate in our supplier program. Currently 9 of the top 30 suppliers ship directly to stock.

Supplier consolidation program—In early 1991 we established an initial goal to reduce the number of our critical production material suppliers by 25% within four years. The supplier selection criteria were quality and delivery performance, level of business activity, supplier's commitment to total quality, level of technology support, and cost. Figure 6.26 shows the results of the program to date. We reduced the number of suppliers from 300 to 175 (a 42% reduction).

Supplier Results Indicator—Area 5.4a described our Supplier Results

Figure 6.26
Supplier Consolidation

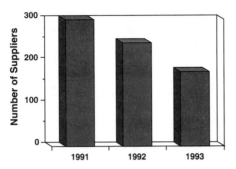

Indicator, designed to measure the combined quality and delivery performance of a supplier. Deployment of the SRI system began in the fourth quarter of 1990. The results are communicated to our critical suppliers quarterly. Results are also reviewed by the TQM Forum. Figure 6.27 shows the improvement trend of the composite SRI for our top 30 suppliers.

At the suggestion of the Supplier Council, we asked suppliers to rate us as a customer. We prepared and

Figure 6.27
Composite SRI

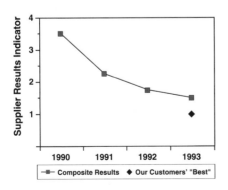

mailed an evaluation questionnaire to our supplier base in December 1992 and 1993. We wanted to determine the suppliers' view of us to identify areas for improvement. We asked our suppliers to rate our personnel normally in contact with suppliers (purchasing, engineering, quality, and accounts payable) on professionalism, responsiveness, and conformance to requirements. Figure 6.28 shows the results. We also asked the suppliers to comment on our communications with them, ease of access, and consistency of requirements. Opportunities to improve our communications were identified and acted upon.

6.4b *Comparisons*

Our most important supplier benchmark is the SRI quality and delivery performance index described in Area 5.4a. To establish world-class benchmarks for these metrics, we conducted a survey of ten of our customers who we consider to be world-class, and with whom we have a highly interactive relationship.

Figure 6.28
Suppliers' Ratings of Collins Technology

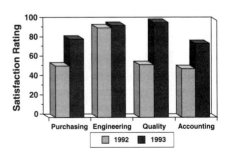

Included were Trubell Industries, Ridgeford Tech, Kenan Corp., and Wrightman Technology. The survey supplied us with their specific quality and delivery criteria of their "best suppliers" which we used to set our Supplier Results Indicator. We are currently developing a benchmark for supplier ship-to-stock. We also attempt to compare our supplier base with the supplier bases of our competitors.

7.0 Customer Focus and Satisfaction *(300 pts.)*

The *Customer Focus and Satisfaction* Category examines the company's relationships with customers and its knowledge of customer requirements and of the key quality factors that drive marketplace competitiveness. Also examined are the company's methods to determine customer satisfaction, current trends and levels of customer satisfaction and retention, and these results relative to competitors.

7.1 Customer Expectations: Current and Future *(35 pts.)*

Describe how the company determines near-term and longer-term requirements and expectations of customers.

☑ Approach

☑ Deployment

☐ Results

Areas to Address

(a) how the company determines *current and near-term requirements* and expectations of customers. Include: (1) how customer groups and/or market segments are determined or selected, including how customers of competitors and other potential customers are considered; (2) how information is collected, including what information is sought, frequency and methods of collection, and how objectivity and validity are assured; (3) how specific product and service features and the relative importance of these features to customer groups or segments are determined; and (4) how other key information and data such as complaints, gains and losses of customers, and product/service performance are used to support the determination.

(b) how the company addresses *future requirements* and expectations of customers. Include: (1) the time horizon for the determination; (2) how important technological, competitive, societal, environmental, economic, and demographic factors that may bear upon customer requirements, expectations, preferences, or alternatives are considered; (3) how customers of competitors and other potential customers are considered; (4) how key product and service features and the relative importance of these features are projected; and (5) how changing or emerging market segments and their implications on current or new product/service lines are considered.

(c) how the company evaluates and improves its processes for determining customer requirements and expectations.

Notes:

1. The distinction between near-term and future depends upon many marketplace factors. The applicant's response should reflect these factors for its market.

2. The company's products and services might be sold to end users via other businesses such as retail stores or dealers. Thus, "customer groups" should take into account the requirements and expectations of both the end users and these other businesses.

3. Product and service features refer to all important characteristics and to the performance of products and services that customers experience or perceive throughout their overall purchase and ownership. These include any factors that bear upon customer preference, repurchase loyalty, or view of quality—for example, those features that enhance or differentiate products and services from competing offerings.

4. Some companies might use similar methods to determine customer requirements/expectations and customer satisfaction (Item 7.4). In such cases, cross-references should be included.

5. Customer groups and market segments (7.1a,b) should take into account opportunities to select or *create* groups and segments based upon customer- and market-related information.

6. Examples of evaluations appropriate for 7.1c are:

   - the adequacy of the customer-related information;

   - improvement of survey design;

   - the best approaches for getting reliable information—surveys, focus groups, customer-contact personnel, etc.; and

   - increasing and decreasing importance of product/service features among customer groups or segments.

   The evaluation might also be supported by company-level analysis addressed in 2.3a.

7.2 Customer Relationship Management *(65 pts.)*

Describe how the company provides effective management of its interactions and relationships with its customers and uses information gained from customers to improve customer relationship management processes.

☑ Approach
☑ Deployment
☐ Results

Areas to Address

(a) for the company's most important contacts between its employees and customers, summarize the key requirements for maintaining and building relationships. Describe how these requirements are translated into key quality measures.

(b) how service standards based upon the key quality measures (7.2a) are set and used. Include: (1) how service standards, including measures and performance levels, are deployed to customer-contact employees and to other company units that provide support for customer-contact employees; and (2) how the performance of the overall service standards system is tracked.

(c) how the company provides information and easy access to enable customers to seek assistance, to comment, and to complain. Include the main types of contact and how easy access is maintained for each type.

(d) how the company follows up with customers on products, services, and recent transactions to seek feedback and to help build relationships.

(e) how the company ensures that formal and informal complaints and feedback received by all company units are resolved effectively and promptly. Briefly describe the complaint management process.

(f) how the following are addressed for customer-contact employees: (1) selection factors; (2) career path; (3) deployment of special training to include: knowledge of products and services; listening to customers; soliciting comments from customers; how to anticipate and handle problems or failures ("recovery"); skills in customer retention; and how to manage expectations; (4) empowerment and decision-making; (5) satisfaction; and (6) recognition and reward.

(g) how the company evaluates and improves its customer relationship management processes. Include: (1) how the company seeks opportunities to enhance relationships with all customers or with key customers; and (2) how evaluations lead to improvements such as in service standards, access, customer-contact employee training, and technology support; and (3) how customer information is used in the improvement process.

Notes:

1. Requirements (7.2a) might include responsiveness, product knowledge, follow-up, ease of access, etc. They do not include product and service requirements addressed in Item 7.1.

2. "Service standards" refers to performance levels or expectations the company sets using the quality measures.

3. The term "customer-contact employees" refers to employees whose main responsibilities bring them into regular contact with customers—in person, via telephone, or other means.

4. In addressing "empowerment and decision-making" in 7.2f, indicate how the company ensures that there is a common vision or basis to guide the actions of customer-contact employees. That is, the response should make clear how the company ensures that empowered customer-contact employees have a consistent understanding of what actions or types of actions they may or should take.

5. In addressing satisfaction (7.2f), consider indicators such as turnover and absenteeism, as well as results of employee feedback through surveys, exit interviews, etc.

6. Information on trends and levels in measures and indicators of complaint response time, effective resolution, and percent of complaints resolved on first contact should be reported in Item 6.1.

7. How feedback and complaint data are aggregated for overall evaluation and how these data are translated into actionable information should be addressed in 2.3a.

7.3 Commitment to Customers *(15 pts.)*

Describe the company's commitments to customers regarding its products/services and how these commitments are evaluated and improved.

> ☑ Approach
> ☑ Deployment
> ☐ Results

Areas to Address

(a) types of commitments the company makes to promote trust and confidence in its products/services and to satisfy customers when product/service failures occur. Describe these commitments and how they: (1) address the principal concerns of customers; (2) are free from conditions that might weaken customers' trust and confidence; and (3) are communicated to customers clearly and simply.

(b) how the company evaluates and improves its commitments, and the customers' understanding of them to avoid gaps between customer expectations and company performance. Include: (1) how information/feedback from customers is used; (2) how product/service performance improvement data are used; and (3) how competitors' commitments are considered.

Note:

Examples of commitments are product and service guarantees, warranties, and other understandings, expressed or implied.

7.4 Customer Satisfaction Determination *(30 pts.)*

Describe how the company determines customer satisfaction, customer repurchase intentions, and customer satisfaction relative to competitors; describe how these determination processes are evaluated and improved.

☑ Approach
☑ Deployment
☐ Results

Areas to Address

(a) how the company determines customer satisfaction. Include: (1) a brief description of processes and measurement scales used; frequency of determination; and how objectivity and validity are assured. Indicate significant differences, if any, in processes and measurement scales for different customer groups or segments; and (2) how customer satisfaction measurements capture key information that reflects customers' likely future market behavior, such as repurchase intentions or positive referrals.

(b) how customer satisfaction relative to that for competitors is determined. Describe: (1) company-based comparative studies; and (2) comparative studies or evaluations made by independent organizations and/or customers. For (1) and (2), describe how objectivity and validity of studies are assured.

(c) how the company evaluates and improves its overall processes and measurement scales for determining customer satisfaction and customer satisfaction relative to that for competitors. Include how other indicators (such as gains and losses of customers) and customer dissatisfaction indicators (such as complaints) are used in this improvement process.

Notes:

1. Customer satisfaction measurement might include both a numerical rating scale and descriptors assigned to each unit in the scale. An effective (actionable) customer satisfaction measurement system is one that provides the company with reliable information about customer ratings of specific product and service features and the relationship between these ratings and the customer's likely future market behavior.

2. The company's products and services might be sold to end users via other businesses such as retail stores or dealers. Thus, "customer groups" or segments should take into account these other businesses as well as end users.

3. Customer dissatisfaction indicators include complaints, claims, refunds, recalls, returns, repeat services, litigation, replacements, downgrades, repairs, warranty work, warranty costs, misshipments, and incomplete orders.

4. Company-based or independent organization comparative studies (7.4b) might take into account one or more indicators of customer dissatisfaction as well as satisfaction. The extent and types of such studies may depend upon factors such as industry and company size.

7.5 Customer Satisfaction Results *(85 pts.)*

Summarize trends in the company's customer satisfaction and trends in key indicators of customer dissatisfaction.

☐ Approach
☐ Deployment
☑ Results

Areas to Address

(a) trends and current levels in key measures and/or indicators of customer satisfaction, including customer retention. Segment by customer group, as appropriate. Trends may be supported by objective information and/or data from customers demonstrating current or recent (past three years) satisfaction with the company's products/services.

(b) trends in measures and/or indicators of customer dissatisfaction. Address the most relevant and important indicators for the company's products/services.

Notes:

1. Results reported in this Item derive from methods described in Items 7.4 and 7.2.

2. Information supporting trends (7.5a) might include customers' assessments of products/services, customer awards, and customer retention.

3. Indicators of customer dissatisfaction are given in Item 7.4, Note 3.

7.6 Customer Satisfaction Comparison *(70 pts.)*

Compare the company's customer satisfaction results with those of competitors.

☐ Approach
☐ Deployment
☑ Results

Areas to Address

(a) trends and current levels in key measures and/or indicators of customer satisfaction relative to competitors. Segment by customer group, as appropriate. Trends may be supported by objective information and/or data from independent organizations, including customers.

(b) trends in gaining and losing customers, or customer accounts, to competitors.

(c) trends in gaining or losing market share to competitors.

Notes:

1. Results reported in Item 7.6 derive from methods described in Item 7.4.

2. Competitors include domestic and international ones in the company's markets, both domestic and international.

3. Objective information and/or data from independent organizations, including customers (7.6a), might include survey results, competitive awards, recognition, and ratings. Such surveys, competitive awards, recognition, and ratings by independent organizations and customers should reflect comparative satisfaction (and dissatisfaction), not comparative performance of products and services. Information on comparative performance of products and services should be included in 6.1b.

7.0 CUSTOMER FOCUS AND SATISFACTION

Customer satisfaction is the hub of our Strategic Business Plan (SBP). All departments continuously strive to increase the quality of products and services offered to the customer. We firmly believe that to compete in today's (and tomorrow's) worldwide business environment, the customers must feel that they are receiving the best product and customer service available. Total customer satisfaction through meeting or exceeding the customers' requirements and expectations is our fundamental business philosophy.

7.1 Customer Expectations: Current and Future

7.1a *Current and Near-Term Requirements*

1. *Market segments*—Market segments are developed based on unique characteristics of the customer groups. Our Commercial customers are pushing the technology envelope and are willing to try new ideas. Our Government and Industrial groups want proven technology that has been used in the field for several years.

2, 3. *Collecting information and determining importance*—The sources of information relating to near-term requirements and expectations of customers come from the following areas: Customer Support Database (CSD); personal contact with customers by the Business Segment Managers (visits with the top 25 customers in their segment at least twice a year); sales coordinators' personal contact with their customers (visits with all their customers at least monthly); internal sources of information including all senior executives and engineers; and information in trade publications about trends in the industry and new technology (all articles about our customers are clipped and maintained in files).

4. *Cross comparisons*—Correlation analysis is done between product quality and customer satisfaction measures to assure that a strong positive correlation exists.

7.1b *Future Customer Expectations*

1. *Time horizon for determination*—Specific plans for each business segment are developed for 1 to 2 years, 4 years, 10 years, and 20 years.

2. *Projections*—Most new technology is only applicable to our Commercial Products business segment since both the Government and Industrial Products groups are often reluctant to try new, not fully proven technology. Most of the opportunities for innovations fall under the Commercial Products group. This is also the area where our

participation on planning and design teams at the customer is most likely to occur. Because we are viewed as a key supplier by all of our customers, they typically are willing to listen to the new ideas we bring to them. The process laboratory (described in Area 5.2b) gives us the ability to develop the new technology and run prototypes for use in our customer's research and development activities while being assured of meeting future environmental requirements.

3. *Potential customers*—With our reputation for accomplishing very difficult technical tasks, potential customers often seek out Collins Technology to determine what our approach would be to accomplish an advanced technical requirement.

4. *Key features*—As a result of the close relationships we maintain and the trust we enjoy with each of our customers, we are often able to participate as a key supplier in their planning process. Thus, we are intimately involved in their knowledge of industry trends and future plans. This information is brought together by the Business Segment Managers when they meet to develop their plans.

5. *Changing market segments*—The most critical area having a bearing on future customer requirements is technological advances in the design and manufacture of

new technology. With this philosophy, many innovators of new technology come to us to test, refine, and introduce their ideas. When new technologies are developed, we approach our key customers and potential customers who would most benefit from incorporating these ideas into their current and future projects.

7.1c *Improved Processes for Determining Customer Requirements*

As referred to earlier, the Business Segment Managers meet annually to develop plans for the SBP. As one of the steps in this process, a review is performed of all customers, potential new products, and applications of new or planned technology. Where opportunities are identified to improve the relationship process with customers, the three managers discuss the merits of the idea and decide whether or not to adopt it for future activities. Some examples of these improvements include the development of the CSD (to collect and consolidate customer data), initiating the customer article clipping service, and the development of the process laboratory.

7.2 Customer Relationship Management

7.2a *Requirements for Maintaining and Building Relationships*

Our products are designed and built to fill the very specific needs of our customers. Our customers' end products could not be produced without high-quality boards. With this basic dependency, it is essential that our product perform flawlessly and to our customers' requirements of quality, delivery, and cost.

Recognizing that each business segment (Government, Industrial, and Commercial) has unique characteristics, we changed our organization in 1989 and established Business Segment Managers. The objective of this change was to create a better understanding of the customer's needs within Collins Technology. At this time, we also established sales coordinators within each business segment assigned to specific customers. In many cases, the customers we deal with have multiple people responsible for purchasing and specifying product from us. By having the Business Segment Manager and sales coordinator structure, we have been able to, in many instances, become a common link between different units of our customers.

Office and home phone numbers of Business Segment Managers and sales coordinators are provided to the customers. A phone directory of all Collins Technology managers is also provided.

The decision to include customers at each new product design review stage was specifically made to enhance our ability to improve our relationships, better understand customer needs, and develop and communicate mutually acceptable expectations. When developing a new product, the sales coordinator creates a cross-functional team to work with the customer (see Area 5.1a). Each sales coordinator also has every resource within Collins available to assist in resolving any customer problem or question.

From these contacts, key indicators have been developed which include response time for:

- quotation
- order entry
- complaint resolution
- technical assistance.

Additionally, the Commercial customers want the ability to change schedules on a 48-hour notice.

7.2b *Customer Service Standards*

The TQM Forum, consisting of senior executives and rotating members, establishes our key customer service standards and goals based on input from benchmark surveys, customer feedback, department managers, Business Segment Managers, and customer-contact employees. The TQM Forum reviews service standards and goals quarterly to increase awareness by all TQM members. Some examples of these standards are listed in Figure 7.1.

Figure 7.1

Service Standards

Service	Standard
Answer Phone	Within 2 rings
Quote Turn Around	Within 24 hours of request
Order Entry	Within 1 hour
Complaint Resolution	Within 24 hours
Technical Assistance	On first contact
Billing Accuracy	3.4 errors per million lines
Warranty Payment	Within 10 days

1. Customer service standards are deployed to the company through the SBP and departmental attainment plans by establishing department goals for all service standards.

2. Employees within each department compile their service standard performance as part of their daily operating process through departmental logs such as complaint or quote response data. All systems have been converted to computerized systems on the LAN.

Each department collects key customer service standard results by business segment and posts them monthly for employees to review and follow the trends toward goals. The TQM Forum and department managers monitor and evaluate our service standard results monthly. We compare the current results to our goals and appropriate action is taken to ensure full company support in meeting our service standards.

Customer-contact employees participate in departmental teams and are represented on the TQM Forum. They play an important role in the process of setting, evaluating, and improving the customer service standards.

7.2c *Information and Easy Access*

The organization of our marketing and sales department provides easy access to the sales coordinators and customer service representatives. This allows for a focused, consistent, personal point of contact between our company and the customer.

Customer service employees are available 16 hours per day on our 800 number to assist customers in placing orders or answering technical, delivery, and price questions. Frequent personal visits, use of fax machines, voice mail, and direct connections between ours and our customers' computers (EDI) allow for a variety of methods of communicating with our customers.

In addition to the above-mentioned methods of contact, all of the customer-contact employees have instant access to Collins' Customer Support Database (CSD). This is software we developed on our LAN. As soon as an employee makes contact with a customer, our requirement is to call up the customer's information on the computer prior to continuing the conversation. This logs the customer contact automatically while, at the same time, connecting the Collins employee to all relevant information relating to the customer. Types of information/systems that can be assessed include order entry; schedule changes; open order inquiry (including status of components and purchased items used in the customer's product); pricing and sales history; quality information relating to each lot shipped and in production; complaint processing; and customer comment/feedback system which allows for recording of all comments (both positive and negative) and provides for automatic reminders of follow-up notes that can be set up for any future date and time.

When a new product is shipped, the system automatically sets up a note to contact the customer on a specific date to follow up on the order. When the customer-contact employee signs on to the system, the first thing they see is a list of all their open notes. When the follow-up is completed and comments recorded in the system, the note is removed automatically from the open file recording the date, time, and person clearing the note.

The use of these systems has greatly improved our ability to quickly provide answers and information to our customers. If a specific customer-contact employee is unavailable when a customer calls in, any other Collins employee has complete access to the customer's full history and current status. It has also allowed us to aggregate information easily on a company-wide basis for analysis. This information is reviewed monthly by the Business Segment Manager and quarterly by the TQM Forum.

Collins Technology is the only supplier in our market with this type of ability to readily access information while at the same time recording data for aggregated company-wide analysis.

7.2d Follow-Up with Customers on Products and Services

The sales coordinator or the customer service representative follows up by telephone within one week of a shipment or quotation to ensure that we have met the customers' requirements. In addition, sales coordinators question the customers during their monthly visits to receive firsthand information on the customers' previous transactions and opportunities for improvement. The customer is asked a series of questions and an open-ended question regarding how Collins Technology could improve the product and delivery. We do not assume silence means satisfaction.

7.2e How Formal and Informal Complaints Are Aggregated and Used

All formal and informal customer complaints are processed through a documented complaint procedure. Figure 7.2 is a flowchart of our complaint process.

The customer service manager trains all customer-contact employees in use of the customer complaint process. To document any formal or informal complaint by a customer, all customer-contact employees use a customer-complaint form or record it in the CSD (Area 7.2c). The customer-contact employee routes the complaint form to the sales coordinator or the customer service manager, who in turn distributes the complaint to the appropriate department manager for corrective action. This process ensures the complaint flow to the manager responsible for the corrective action. Direct contact with the customer accomplishes the final resolution of the complaint. We inform the customer of the cause of the problem and what steps have been taken to prevent the problem from occurring again. At that point if the complaint was on a manual form the customer service manager logs the complaint resolution, the nature of the complaint for Pareto analysis, and the data for total response time into the LAN system. The complaint data are aggregated by the Business Segment Manager and presented to the TQM Forum and department managers for evaluation.

The form is readily available and is used by everyone who has any customer contact. This form assures that all information needed to understand the customer's complaint is collected up front and makes it easier to be entered in the database. The customer complaint process was added to the CSD in November 1991.

Figure 7.2
Complaint Process Flow

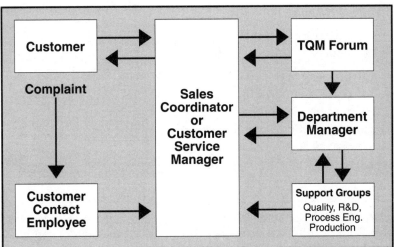

7.2f *Customer-Contact Personnel*

Customer-contact personnel are our frontline employees and must be the best available within our company. We consider senior executives, Business Segment Managers, sales coordinators, department managers, customer service representatives, design engineers, quality engineers, and the receptionist all to be customer-contact personnel.

1. *Selection factor*—Management normally selects customer-contact personnel from within our company. This assures firsthand assessment of the employee's personal attributes and knowledge of products and services. Prioritized personal attributes that we consider imperative for customer-contact personnel are the following:

   - customer-oriented attitude/personality

   - decisiveness

   - communication skills

   - education/technical understanding

   - planning and organizational skills.

The most important factor in the selection of customer-contact representatives is their ability to act for and represent our company on a stand-alone basis. If so, then with our company-wide customer orientation and TQM priority for customer support, they will excel in their responsibilities.

2. *Career path*—We promote customer-contact employees from within based on the attributes described above. It is important for employees to recognize that a career path is available and to strive for continuous learning and improvement. To enhance the philosophy of promotion from within, we encourage higher education through tuition reimbursement. To date, customer-contact employees have received three MBAs and four BAs. Another benefit of promotion from within is that internally selected personnel are cross-trained in more than one job function within the company, thus creating a more capable team of employees. The following is an example of a career path within the marketing department.

   Product Design Engineer

   ↓

   Customer Application Engineer

   ↓

   Sales Coordinator

3. *Special training*—Department employees train customer-contact personnel in the operations of each functional area so that they understand the design

and manufacture of the product. Personnel actually produce boards to understand the intricate details of the process and to appreciate the technology used to produce the product.

Customer-contact personnel receive ongoing training through internal and external courses, such as marketing and sales processes and policies, product specification and general capabilities, sales techniques, team building, listening skills, customer service skills, skills in customer retention, proposal writing, and negotiation skills. Customer-contact personnel have received over 2000 hours of outside training and 3000 hours of internal training. Outside training consists of courses such as "How to Increase Your Customer Satisfaction," "Implementing a Strategic Business Plan," Tom Peters' seminar on "Thriving on Chaos," Quality Consortium seminars on "SPC," and "How to Improve Customer Satisfaction." Internal training includes such courses as product documentation, terms and conditions, and market definition.

4. *Empowerment and decision-making* —To make doing business with our company easier for the customer, we have empowered our customer-contact employees with the authority to make customer-related decisions. As an example, customer service representatives

have the authority to make customer-related decisions up to $10,000. They can authorize the return of products, replace or repair products, or credit the customer entirely on their own using this empowerment.

Senior executives fully support customer-related decisions by the customer-contact employee. The basis for this empowerment is the selection of the best personnel, as described in Area 7.2f(1) and extensive training described in Area 7.2f(3).

5. *Satisfaction determination*— The TQM Forum conducts an employee quality survey once a year to determine the attitude and morale of the employees within each department. We group our customer-contact employees together to analyze the results of their surveys and how they affect our customer relations. Senior executive and Business Segment Manager visits to the customer are also used to gather information and to cross-check the results of our internal employee quality survey. The results of the employee quality survey show our customer-contact employees scored 95 out of a maximum 100 points, which indicates a good attitude and high morale. The employee quality survey results also pinpoint service areas that need the attention of the TQM Forum to ensure that we take appropriate

action for continuous attitude and morale improvement.

6. *Recognition and reward*—We recognize employees for special achievements through a formal recognition program every month at our company meetings. In addition, we empower various levels of management to give bonuses to employees who have "gone the extra mile" for the customer. "Management by walking around" is practiced by our company. Employees receive praise on a routine basis. There is also the internal prestige of being one of the recognized employees, which is a reward in itself. (See Item 4.4 for other forms of recognition.)

7. *Turnover*—Turnover is an expense that we continuously track. We make every effort to keep an employee, not only to reduce the cost of training, but to maintain a team of highly motivated employees, which results in a good continuous customer relationship. Our customer-contact employee attrition rate has averaged only 8% since 1988, as compared to a 25% industry-wide average.

7.2g How Company Evaluates and Improves Practices

1. The TQM Forum uses customer satisfaction results to monitor and evaluate customer relationship management. Additionally, Collins Technology is benchmarking our sales and service

organization against the best as identified by *Sales and Marketing* magazine. The organizations include Digital Equipment, Northwestern Mutual Life, Abbott Laboratory, Tektronix, and 3M.

2. Customer service standards results are tracked. These include:

- quotation response time
- technical assistance response time
- customer service response time
- telephone responsiveness
- complaint resolution time
- sales order entry errors
- cycle time
- customer feedback
- input from customer contact employees
- input from Business Segment Managers.

The TQM Forum uses these indices and information to establish goals through the SBP for each company department. During the SBP review cycle, these indices are reviewed for adequacy and success in dealing with customer expectations. Departmental attainment plans are structured for elements such as additional training and application of new technology. As a result, additional 800 lines and the Customer Support Database were added.

7.3 Commitment to Customers

7.3a *Promoting Trust and Confidence*

Our quality policy, our company's business philosophy, states (see also Area 1.1b):

The company will meet or exceed the customers' expectations in every product or service we provide, without exception.

All company employees make this commitment to our customers. We wrap our SBP, departmental attainment plans, and quality plans around meeting this commitment.

1. *Principal concerns of customers—* The concern of all customers is very simple: Can we supply a quality product, on time, at a competitive price? All three of these concerns blend into our TQM System so that every aspect of our plans, decisions, and resource commitments are directed toward these interests. This commitment is responsible for our success to date and the focus of our future business.

2. *Conditions—*We have no contractual conditions that weaken our commitment to our customers. We accept our customers' requirements with no hidden conditions associated with our products and services.

3. *Clear communications—*Our terms are concise, simple, and easy to read. A brief, one-page document defines our contractual conditions and product warranty, which is a four-year warranty.

7.3b *Improvements of Commitment*

The commitment to customers is reviewed with the SBP on an annual basis in the areas of competitive analysis, strengths assessment, customer analysis, and business segment as shown in Figure 3.1. With our corporate goal of exceeding our customers' expectations, the commitment area is very carefully reviewed to be sure that we lead our industry and have no gaps.

On an annual basis, the business segment managers meet with the express purpose of identifying needs, for each of the segments, to enter into the SBP. Part of this analysis includes finding opportunities, based on customers' needs, which could result in improving Collins' tangible commitments to those customers. The TQM Forum, then correlates these ideas with the current results and incorporates them into the SBP.

1. In addition to our written warranty, our implied warranty is to repair or replace any product the customer believes is our responsibility. The cost of this "stretch" effort to quickly and fairly resolve a complaint further demonstrates our service commitment to the customer. We are the only multi-layer board

manufacturer in the industry to provide product reliability data to our customers. We request customer help in establishing our test criteria to ensure that the reliability test data will be useful in their applications. The reliability program is a direct result of the TQM System and has solidified a closer relationship with our customers.

Rapid response to our customer problems is another example of our commitment to the customer. It we cannot solve a problem over the telephone, then we go to the customer the next morning. As an example, a customer had production line dropouts attributed to our product. We responded so quickly that the customer was initially dubious of our intentions. The problem resulted from a change in the customer's processes, and we received a personal statement of gratitude from the division manager for our rapid response.

In general, we want our dealings with our customers to be "win–win." Consequently, we share the advantages achieved from our quality improvements.

2. As a result of our TQM System, our product reliability has increased so that we now offer a four-year product warranty.

3. This warranty period is four times the industry norm for an electronic component and is proof of our confidence in the quality of our product.

Our increased commitments to the customer are direct results of continuous improvements made over the past four years through our TQM System.

7.4 Customer Satisfaction Determination

7.4a *How Customer Satisfaction for Customer Groups Is Determined*

1. *Methods and market segments—* Our sales are segmented into three groups: Commercial, Government, and Industrial. The information gathered from our customer satisfaction, future requirement, and benchmark surveys ensures that the needs of individual business segment products and services are considered.

Each year we send a customer satisfaction survey to the top 40 customers in each of our three business segments. These customers represent over 85% of our sales. Starting in the third quarter of 1991 this survey included all customers.

How objectivity is assured— Internally, we monitor key

customer satisfaction indices which affect the customers' perception of our company. Each department, including manufacturing, quality, engineering, and marketing, updates these indices monthly. Internal key customer satisfaction indices monitored are:

- on-time delivery
- returned material
- warranty cost
- complaints
- market share by business segments
- gains and losses of customers.

Our customers set their service preferences through our annual customer benchmark survey. In addition to providing benchmarks for service standards, the customer is requested to prioritize the products and services that they consider to be important in our business relationship. As an example, our customers prioritized the following product and service standards to aid us in understanding their requirements and to ensure that we capture key internal and external customer satisfaction data in our customer surveys:

- quality of products
- on-time delivery
- competitive price

- technical/design assistance
- direct sales
- order entry support.

2. *Key information*—The customer satisfaction survey asks the customer to rate Collins Technology using the measurement scale of outstanding (5) to poor (1) in the following categories as compared to other multi-layer board manufacturers.

- customer service responsiveness
- ease and assistance in placing orders
- technical assistance
- new product design
- product quality
- product reliability
- delivery performance
- conformance to specifications
- packaging and labeling
- complaint resolution.

Multi-level departments within the customer organization (i.e., purchasing, quality, engineering, and manufacturing) all receive the customer satisfaction surveys to increase the objectivity of the data. When we send surveys to multi-level departments within customers, we encode the surveys by department. This allows us to cross-tabulate the information by our customer's function and evaluate our performance within each area. Additionally, we visit every major customer at least every month, where we spend several hours reviewing different customer satisfaction survey questions and soliciting ideas for improvements.

7.4b How Customer Satisfaction Relative to Competitors Is Determined

1. *Company-based studies*—Sixty percent of our business is single-sourced, and in these cases, we ask our customers to compare us to their leading suppliers. In those cases where other multi-layer board manufacturers supply the customer, we ask the customer to compare us to our competitors and to their best suppliers regardless of industry. We use the ultimate test of customer satisfaction relative to competitors by measuring our 65% market share against their market share.

2. *Customer-based studies*—We feel that the best information from independent sources is derived from the following areas: retention of customers, successful quotations, results of customer audits (see Area 5.5a), and market share analysis. All of these data indicate customer satisfaction relative to our competitors.

External customer satisfaction data and internally generated customer satisfaction indices are cross-compared to verify the accuracy and validity of the data. A customer satisfaction team uses external and internal indices that do not correlate or negative trends as areas for improvement (i.e., improvement in the customer satisfaction survey questions, the addition of another question in the survey, or possibly the addition of another measurement category in the survey). This analysis process resulted in an improvement to the 1992 customer satisfaction survey process by sending the surveys to multi-department personnel within the customer's organization, and by asking customers for comparisons to their leading suppliers regardless of industry. This improvement increased the overall accuracy of the customer's response and broadened our scope beyond our industry.

7.4c Evaluation and Improvement

The customer satisfaction survey questions and measurement technique were formulated based on feedback for product and services that the customers considered important. We asked our customers what product and service categories they considered important in judging customer satisfaction and how to score the questions to make them easy and accurate.

During personal visits, we ask customers to review our surveys in detail, and comment on the effectiveness of

Figure 7.3

Customer Satisfaction Survey Results

the questions and measurement scale and to make suggestions for improvement. Most customers do not like to spend hours filling out complicated surveys; therefore, it is critical to develop a quick, comprehensive survey that addresses our performance in key product and service areas compared to competitors and other similar suppliers. We have found the best approach is to ask the customer to help with the development of the survey. The result was an 85% first-time response to our new survey.

Each review and update of the SBP monitors our customer satisfaction. We have not lost a customer in the last 12 years because of competitive reasons, and thus we must rely on satisfaction levels. Our data system contains all customer contacts with any employee. The data are continually monitored and analyzed for any adverse trends in customer comments.

Annually, the Business Segment Managers meet to review and discuss ways of improving their measurements of customer satisfaction. The annual customer survey was initiated as a result of one of these meetings. The subsequent improvements to the survey were also results of these meetings.

7.5 Customer Satisfaction Results

7.5a *Trends of Customer Satisfaction*

Figure 7.3 illustrates our improvement trend in overall customer satisfaction for our largest customers, representing 85% of our total volume. The overall customer satisfaction scores are composed of many indices and are based on the customer's perception of our performance. Our overall score of 4.3 (excellent) points out the overall effectiveness of our TQM System. Our goal is to improve this overall score to 4.8 in 1994.

Figure 7.4 shows that board quality for our major customers is now at

Figure 7.4

Quality Rating—Major Customers

100%. Prior to the implementation of our TQM System, quality was our greatest cause of customer complaints. Implementing internal quality programs such as stabilizing and controlling processes, and implementation of modern production techniques, increased our quality to 100% as measured by our major customers.

7.5b *Adverse Indicators*

Collins Technology has not experienced any product or service claims, refunds, recalls, litigation, or downgrades in the 20 years we have been in business, nor have we received any state or federal sanctions in the history of our company.

Warranty cost for the last three years has averaged less than ¼% of our total sales.

As shown in Figure 7.5, our field quality has increased over the previous three years. Field returns, replacements, and repairs combined into a field quality index to give an overall quality score. All three of the indices are individually tracked. This overall rating is 2 ½ times as good as our major competitor's. The total field quality score also includes our informal warranty policy of replacing boards the customers believe are our responsibility.

Figure 7.6 shows that our complaints have decreased significantly over the past three years as a result of our increased awareness and preventive actions taken through our TQM System. The 1993 complaint level represented ⅓% of our total products shipped.

7.6 Customer Satisfaction Comparison

7.6a *Comparison of Customer Satisfaction*

We compare our market share to our competitors as our measurement of customer satisfaction. As shown in

Figure 7.5

Field Quality

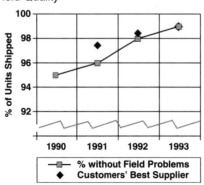

Figure 7.6

Customer Complaints

Figure 7.7
Market Share Comparison

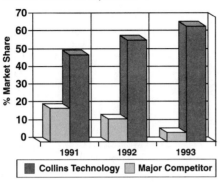

Figure 7.8
Worldwide Market

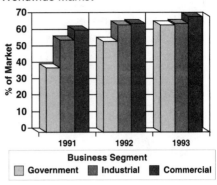

Figure 7.7, we presently have 65% of the market versus our major competitor's 5%.

Table 7.1 illustrates our ratings as compared to our major competitor for individual indices used to determine customer satisfaction. In all measurement areas we exceed our major competitor's performance.

Figure 7.8 shows our market share in our three business segments.

7.6b *Trends in Gaining or Losing Customers*

As described in the overview, in 1986, we made a strategic decision to reduce our customer base. The customers we continued to do business with were the ones who we could best serve based on our ability to produce product. As shown in Figure 7.9, the number of customers varies from year to year. It is significant to note that these variations are due to mixes of products within customers, not as a result of placing business with competitors. Our most outstanding trend is the fact that we have not lost a single customer to a competitor in the last 12 years.

Our customer retention record is the direct result of providing customers with products and services that meet or exceed their expectations.

Table 7.1
Competitive Comparisons

Index	Collins	Major Competitor
On-time Delivery (*% of shipments*)	99.5%	70%
Customer Complaints (*% of products*)	0.3%	12%
Warranty Costs (*% of sales*)	0.25%	5%
Product Quality (*ppm units shipped*)	3.4 ppm	150,000 ppm

7.6c *Trends in Gaining and Losing Market Share*

We have increased our market share for the third consecutive year in each of our three business segments. Figure 7.7 shows the total increase for our three business segments combined. Figure 7.8 illustrates our market increase for each of our business segments.

This continuous market share increase is a direct result of deploying the TQM philosophy throughout all company departments.

In the Far East we have increased our specialized board market share by offering products and services with higher quality than our Far Eastern competitors. As shown in Figure 7.10 we have actually increased our market share to 50% in spite of pressure from Far Eastern competitors. Even with a premium-priced product, we maintain a dominant share of the Far East market.

Figure 7.10

Market Share in the Far East

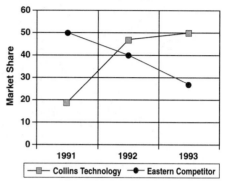

In both the domestic and international markets, our high market share and our continued growth can be directly attributed to our commitment to quality excellence through our TQM philosophy.

Figure 7.9

Number of Customers

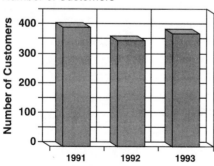

ACRONYM LIST

Acronym	Name
A/P	Accounts Payable
ASQC	American Society for Quality Control
BoD	Board of Directors
CAD	Computer-Aided Design
C.C.	Customer Contact
CEO	Chief Executive Officer
CET	Collins Employee Team
CFT	Cross-Functional Team
COQ	Cost of Quality
Cp, Cpk	Capability Indices
CSD	Computer Support Database
DAP	Department Attainment Plans
DFT	Department Functional Team
DoD	Department of Defense
DTC	Design to Cost
EAP	Employee Assistance Program
ECI	Error Cause Identification
ECN	Engineering Change Notice
EDI	Electronic Data Interface
EEOC	Equal Employment Opportunity Commission
EPA	Environmental Protection Agency
HR	Human Resource
IIE	Institute of Industrial Engineers
IIR	Incoming Inspection Requirement
IPC	Institute of Printed Circuits
IRS	Internal Revenue Service
ISO	International Standardization Organization

Acronym	Name
LAN	Local Area Network
MBO	Management by Objectives
MBWA	Management by Walking Around / Management by Wandering Around
MIL-STD	Military Standard
MRP	Material Resources Planning
NASA	National Aeronautics and Space Administration
NIST	National Institute of Standards and Technology
NSA	National Security Agency
OSHA	Occupational Safety and Health Administration
PCB	Printed Circuit Board
PDR	Preliminary Design Review
PM	Preventive Maintenance
P/O	Purchase Order
QA	Quality Assurance
R&D	Research and Development
Red Book	Data reviewed by the TQM Forum
RFQ	Request for Quote
SBP	Strategic Business Plan
SCD	Specification Control Drawing
SIC	Standard Industry Classification
SPC	Statistical Process Control
SQR	Supplier Quality Rating
SQS	Station Qualification System
SRI	Supplier Results Indicator
TQM	Total Quality Management

8 | Evaluation Process: Comments

INTRODUCTION

In the first stage of the evaluation process, several Examiners complete an independent review of a written response. The output from this review is comments and a score for each Item. The comments identify both the strengths and the areas for improvement. This evaluation process is necessary for self-improvement. It provides the yardstick to identify where you are starting and measures the impact of your improvement activities on the total system. The evaluation process also identifies the strengths of your system, which should be nurtured, and the areas that need improvement, which should be used in your business and quality planning cycles to achieve continual improvement.

This chapter begins with the purpose of comments from independent Examiners, and then provides guidelines for writing comments. The second stage of the evaluation process requires the Examiner team to reach consensus on comments and scores. The consensus process is introduced in the second half of this chapter and discussed more fully in Chapter 9.

 ## PURPOSE OF COMMENTS

A major benefit to an organization that decides to develop a written response to the Criteria and have it independently reviewed is the comments recieved from the Examiners, which can be used to drive the improvement process. The identification of strengths as well as areas for improvement is important. The strengths indicate what is going well and should be continued. The areas for improvement should be studied and prioritized to select key processes for improvement. Both types of comments should be assessed carefully and considered as input to the business planning process.

The comments serve several purposes. First, they are the basis for the feedback report. Organizations take action based on the feedback report. If the comments are not accurate, specific, understandable, and action-oriented the review process loses credibility. An Examiner has an obligation to study the written response and write comments that reflect a detailed review. Another use of comments is within the review process itself. Comments act as the primary means of communication during the consensus process and between the examining team and the Judges. Further, they are the basis of the score.

The score needs to be consistent with the comments. It is not appropriate to write a comment that states "It does not appear that the planning process is evaluated and improved" and then score the response 80% for that Item. That comment would typically be associated with a score of 30% or less. The score may be in the 40% to 60% range, which requires that an improvement process be in place, if all other aspects of the response were strong. (See Chapter 9 for more details on scoring.)

The comments are also used in the preparation for site visits. Major initiatives that serve as the rally points for the organization need to be verified on the site visit. Issues that were not clearly communicated in the response need to be clarified. (See Chapter 11 for more details on site visits.)

COMMENT GUIDELINES

In the independent review, each Examiner writes comments for each Item. These comments should describe either a strength or an area for improvement. A strength is some feature, activity, or result of the organization's system that the Examiner feels has a positive effect on their improvement efforts. When the comment is written down, the Examiner indicates a positive comment with the plus (+) sign. Areas for improvement, indicated by a negative (−) sign, are deficiencies identified by the Examiner. Because the Examiner uses only the written response to obtain information on the organization's total system, the deficiency may be in the development of the response as opposed to in the system itself. For example, a comment to Area to Address 3.2a may be

> (−) It is not clear how the overall requirements are deployed to suppliers. Reference is made to Item 5.4, but no direct answer was found there regarding this issue.

The reason for the lack of information in the written response may be (1) there is no deployment process, (2) there is a deployment process, but because of miscommunication among the response development team it was left out, or (3) a combination of both. It is not the job of the Examiner to improve the organization's system but simply to provide feedback on what is contained in the written response.

Typically there would be three to five comments per Item. Note that this is per Item not per Area to Address. Because there are 24 Items in the 1995 Criteria, a total of 72 to 120 comments would be written for the complete written response. For certain lengthy Items, the number of comments could exceed five. Some Examiners, especially new ones, want to provide a detailed review of the response and write 10 to 15 comments for each item or three to five comments for each Area to Address. This could result in 200 or more comments—an overwhelming number of comments that could leave the organization confused. They would not know which comments dealt with systemic issues and which with smaller, detail issues. It is important that the comments reflect the most important observations rather than all observations.

As the team goes through the consensus process, the comments of all the Examiners are consolidated. It is inane to believe that every Examiner will have the same comments. Yet, the final number of comments should remain about the same as that just recommended—three to five comments per Item. Since different people have different backgrounds and focuses, the team members need to reach agreement on the issues and their priorities and decide which three to five comments adequately reflect their key consensus issues.

The comments are to be nonprescriptive. That is, the comment should not prescribe or even recommend the use of a specific tool, technique, or strategy—even if the organization could benefit from its use.

It would be inappropriate to have a comment that stated "quality function deployment (QFD) should be used in the design process." The Criteria do not require the use of specific tools. If an organization used QFD, it would be recognized as a systematic way to translate the voice of the customer. But it would also be acceptable if another method were used. Some articles and books that have been written on the MBNQA claim that Examiners look for cost of quality, process mapping (flowcharts), QFD, quality circles, and control charts. But no specific tools or methods are required by the Criteria. There is an underlying paradigm contained in the core values and concepts; but no specific tools are specified. The paradigm is one of customer-focused quality, leadership involvement, continual improvement and learning, prevention, full participation and development of employees, design and process quality, and the use of data and information for improvement and awareness of the continuity of improvement. If the organization's philosophy and implementation is consistent with this paradigm, then the internal improvement processes will identify the tools, techniques, and strategies that work best for a particular organization.

Being nonprescripive is, perhaps, the most difficult part of writing comments. Examiners, especially internal Examiners, want to help the organization on its improvement journey. If they see an area for improvement which they feel could be improved by using a tool with which they are familiar, most Examiners want to tell the organization about it. But this is not fitting because it could stifle the innovation and diversity that the Award process wants to foster. Again, the Examiners' job is simply to provide an independent and unbiased review and feedback of the information contained in the written response.

The comments written by the Examiners should state observations and evaluations based on the written response. No other source of information about the organization should be used. Examiners should give credit, by acknowledging a strength, for what has been accomplished—even if it falls short of perfection. However, Examiners should avoid mixed comments—that is, comments that contain both a strength and area for improvement (see Figure 8.1). For clarity, two comments should be written. The comment crediting the strength should be balanced with an area for improvement comment to identify the shortcoming. For example, consider Item 1.1, Senior Executive Leadership. A comment indicating a strength might be "(+) The senior executives attended a five-day training class on quality." Under the areas for improvement there could be another comment stating "(–) The senior executives' involvement in quality seems limited to attendance at a five-day training class." These two comments together give credit for what has been done but also point out the shortcoming.

For emphasis, Examiners may use the notation (++) or (– –) for select comments. This provides a quick way for Examiners to highlight the written response's key points, that is, elements that permeate and are critical to the organization's philosophy and strategies. Elements which, if missing, will lead to a downward spiral in quality and performance excellence. These would include the core values and concepts (especially if missing) but may also include elements specific to the organization. There should be few, if any, comments with the double notation.

Identification of these key points is useful in the consensus process. Because the organization receives comments under two headings—strengths and areas for improvement—the organization will not be aware of the double plus or minus designation. The words in the comment need to convey this emphasis.

Examiners should write short, clear comments of one or two lines. The comments are not intended to be a tutorial, but they should convey a complete thought. If the comment needs a verbal explanation, then it will not be useful to the organization since there will be no additional explanation given in the feedback report to clarify the comment. Examples of comments that are not very useful are shown in Figure 8.1. A rewrite of those comments is given in Figure 8.2. The reason *why* the Examiner thought an activity was good or was limited is now contained in the comment. Note that the comments are still short and to the point. The mixed comment is now written as two comments. A strength and an area for improvement are both identified.

The comments should be constructive, with a positive tone. Preferred wording would be to state "It appears …," or "It is not evident …" rather than say "The organization does not do…." The organization may have a process in place but did not do a good job of describing it. Perhaps they misinterpreted the Criteria and did not realize they should have described a particular process at that time.

Since the consensus meeting and/or feedback report development usually occurs several days or weeks after the initial reviews, an Examiner may not remember which Area to Address was the focus of a particular comment. For the internal evaluation process, it can be a significant timesaver during the consensus process if the Examiners note the Area to Address (a, b, c, etc.) associated with each comment. In writing the feedback report or reaching consensus on comments, the team should use an agreed-on format. Perhaps all of the comments for

Figure 8.1

Examples of Not-Very-Useful Comments

Comment	Reason
+ Collins' reward and recognition program is good.	Comment does not state why the Examiner concluded the program was good.
– Limited benchmarking.	Comment does not explain what is limited about the benchmarking. Does the Examiner think it is limited because it only applies to key products or because it is at the early stages?
– The senior executives attended a five-day training session on total quality leadership but it was not clear if they participated in other quality activities.	Comment is mixed. Both a strength and area for improvement are combined into one. For clarity, it is best to keep them separate.

Figure 8.2

Comments Rewritten to Be Useful

+ Collins' reward and recognition program emphasizes teams and the use of quality tools and techniques.

– The company only benchmarks other electronic companies.

+ The senior executives attended a five-day training session on total quality leadership.

– It is not clear what quality activities have senior executive participation.

"a" are discussed, then those for "b." Alternatively, all strengths might be discussed first across all Areas. Whatever format is used, having a reference to the Area is a timesaver.

The comments should be specific to the written response. If the organization refers to their employee involvement teams as the "Eagles" then the comment could state "The activities of the Eagles cover ...." This is in the language of the organization. If the response refers to concurrent engineering then the comments could reflect the term "concurrent engineering."

The score needs to be consistent with the comments. Examiners *cannot* count the number of comments and derive the score. It is the content of the comment that is essential. The scoring guidelines require certain characteristics to be achieved to reach a scoring range. (See Chapter 9 for more details on scoring.) The content of the comments should reflect the requested characteristics. For example, the presence or absence of the evaluation and improvement of a process could be reflected in a comment. The breadth and depth of usage of a process described would also be appropriate for a comment.

Some examples of useful comments are shown in Figure 8.3.

CONSENSUS

In the national evaluation process, as well as in many state evaluation processes, four to six Examiners complete an independent review of the application. On the basis of these scores, the applicant may be selected to go to the next stage—consensus. Many organizations that use the Criteria for internal assessment have all

Figure 8.3
Examples of Useful Comments

(a)++ Excellent integration of human resource improvement strategies with quality requirements and business plans.

(a)– Trends in employee satisfaction indicators were omitted for 1991 and 1993.

(b)+ Company collects data on all key areas with a majority of their systems operating in real time. Customer, supplier, and associated databases seem quite good.

(a)+ All customer satisfaction trends presented show significant improvement during the last four years.

(a)– Customer satisfaction data show company behind top competitor.

(a)++ Company uses extensive means to determine customer requirements: focus groups, customer panels, mail and telephone surveys, in-person interviews, etc.

(a)– No mention was made of quality improvement efforts in the support areas of the company.

"applicants" go to consensus. The purpose of consensus is to pool information and viewpoints. Diversity is welcome here. The opinions of Examiners from marketing, manufacturing, engineering, finance, and so on may be different. Each views the written response from their own background, experience, and training. This impacts how the Criteria and answers given are interpreted. In the consensus process, the Examiners should be exposed to the various views. The Examiners need to achieve basic agreement on the comments: strengths and areas for improvement. The consensus comments are used for the feedback report. (See Chapter 9 for additional information and guidelines on the consensus process.)

SUMMARY

The multiple role that comments play was presented. The chapter began with the purpose of comments, and then gave guidelines for writing comments. Examples of useful and not-very-useful comments were presented and discussed. The second stage of the evaluation process requires the Examiners to reach consensus on comments and scores. The consensus process was introduced in the second half of this chapter.

CHAPTER

9 | Scoring

INTRODUCTION

The system for scoring written responses is discussed in this chapter. An organization using the Criteria for self-evaluation and improvement will find that this scoring process adds a new depth to the Criteria and adds the details to their yardstick.

Three evaluation dimensions are used in scoring the Items: approach, deployment, and results. These three dimensions are presented here along with the scoring guidelines. The guidelines provide mile markers to indicate progress on the journey toward customer delight. The chapter ends with a summary of the scores of previous MBNQA applicants and general findings on the relationship between scores and characteristics of the organization.

 ## THE THREE DIMENSIONS OF SCORING

Whenever a written response is reviewed for scoring or feedback, each Item is critiqued based on the three evaluation dimensions just mentioned: of approach and deployment, or results. The Criteria require one or more of these dimensions to be addressed in each Item. In concert with the core values, the *approach* is to be a prevention-based system, showing continual improvement and excellent integration. *Deployment* refers to the extent that the approaches are applied by all work units to all processes and activities. The results derived from the approach need to be sustained and show excellent levels and trends in all areas deployed.

Items that request information pertaining to an approach also require information on deployment. Approach and deployment are always linked together in the Criteria. An organization is never asked to describe a process without showing its breadth of use. The response to these Items would necessarily include a description of processes, tools, and techniques as well as identification of users.

Other Items solicit only results. The key word used to indicate a request for results is *trends*. When the word *trends* appears in the Criteria, analysis of data indicating sustained use is expected. No minimum period of time is specified for trend data. However, this is usually interpreted to mean that an analysis of three to five years of data is to be part of the response. In cases where the deployment is recent or trend data are unavailable, the Examiner can only focus on the current level(s) of performance and not sustained improvement.

A grouping of Items by the three dimensions is shown in Figure 9.1.

Approach

Approach refers to how the requirements defined in the Criteria are met. Some of the key points looked for in the responses are prevention, appropriate use of tools and techniques, integration, and improvement cycles. The techniques used should demonstrate a prevention-based approach. For example, the use of supplier certification to assure that incoming material will be acceptable would be a prevention-based approach as opposed to the use of an inspection upon receipt, a detection approach. The methods used should fit into an overall strategy. Unfortunately, this is not always the case. An executive at a Baldrige recipient company said that, in the past, management involvement use to be a flurry of activity. He referred to this as "ricochet management." There was no overall integration or focus. Improvement came in isolated pieces. This was a major challenge that his organization had to overcome.

Many of the Items ask for a description of the evaluation and improvement process *as applied to the approach just described* in the response. This is probably the hardest area for an organization to address. It does not seem to be the American

Figure 9.1
Approach, Deployment, and Results by Item

Focus on approach and deployment

Category	Items
1.0	1.1, 1.2, 1.3
2.0	2.1, 2.2, 2.3
3.0	3.1, 3.2
4.0	4.1, 4.2, 4.3, 4.4
5.0	5.1, 5.2, 5.3, 5.4
7.0	7.1, 7.2, 7.3

Results are specifically requested in

Category	Items
6.0	6.1, 6.2, 6.3
7.0	7.4, 7.5

way to go back to a process or technique and review its success. Further, if it is working, the thought of making it even better contradicts the common philosophy of "If it ain't broke, don't fix it." Perhaps we need a new saying. At Federal Express, the saying is "If it ain't broke, improve it."

Deployment

Deployment refers to the extent to which the organization's approaches are applied to various work units. This examines the breadth of use of the approach. The question asked is "Are the methods used with all transactions with all customers and all suppliers, all operations and processes, all products and services, and all employees?"

Approach and deployment are interwoven. Without deployment, an approach would only be an idea. Deployment addresses how broadly or narrowly the approach is applied. For example, do all employees have access to quality-related training or just those at management levels?

Results

One of the things that separates the Baldrige process from other quality road maps is its inclusion of results as well as approach and deployment. In those Items requesting results, an analysis of data is expected. The word *trends* indicates that data showing sustained use and improvement is expected in a graph, table, or chart format. This information should demonstrate the results achieved from the approaches described in the response. From this information, the current quality levels and the rate of improvement should be obvious to the reader of the response—if it isn't, the response needs to be rewritten.

Multiple years of data are used to illustrate that the successful results have been sustained and are due to the documented approach. A single indication (point) of "improvement" may be due to the normal variation of the system or an outside influence and not because of the activities in place. It is up to the developers of the response to show that the results are due to their approach, and were not caused by some outside or random event.

But even ongoing improvements, demonstrated with data, are not sufficient. Comparisons are also needed. We must remember that one of the purposes of the Criteria is to identify excellence. An organization may have very good, even dramatic, improvement trends, and yet not be a model of excellence. It all depends on their starting point. To provide the organization, as well as the Examiner, with an external measure, the Criteria requests comparisons. The comparisons may be with an industry average, industry leaders, or with benchmark data from world-class organizations. An organization needs to demonstrate they are one of the leaders. They do not need to be number one, but must be clearly one of the best. Some of the comparisons should be outside the organization's business focus to reflect performances of best practices.

 ## SCORING GUIDELINES

A numeric rating is given to each of the 24 Items. This rating is used to calculate the score for the seven Categories. For many years the point allocations were modified a small amount, shifting points from approach and deployment Items to results Items. Some people want all of the Criteria and all of the points to reflect results. They have no interest in knowing the process used to achieve the success. They just want to know the bottom line—who is the "winner." Others want more emphasis on approach and deployment. This reflects an attitude that acknowledges the important role the process plays. Presently, 59% of the points are earned through approach and deployment Items and 41% of the points are for results Items.

The points for each category are shown in Figure 9.2. Twenty-five percent of the points are given to Category 7.0, Customer Focus and Satisfaction. This reflects one of the core values of the Criteria—customer-driven quality.

Figure 9.2

Relative Impact of Each Category

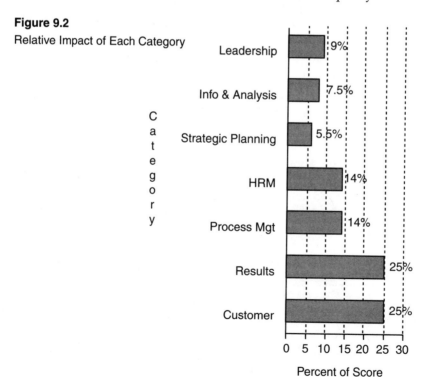

Percent of Score

Scoring Steps

The scoring process involves several steps (see Figure 9.3). These steps would be followed whether you are reviewing an application for the Award process or a written response for your own internal improvement process. First, read the entire response, cover to cover, including the overview section. For each Item, reread the Criteria, even though you may be familiar with the Criteria. Any evaluation should be focused on what the Criteria actually request, not what you remember them saying. Then read the response to the Item. Write comments that reflect the strengths and areas for improvement that were identified in comparing the organization's answer to the Criteria. Keep the overview and the rest of the written response in mind as you write comments. Although the organization should respond to the Item fully, it is our experience that this does not always happen. It is the obligation of the Examiner to keep the total system in mind. Although the response should include cross-referenced material, you should not penalize the organization if the required information is described in another Item. (See Chapter 2 for interrelationships between the Items.)

After the Item has been reviewed and comments written, assign a score (details on how to assign a score are given later). Review the score and comments for consistency; that is, a high score should have meaningful strengths and a lower score should have significant areas for improvement. Go to the next Item and continue the process.

Assigning Scores

The scoring guidelines are defined in the Criteria and are shown below in Figure 9.4. Sometimes the scoring guidelines are referred to as the "Green Sheet." In the initial years of the Award process, the guidelines were printed on green paper in the notebooks used in Examiner training. This allowed the Examiner to find it quickly since it was used throughout the training. Hence the name Green Sheet.

Figure 9.3
Scoring Steps

- Read entire response
- For each Item:
  - Reread the Criteria
  - Read the response
  - Write comments
  - Assign score
  - Review score and comments for consistency

There is no specific number of favorable or unfavorable comments associated with a given score. That is, the score does not reflect the proportion of favorable comments given. A response may receive three comments pertaining to strengths for a given Item and two comments identifying areas for improvement. If the two (–) comments relate to core concepts, the score may be quite low. For example, consider the comments: "(–) It appears the organization reacts to problems but does not prevent them" and "(–) The evaluation and improvement process were not described." Because prevention and improvement are two of the core concepts, a low score would be given. In a different case, if the two comments under areas for improvement were more minor in nature, the score could be quite high.

Even Award recipients receive comments on areas for improvement. No organization is perfect. There is always room for improvement. The areas for improvement comments for high-scoring responses typically demonstrate a need for some fine-tuning and do not reflect that core concepts are being overlooked.

To assign a score to the response of an Item, you should always begin at the 40% to 60% range on the guidelines. For an approach and deployment Item, this corresponds to the following:

- a sound, systematic approach, responsive to the primary purposes of the Item

- a fact-based improvement process in place in key areas; more emphasis is placed on improvement than on reaction to problems

- no major gaps in deployment, though some areas or work units may be in very early stages of deployment

If you feel the response satisfies these guidelines, the next higher scoring category should be checked to see if the additional requirements of that scoring range are also met. Assuming that none of the guidelines for the higher score, 70% to 90%, are satisfied, you would assign a score from the 40% to 60% range depending on your overall impressions (which should be supported by your written comments). Since scores are assigned in increments of 10, a 40%, 50%, or 60% would be assigned for this scenario. Correspondingly, if the guidelines for the 40% to 60% were not met, you would continue with the next lower scoring range—10% to 30%.

The 40% to 60% scoring range should always be the starting point. This technique is called *anchoring*. It is a technique used to reduce variation when dealing with subjective probabilities or, in this case, subjective scoring.

Figure 9.4

Scoring Guidelines

SCORE	APPROACH / DEPLOYMENT
0%	• no systematic approach evident; anecdotal information
10% to 30%	• beginning of a systematic approach to the primary purpose of the Item
	• early stages of a transition from reacting to problems to a general improvement orientation
	• major gaps exist in deployment that would inhibit progress in achieving the primary purposes of the Item
40% to 60%	• a sound, systematic approach responsive to the primary purposes of the Item
	• a fact-based improvement process in place in key areas; more emphasis is placed on improvement than on reaction to problems
	• no major gaps in deployment, though some areas or work units may be in very early stages of deployment
70% to 90%	• a sound, systematic approach responsive to the overall purposes of the Item
	• a fact-based improvement process is a key management tool; clear evidence of refinement and improved integration as a result of improvement cycles and analysis
	• approach is well deployed, with no major gaps; deployment may vary in some areas or work units
100%	• a sound, systematic approach fully responsive to all the requirements of the Item
	• a very strong, fact-based improvement process is a key management tool; strong refinement and integration—backed by excellent analysis
	• approach is fully deployed without any significant weaknesses or gaps in any areas or work units

SCORE	RESULTS
0%	• no results or poor results in areas reported
10% to 30%	• early stages of developing trends; some improvement *and/or* early good performance levels in a few areas
	• results not reported for many to most areas of importance to the applicant's key business requirements
40% to 60%	• improvement trends *and/or* good performance levels reported for many to most areas of importance to the applicant's key business requirements
	• no pattern of adverse trends *and/or* poor performance levels in areas of importance to the applicant's key business requirements
	• some trends *and/or* current performance levels—evaluated against relevant comparisons *and/or* benchmarks—show areas of strength *and/or* good to very good relative performance levels

Figure 9.4

Scoring Guidelines (*continued*)

70% to 90%	• current performance is good to excellent in most areas of importance to the applicant's key business requirements • most improvement trends *and/or* performance levels are sustained • many to most trends *and/or* current performance levels—evaluated against relevant comparisons *and/or* benchmarks—show areas of leadership and very good relative performance levels
100%	• current performance is excellent in most areas of importance to the applicant's key business requirements • excellent improvement trends *and/or* sustained excellent performance levels in most areas • strong evidence of industry and benchmark leadership demonstrated in many areas

Factors Affecting the Score

In the Baldrige review process, the team members assigned to an application initially do independent reviews. It is recommended that those using the Criteria for self-improvement also follow this policy. This will assure the widest possible review since an individual reviewer will not bias other reviewers.

When an Examiner is first given a written response, other Examiners reviewing the same response are not identified. This is done so that an Examiner cannot call and discuss any questions or get other opinions. The initial review is to be an independent review by the Examiner using just the material in the written response. Library searches or reports from stockbrokers should not enter into the evaluation. When organizations use the Criteria for their own self-evaluation and improvement, it is recommended they also use multiple Examiners for feedback. There will be variation in the comments and scores given by Examiners. Even if they all are seasoned Examiners, there will still be differences in scores because they have different experiences and backgrounds. These differences could affect the scoring in the positive or negative direction. That is, an individual Examiner may generally score the Items consistently higher or lower than the rest of the group. Some of the factors that affect scores in the negative direction include

- Overreading of 50% point

- Examiner's outlook

- Expecting too much evidence

- Failure to weigh business factors

- More acceptable to look tough

- Weighing areas for improvement heavier than strengths.

Consider the first factor. When "overreading the 50% point" occurs, the typical cause is the Examiner is using the Criteria as a checklist. The Criteria are not and should not be viewed as a checklist. An Examiner should not expect to find details supporting every word or phrase in the Areas to Address. In scoring a written response, the Examiner needs to determine if the overall *intent* of the Criteria Item is being met.

The impact from the Examiner's outlook is significant. Examiners bring together their training, job experience, industry norms, and other factors. Suppose an Examiner comes from an industry that is using techniques that were state of the art 20 years ago but are not considered among the best practices today. Further, this person tried to implement some new methods but met resistance and was not successful in his/her own organization. This person may be very impressed with a specific organization that has successfully implemented some of those new methods. On the other hand, an Examiner having different, and in this case more positive, experiences may not consider the same methods as innovative. The Examiners' outlook influences their interpretation of the Criteria, the scoring guidelines, and the written response.

The third factor is that some Examiners expect too much evidence in the written response. The page limitation needs to be kept in mind. To include everything within the page limits, the organization will necessarily consolidate and summarize discussions supporting their strategies, activities, and results. The written response should be accepted as being truthful. If a company claims to have employee involvement teams, then the Examiner is expected to accept that claim. If a later response seems to contradict this statement, then a comment could be written stating the perceived contradiction. Additionally, failure to weigh business considerations, such as size of the company, could also lead to variation in scores. A smaller company may have less formality and less complex systems. They are not expected to have the same level of community involvement. But they should still have clearly defined practices.

Another factor affecting scoring in the negative direction is the notion that the review is an audit. Some "stereotypes" are typically associated with audits, for example: An audit is not successful unless every problem is found; a good auditor will be tough and not fooled; the company is guilty until proven innocent. With these common clichés of an audit, the focus is on the negative, with little concern for the positives. This links to the last factor given earlier in the list of negative factors: weighing areas for improvement heavier than strengths. In the Award process, the Examiners are required to provide feedback on the strengths as well as areas for improvement. They are to look at the total system. If an Examiner weighs the areas for improvement heavier than strengths, the score will be lower than it should be.

If someone tends to consistently rate responses low, they may want to recalibrate themselves. One way to do this is to write all the strengths for an Item before thinking about the areas to improve. By focusing on the strengths first, an Examiner is more apt to give partial credit, that is, to recognize the progress the organization has made. Consider these comments:

(+) The senior executives attended a two-day training class on quality and held an all-employee meeting to discuss the company vision.

(–) The involvement of the senior executives seems limited to an occasional special event.

These two comments could appear in the feedback report. The first comment is a strength. The executives are given credit for what they have done. The second comment would be under the areas for improvement. Someone who tends to rate responses lower than the rest of his or her team focuses on the second comment. By writing strengths first and (sometimes) struggling to write something positive, perhaps the first comment would also be written.

There are also factors that affect the score in the positive direction. When there is a score that appears to be an outlier from the rest of the group, it oftentimes is on the low side but not always. Outliers on the high side are typically due to an Examiner accepting vague statements as evidence, reading more into the response than is actually there, and/or looking at the score in a relative or competitive manner (i.e., give a high score because the organization is so much better than the other organizations he/she knows). If someone tends to consistently rate responses higher than the rest of the team, they may want to review the scoring steps.

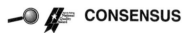 ## CONSENSUS

Because we recognize that natural variation in comments and scores will occur, a consensus process is used to select candidates for the second stage review in the national model. Many organizations that use the Criteria internally have all written responses undergo the consensus process. The purpose of consensus is to have a pooling of information so that Examiner knowledge is additive. All of the team members are exposed to this pool of knowledge by having an open discussion. First, there is a need to achieve agreement on major strengths and areas for improvement. Once comments are agreed on, the scoring guidelines are used to determine which scoring category is consistent with the team's comments.

In the national model, the consensus scores are used to determine site visits, and consensus comments are used in the feedback report (see the next chapter).

NIST has prepared material addressing the consensus process.[1] This material discusses the process as well as logistics. It has been adapted, as follows, assuming use by a large organization with a central Board of Examiners.

Note: In the national model there are references to NIST observers and to the administrator. For internal use, the role of the observer would be assumed by the person(s) who has the responsibility to monitor the constancy of purpose among the Examiner teams. The administrator would be the support person who is responsible for the clerical and logistical activities associated with this process. The following text uses the terms *observer* and *administrator* in this context. Also, NIST advocates the use of phone call(s) among the Examiner team members for time and cost savings since in the national model it is rare for the Examiners to be located near each other. For state and organizational models, the phone call(s) could be effectively replaced by a meeting.

Achieving Consensus

Applications that reach the consensus review step will undergo consensus scoring. (Remember that the term *scoring* refers to the total process of identifying strengths, areas for improvement, and site visit issues for each Item and to the process of assigning a numerical score value to each Item.) Consensus scoring is a particularly important step in the Award process because consensus numerical scores play a major role in determining which applicants will receive site visits; consensus issues are used in site visits; and consensus comments are used in the feedback report. All members of the examination team must plan and work together to ensure that consensus is reached in a fair and effective manner.

Some Frequently Asked Questions on Consensus

1. *What is consensus?* "Consensus" is a rationally derived decision on a feedback comment, a site visit issue, or a numerical score, based on the contributions of *all* team members. Numerical consensus, for example, is not simply the mean value of all numerical scores of team members. It *may* be the mean value if there is high unanimity of scoring from the individual scoring by team members and the team leader elects to use the mean value. Otherwise, it will be the value agreed on by the team after discussion of issues that led to any major disparity in individual numerical values.

2. *Who takes the lead in achieving consensus?* Consensus is initiated by and implemented through the leadership of the team leader, one of the Senior Examiners on the team.

---

[1] *Handbook for Examiners*, NIST, Gaithersburg, MD, 1994.

3. *How is consensus achieved?* All applications that reach the consensus review stage are the subject of a consensus discussion to resolve issues and to improve scoring and feedback. Consensus of specific Items can be achieved by either of two methods:

- *Method 1:* The team leader elects to achieve consensus through a meeting or a conference telephone call in which all team members discuss the examination issues that need resolution, or

- *Method 2:* The team leader assigns a numerical score for a specific Item and prepares written comments on strengths, areas for improvement, and site visit issues, taking into account the written input of all team members, but without addressing the specific Item in the consensus discussion.

4. *How does the team leader choose which method to use?* The team leader must first determine which of the preceding methods to use in achieving consensus. The following steps are recommended to aid the team leader in decision-making:

- Review the scoring worksheets of all team members and assess the degree of unanimity that exists as a result of the independent scoring.

- Look for significant differences among the written comments on strengths, areas for improvement, or site visit issues.

- For each of the scored Items, calculate, using the following equation, the difference between the highest "percent score" and the lowest "percent score" assigned by the individual team members:

$$(S_H - S_L) = \text{difference between "percent score" values}$$

where S_H = highest "percent score" value and

S_L = lowest "percent score" value.

If the difference in "percent score" value obtained is 0, 10, 20, or 30, the team leader may use Method 2. In the national model, if the difference in "percent score" value is 40 or greater, the team leader must use Method 1.

Resolution of Differences

In the NIST-developed guidelines for resolving scoring differences in the consensus process, the difference is determined between the highest and lowest score for an Item. If that difference is 20% or less, average the scores. If it is 40% or greater, then the team needs to confer until the scores are closer. If the difference is 30%, then either average or confer depending on the general variation of the scores. If a team had general agreement on most scores but had a couple of spreads of 30% or 40%, then it would be likely that the team would confer on those Items. If a team had wide differences on many Items, the Items with a 30% range may be averaged.

The Examiners assign a score in percent in increments of 10. Consider a case where there are five Examiners and they have the following scores for Item 7.1: 40%, 30%, 40%, 50%, and 60%. The range of their scores is 30% (max 60 – min 30). The team could average the score and report 44% or could confer to reach a consensus. An average is *not* a consensus. Also, note that the agreed-on score is 44% and was not rounded to an increment of 10.

5. *Who participates in a consensus discussion?* The consensus discussion includes all Examiners and Senior Examiners assigned to the particular application, plus an observer. Normally, there will be a total of six Board members, including the team leader.

6. *What are the responsibilities of Examiners in the consensus stage?*

   - Provide a list of your availability (dates and times) to participate in the consensus discussion.
   - Clear your calendar, as necessary, to participate in the consensus discussion.
   - Review the written comments of all team members and consider all points of view presented.
   - Prepare to lead or participate in the discussion of all Items requested by the team leader.
   - Send written assignments, agreed to with the team leader, to other team members at least two days prior to the consensus discussion.
   - If the discussions are via phone, provide your phone number for the conference call to the administrator.
   - Participate in the consensus discussion.
   - Provide any follow-on written comments requested by the team leader in a timely manner.

7. *What are the suggested steps in achieving consensus?* Prior to the consensus discussion, the team leader (one of the Senior Examiners) must take these steps:

   - Identify the tentative issues to be resolved.
   - List the issues clearly and concisely for your own use in developing a proposed strategy. Include strengths, areas for improvement, site visit issues, and numerical scores.
   - Schedule the consensus discussion.
   - The team leader must determine a tentative date acceptable to each team member in a timely manner. Encourage the team members to adhere to the tentative date and time, but, if absolutely necessary, the date/time can be changed. Also identify one or two dates/times that could be set aside for a follow-up consensus discussion, if needed.

- Call the observer and the administrator to report the findings of the contacts with the team members (regarding the scheduling of the consensus discussion). In conjunction with the observer and the administrator, finalize the dates and times.

- Set a time limit for the first consensus discussion. A maximum of a full day for meetings and three to four hours for calls is suggested per session.

- The administrator will make all the physical arrangements for setting up the consensus discussion (meeting or conference call).

- Plan for the consensus discussion.

- Discuss your tentative plans for the consensus discussion with the observer and develop a tentative agenda.

- Develop assignments for all team members and clearly communicate them to the team.

- Develop the agenda and communicate it to the team and observer.

- The administrator will have sent copies of all team member scorebooks to all Examiners participating in the consensus discussion and to the observer. The team leader should confirm that all team members have received all scorebooks.

- Follow up the initial contact(s) with each team member with a letter (via fax) at least one week before the consensus discussion. In the letter, remind the team members of the date and time for the discussion(s). Remind them that they should be receiving a copy of all application scorebooks from the administrator. Ask them to call you if the application scorebooks have not arrived by a certain date. Remind them that it is essential that each team member participate in the entire consensus discussion. Ask them to contact you immediately if a problem arises with the proposed times/dates. Remind them that it is essential that each team member fully prepare prior to the consensus discussion. Describe the procedures you intend to use during the call for resolving issues and arriving at a consensus. Include the agenda with the letter. Ask for comments on the procedures and the agenda and give a firm date for a response. Take all comments and suggestions into consideration, but remember that you may not be able to please everyone. Ultimately, the team leader must select a process and a set of ground rules with which he or she is comfortable.

- Finalize, with the observer, the arrangements for the consensus discussion.

- If it is necessary to alter the procedures or the agenda based on team member comments, communicate to all team members and the observer the final agenda and a description of the procedures to be used during the call for resolving issues. This must be done at least two days before the consensus discussion.

Notes

1. The administrator will provide detailed instructions on the logistics of the discussions (either meeting or conference call), including any contingency plans; for example, for conference calls, the number each Examiner should call in the event he or she is cut off during the call.

2. If a new Examiner is added at the consensus stage, he or she *must* complete all individual scoring before being sent copies of the score sheets from other team members.

3. The application scorebooks of any Examiners who will not participate in the consensus discussion will be sent to the team leader but not to all team members. These application scorebooks are provided so that any useful information can be extracted by the team leader for use in the consensus process. (Ideally the original Examiner team would conduct consensus but occasionally there is an urgent situation that requires an Examiner to be replaced.)

Conducting the Consensus Discussion

Begin the consensus discussion when all team members and the observer are in attendance. The team leader or designated team member must record notes that:

1. Capture the *main* ideas or issues discussed,

2. Capture how significant issues (the reasons) were resolved, and

3. Capture the consensus scores agreed on.

Review briefly the agenda and the procedures to be used in reaching consensus. For example, it may be useful to have a specific order in which the leader addresses team members to determine if they have a response to an issue being addressed; it may be useful to begin discussions of each issue to be addressed by having one or more team members whose numerical score was near the midrange speak first (so that those who gave higher or lower scores will not feel placed in a defensive posture); and with a conference call, it may be useful to have each member state his or her name when beginning to speak. It is also essential to refrain from isolating or polarizing one or more team members; the team leader must be particularly sensitive to "ganging up" on a high or low scorer. Some previous teams have found it helpful to use the stepwise approach used in the Examiner Preparation Course:

1. Agree on the meaning of the Criteria.

2. Agree on the applicant's response.

3. Agree on the meaning of the scoring guidelines.

4. Agree on the score.

The team leader must ensure that the reasons for the score/comments are brought out and that each team member (even a quiet one) has an opportunity to contribute to the consensus discussion. It has been found useful in the past for the team leader to present a closing summary/closure statement of each issue on the agenda at the conclusion of discussion of the agenda item and that concurrence on the summary be achieved before proceeding to the next agenda item. The summaries by the team leader help ensure that all team members are kept up to date with action items/resolutions. It is strongly encouraged that the team leader achieve formal closure of each agenda item before moving on to the next item. Remember to adhere to the agenda, and to adhere to the previously agreed-on time schedule.

Close the discussion. If all agenda items have not been completed at the end of the time, reaffirm the schedule for the follow-up conference call. If appropriate, summarize the key resolutions and action items. Thank all team members for their cooperation and effort.

Follow up on the consensus discussion by completing the resolution of any remaining agenda items as agreed on, and by completing the worksheets as directed in the application scorebook. Send the worksheets to each team member and to the administrator as requested.

IMPACT OF MISSING INFORMATION ON SCORING

If information is missing in the written response that reflects the central issue of the Item, the impact to the review of the response is major. If the missing information is one detail among many in the Item, the impact is slight. If the missing information is not identified in the Criteria, then there should be no impact. This last point seems obvious. But there have been times when a review team agrees that some needed information is missing only to realize that, although they would like the information, it was not requested in the Criteria. Most often this occurs in reviewing an approach/deployment Item. The team may want evidence of success, but this evidence is part of another (later) Item that requires results. Examiners are cautioned to avoid long lists of missing information. They should only identify the key issues that are missing.

SCORING APPROACH AND DEPLOYMENT ITEMS

It is often difficult to discern the extent of deployment, refinement, and maturity from a written response. For this reason, primary emphasis in evaluating written responses and determining appropriate scores is placed on the approach, provided that good evidence is presented that the approach is being implemented. Site visits focus heavily on deployment, refinement, and maturity using a variety of techniques and indicators.

Scoring problems associated with approach and deployment Items include the following:

- Not evaluating the strengths and areas for improvement in the context of their significance to the company's business, size, customers, etc.—information identified in the business overview

- Searching for results in Items that do not ask for results

- Using the Criteria as a checklist, thereby expecting organizations to address every issue rather than the basic intent of the Items and areas

- Expecting too much evidence rather than accepting facts at face value

- Treating opportunities for improvement as weaknesses.

SCORING RESULTS ITEMS

The page limit restricts organizations from reporting all results and details that may be relevant in scoring. If an organization's written response justifies a site visit, several factors will need to be considered during the visit:

- Relationship between the documented process and results

- Overall scope of data

- Bases for comparisons and benchmarks

- Validity of data

- Unusual features in results or trends.

Scoring problems associated with results Items include these:

- Failure to focus on the key business factors and relevant data

- Setting the requirements for the 40% to 60% range much too high, expecting that this represents excellent results in all areas

- Overreacting to an individual data point in a trend line

- Not making the distinction between anecdotes and illustrations.

ALLOCATING POINTS

Once a score is agreed on (generally referred to as a consensus score but, in fact, it may be an average), it is multiplied by the maximum number of points for that Item. For the example previously discussed involving Item 7.1, the score of 44% is multiplied by 30 points, from Figure 9.5. This results in 13.2 points (rounded to 13 points) for Item 7.1. The maximum number of points for a single Item is 130 points for Item 6.2.

Even if the point allocation for each Item were to change, the percent scores given by the Examiner would not change. If the consensus score is 60% for Item 1.1, it remains at 60% whether the point allocation is 45 points or 100 points.

 Scores are developed, recorded, and used *prior* to a site visit. A high score results in a request to the organization for a site visit. The scores are *not* adjusted based on the site visit. The purpose of this is to keep the focus on the comments and not the score. By the time there is a site visit, the comments are the key communicators, not the scores. An organization is told what range their score was in but not the actual number (see Figures 9.6 and 9.7).

Figure 9.5
Points for Each Item—1995

Points	Items
15	2.2
20	1.3, 2.1, 3.2, 4.1
25	1.2, 4.4
30	5.3, 5.4, 7.1, 7.2, 7.3
35	3.1
40	2.3, 5.1, 5.2
45	1.1, 4.2, 6.3
50	4.3
55	—
60	7.5
65	—
70	—
75	6.1
100	7.4
130	6.2

Figure 9.6

Results Prior to Site Visit

Scoring Range	Number of Companies						
	1988	1989	1990	1991	1992	1993	1994
0–125	0	0	0	3	0	2	1
126–250	0	1	7	14	11	6	7
251–400	1	8	18	38	27	18	20
401–600	31	15	51	36	36	36	36
601–750	23	12	19	15	16	14	7
751–875	11	4	2	0	0	0	0
876–1000	0	0	0	0	0	0	0
Number of Applicants	66	40	97	106	90	76	71

Figure 9.7

Graph of Results Prior to Site Visit

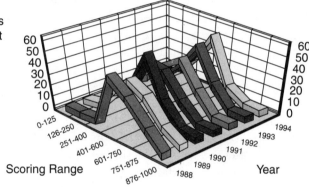

GENERAL FINDINGS

Characteristics of Higher Scoring Companies

Higher scoring companies have some common characteristics. Many of these characteristics have been identified as the Key Excellence Indicators for each Category of the Criteria (see Chapter 4.) High-scoring organizations focus on aggressive quality goals and plans, instead of just fine-tuning their current systems. Their leadership is proactive and involved in the quality journey as well as creating an environment that encourages involvement and growth of employees. These organizations are open to new ideas and search out best practices through benchmarking. Response time is a universal driver for these companies. Processes are streamlined to accomplish any and every task in a shorter period of time. They cut the time it takes to bring out a new product or service. They reduce the time to respond to a customer or employee by streamlining and simplifying processes.

High-scoring companies have proactive customer systems to establish and build a relationship that delights the customer. Facts and data drive their decision-making. A major investment is made in the people of the company since they are recognized as the greatest resource a company could have. And, in all these companies, there is a philosophy of continual improvement. The improvements may be incremental or giant leaps, but the focus is to prevent problems not just detect and remove problems.

Characteristics of Lower Scoring Companies

 The characteristics of high-scoring companies can be contrasted with some of the characteristics found in organizations that score low with the Baldrige scoring guidelines. Here the leadership tends to be passive regarding the transformation to a customer-driven company. The leadership may say the right words, occasionally, but does not get involved. The customer systems tend to be reactive. Lower scoring companies fix the problem but do not fix the system that caused the problem. They have limited measures and benchmarks. In general, data are either not easily available or not in a format that is easy to use. The narrow focus in these companies has resulted in quality being applied only to the key product/service area.

SUMMARY

The system for scoring written response was discussed in this chapter. The three evaluation dimensions used in scoring the Items—approach, deployment, and results—were presented with a description of problems associated with scoring accurately. The scoring guidelines, which provide mile markers to indicate progress on the journey toward customer delight, were reviewed. The consensus process was detailed including the logistic issues of consensus. The chapter ended with a summary of the scores of previous applicants and general findings on the relationship between scores and characteristics of the organization.

In conclusion, organizations that are successfully using the MBNQA Criteria for self-improvement have identified their strengths and areas for improvement. Then they build on their strengths and improve their weaker areas.

10 | Feedback Report

INTRODUCTION

The feedback report is a written summary of the Examiner team's assessment of the organization. It contains the strengths and areas for improvement based on the information contained in the written application and the site visit (if conducted). The feedback report is considered by many to be the most valuable part of the Baldrige review process. If it is rated as the second most important, then first place is held by the organizations reviewing themselves against the Criteria.

However, all the documents in the world will not improve the processes in an organization, no matter how well they are done. Improvement, the continuation of the quality journey, requires that the organization take well-defined actions based on this information. This chapter identifies what an organization should expect from a feedback report and provides guidelines for its use. It discusses the purpose, elements, and development of a feedback report.

A feedback report for Collins Technology, whose application is given in Chapter 7, is included as an example for the general reader and a comparative "strawman" for those readers who developed their own comments for the case study.

 ## PURPOSE

The feedback report is the only mechanism Examiners have to communicate their evaluations to the applicant (in the national Award process.) Each applicant for the Baldrige Award receives a written feedback report. It contains the strengths and areas for improvement for each Item for those applicants whose score is over 250 and by Category otherwise. The feedback report is based only on the information contained in the written application and the site visit (if conducted).

This feedback is probably the most important aspect of the entire process because it identifies the key strengths and areas for improvement of the organization when compared to the Criteria. It is the measure of the Baldrige process yardstick applied to the organization. This is vital to the improvement process because it provides an unbiased review of the organization. Without this feedback, an organization could "miss the forest for the trees." An organization using the MBNQA Criteria for self-improvement must decide how it will obtain this necessary feedback (see Chapter 5).

Valid feedback requires several elements. First, the written response must effectively communicate the organization's situation. The response is the only written source of information about the organization available to the Examiners. If the organization does not provide a coherent, understandable, and complete response, the Examiners cannot "fill in the blanks." Once they have received the written response, the Examiners make a thorough examination of the response relative to the requirements of the Baldrige Criteria. Observations about key strengths or areas for improvement are based only on the written response and site visit (if any). These findings must then be clearly communicated to the organization through the feedback report.

To assure effective communication of this feedback, some organizations with an internal Baldrige evaluation process expand on the national model by allowing a "feedback visit" by the Examiner team. This visit typically must be requested by the business unit. This is not a "second chance," "bitching," or a "what do I need to win" session. The Examiner team was unbiased and nonprescriptive in their comments and continue to remain so during this follow-up visit. The sole purpose is to allow the business unit to get additional details about and insight into the feedback report.

USING THE FEEDBACK REPORT

The feedback report should be used by the organization in its strategic planning process. The focus should be on improving the processes to delight the customer, productivity, and asset utilization. Organizations that focus on winning an award may become defensive when they receive their feedback report. After all, no one likes to hear negative things about their pride and joy; this is the "ugly baby" syndrome. However, no matter how good the company is, there is always room for improvement. Even recipients of the Award have received reports identifying improvement opportunities. So if an organization applies for the Baldrige with the intent of getting the "gold" and expects only praise, they are in for a surprise.

Because the feedback report is an important part of the improvement process, it is not surprising that the organization would like to get as much direction out of the report as possible. However, the focus of the improvement process should be using and understanding the Criteria, that is, the road map, and not the feedback report, the results of applying the yardstick. The organization should not expect the report to tell them what to do, only to identify strengths and areas for improvement as compared to the Criteria. Some organizations focus too narrowly on the Criteria; they need a broader view. To this end, the feedback report does not indicate the specific Area to Address for individual comments. The focus of the comments should be on the overall intent of the Item.

The use of the information in the feedback report is a crucial step in the Quality Journey. The conversion of the assessment into action plans should be an integral part of the overall business planning process. Organizations that set up a separate process to identify action plans to "improve the application" will generally end up with a good written response but short-lived and costly successes. Improvements achieved may be "in spite of the system" rather than "because of the system."

⦿᎒ PDSA

There is no single approach to this improvement process. Many excellent organizations are using strategic planning and improvement processes based on the PDSA (Plan-Do-Study-Act) cycle[1] (see Figure 10.1), which is simply a variation of

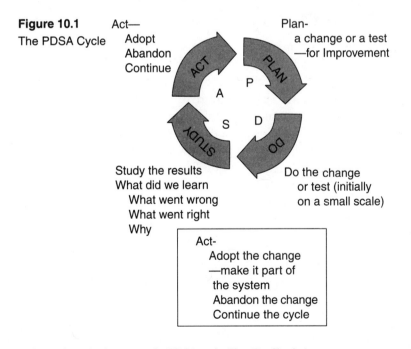

Figure 10.1
The PDSA Cycle

Act—
Adopt
Abandon
Continue

Plan-
a change or a test
—for Improvement

Study the results
What did we learn
What went wrong
What went right
Why

Do the change
or test (initially
on a small scale)

Act-
Adopt the change
—make it part of
the system
Abandon the change
Continue the cycle

---

[1] This cycle is also known as the PDCA cycle: Plan-Do-Check-Act.

the scientific method. Many people call this the Deming Cycle, after Dr. W. E. Deming; Dr. Deming referred to it as the Shewhart cycle since Dr. W. Shewhart was one of the pioneers in applying the scientific method to industry. Dr. Deming's name became attached to this cycle by his Japanese students in the 1950s. Dr. Deming did expand on the original application of Dr. Shewhart by explicitly including the future process related "Act" step.[2]

In Chapter 3, Using the Criteria for Self-Improvement, we used the PDSA cycle to demonstrate an improvement and nonimprovement approach to using the MBNQA process. Organizations that use the PDSA approach adapt it for their own systems. For example, Eastman Chemical uses the PDSA as the basis for their Quality Management Process (QMP) (see Figure 4.4 in Chapter 4). The QMP is used to address performance improvement at all organizational levels. The *assess organization* box in Figure 4.4 is where the information from a feedback report would enter the process.

Application of the PDSA cycle to converting the feedback report information to action plans will depend on the organization's approach to strategic planning, but would include the following elements.

Plan

Identify Improvement Opportunities

Review the strengths and areas for improvement identified in the feedback report to determine if these match the organization's view of itself. If not, establish if this is because the organization is interpreting the Criteria differently, the response development process needs to be improved, or some other reason. Consolidate all information and determine which strengths should be reinforced and which areas for improvement acted on. One way of doing this is by prioritizing the processes, recognizing their causal nature as system or local, establishing team ownership, assigning a management sponsor, and defining the scope of the team's activities (i.e., establish the team's charter). Be realistic. If most of the improvement plans deal with system-related processes, then restrict yourself to 6 to 8 key issues. If most are of a local nature, then 10 to 12 key issues may be handled.

Initiate Team Activities for Process Improvement

Organize a cross-functional team if it does not already exist, with an established mission and boundaries. Allocate the necessary resources, including a large dose

---

2 Based on a presentation to the Deming Study Group of Greater Detroit by R. Moen, June 1994.

of patience for management. The team should then identify internal and external customers and suppliers, envision process breakthrough by visualizing an improved process adapted to customers' needs and desires, determine appropriate measurements, and plan for data collection and analysis. These activities should interface with established processes within the organization; for example, the processes described in Category 2, Information and Analysis.

Do

Pilot and Verify the New Process

Communicate changes to all affected employees, suppliers, and customers, provide necessary training, implement the suggested changes, and initiate data collection and analysis. If the planning stage was successful, there should be no surprises.

Study

Study the Results

Determine what can be learned from all the activities, what went wrong, what went right, and, most importantly, why.

Act

One of three things can happen at this stage:

1. *Implement system changes:* Standardize and integrate the new process throughout the system considering other system factors, update all related process documentation as well as procedures and forms, communicate changes to all affected employees, suppliers, and customers, provide necessary training. In addition, the leadership team should look for replication opportunities.

2. *Abandon the pilot changes:* Communicate the removal of the changes to all affected employees and suppliers identifying why such actions are necessary, restore the original process, update all related process documentation so that such changes will not be tried under the same or similar circumstances in the future.

3. *Continue:* If the data are inconclusive, continue with the PDSA cycle.

In actual practice, use of the PDSA is an iterative process since the Act step immediately leads to the Plan step. This leads to several "turns" of the cycle before the total improvement process is complete. In the preceding steps, it was assumed that information on the existing process was available to the improvement team. If this is not true, then the team should first document and quantify the process before determining or making any changes. In addition to this, the team should determine why the information was not available—the reasons may point to an opportunity with more impact than the original project.

THE FEEDBACK REPORT

Because the feedback report is an important part of the improvement process, it is the obligation of the Examiner team to provide the organization with a document that is of value to them. The team must assume that the users of the report understand the Criteria and the links among the Items. The report is not a tutorial on the Criteria. However, the report must be specific to the written response—specific is not prescriptive. All the guidelines of writing good comments apply to the feedback report. The primary difference between the feedback report comments and the consensus comments (often called the Senior Examiner's report) is that the feedback report does not identify the Item scores and the Area to Address with the comment. This means that the comments (initial and consensus) should be rephrased to include the specific Item intent.

A feedback report has five sections:

1. Introduction

2. Background: application review process

3. Scoring: distribution of numerical scores for all applicants

4. Scoring: summary of applicant

5. Strengths and areas for improvement—three to five comments per Item

In the national process the first two sections are written by NIST and are standard for all applicants. The introduction contains a few sentences explaining what is contained in the report. The background section explains the application review process and provides summary statistics pertaining to the number of applicants that completed each step. For example, in 1994 there were 71 applicants; 29 of them reached the consensus stage and 14 received a site visit. (See Figure 10.2 for the number of applicants for each year, from 1988 to the present.) The process flow of the evaluation process is included in the background section.

Section 3 of the report contains the scoring distribution for all applicants for the year. The actual scores are not revealed, only the range in which the score fell. The predetermined ranges used each year are shown in Figure 10.3. Note that the ranges are not of equal size with the largest spacing in the center. Experience has shown that the typical "good" U.S. organization will fall in the 251–400 range.

Figure 10.2

Number of Applicants Selected for Consensus Review

	1988	1989	1990	1991	1992	1993	1994
Number of applicants to consensus	all	23	26	28	30	29	29
Total number of applicants	66	40	97	106	90	76	71

Section 4 of the feedback report tells an organization the range of its score. The organization does not receive its own score in absolute terms, only the range of occurrence. For example, an organization may be told they scored in range 5, which is 601–750. This score was arrived at through the consensus review process. This score is determined prior to any site visit, if the organization received one. The scores lead to an invitation for a site visit. The scores are not changed after a site visit. The focus is to be on the comments not the score. One way to attempt to make this happen is to ignore (not change) the score based on a site visit. The Examiners communicate their findings through their comments and indicate if the score could be raised, lowered, or stay the same. But the score is not changed to reflect the findings of the site visit. In addition to the score, the fourth section contains a summary of the key findings. The section is a maximum of one page in length.

The bulk of the feedback report is contained in Section 5, strengths and areas for improvement. For this reason, there is a strong link between this chapter and Chapter 8, Evaluation Process: Comments. An Examiner needs to write clear, concise, and accurate comments in order for the feedback report to be useful. If the organization has received a site visit, then comments stating "It is not clear . . ." should not be present in the feedback report. The site visit should have clarified any significant issues.

Figure 10.3

Range of Scores

Range	Range Number
0–125	1
126–250	2
251–400	3
401–600	4
601–750	5
751–875	6
876–1000	7

If an organization did not proceed past stage one of the evaluation process—the independent review—then one of the Examiners that did the independent review will be asked to write the feedback report. This Examiner will be given the comments and scores from the other Examiners to consider as input. In this case, the Examiners do not discuss the written response. The submitted comments and score are the only input received from the other Examiners. If the organization received a score of less than 250, the comments would be given on a Category, not an Item, basis.

If an organization did reach the consensus review stage, then the comments developed through the consensus process become the basis for the feedback report. One of the Examiners is asked to write the document based on the consensus comments. This would involve some fine-tuning of the comments but the essence of the report would remain as developed through consensus.

For those organizations receiving a site visit, the feedback report contains information learned by the Examiners from the written response and the site visit.

GUIDANCE

 NIST provides written guidance to the Examiners in addition to covering the topic in the Examiner training session. A portion of this information is shown here:[3]

Making the Most of Your Written Comments

 "Feedback Reports should be relatively short (30 pages or less). Thus, the comments of Board members must be condensed and focused upon key issues. The following guidelines are intended to aid the Examiner in identifying and composing comments for strengths and areas for improvement, and in writing Feedback Reports.

- ☑ Use bullet form.
- ☑ Use complete sentences.
- ☑ Address only the most important issues.
- ☑ Try to provide two to five comments per Item.
- ☑ Be constructive; use a positive tone.
- ☑ Be specific and clear.
  - Remember: The applicant will want to begin improvements based upon the observations contained in the Feedback Report.

---

 3 *Handbook for Examiners*, NIST, Gaithersburg, MD, 1994.

☑ Avoid jargon or acronyms.

☑ Be nonprescriptive; state observations and evaluations; refrain from the use of "could" and "should."

☑ Reflect the numerical score.

- Low scoring applications lack basic information.
- Higher scoring applications contain basic information but may lack information on finer points.

☑ Comment only on areas contained in the Criteria.

- A number of techniques, tools, and approaches advocated by authorities in the field go beyond the statements of the Award Criteria. If terms referring to such techniques are used in the application, the writer must be careful not to allow his/her own background or opinions to affect statements in the Feedback Report.

☑ Take special care in identifying areas for improvement.

☑ Be sure that each statement is clearly justified. Areas for improvement should be traceable to omissions and problems noted in the review of the application.

☑ Write legibly.

☑ Check for inconsistencies and conflicting statements in the comments.

☑ For site-visited applications do not use such phrases as:

"It is not clear that. . . ."
"It does not appear that. . . ."
". . . is unclear."
(If the issue cited is important and you visited the company, it should have already been "clarified.")

 ## SAMPLE FEEDBACK REPORT

A feedback report for Collins Technology, which follows, is a sample of what an organization would receive. Examples of all five sections are included.

The Collins feedback report is written as if the Collins written response reached the consensus stage, but did not receive a site visit. At the scoring level achieved by Collins, a real application might have been selected for site visit. However, from a case-writing point of view, the creation of the kind of detail normally uncovered in a site visit was beyond the scope of a case study. If this feedback report were developed after a site visit, phrases like "it is not clear that" would not be included.

An actual feedback report would contain the numbers of applications reaching each stage of evaluation and the distribution of written scores. These numbers can be obtained from the Office of Quality Programs (National Institute of Standards and Technology). See the appendix for contact information.

Note: An actual feedback report would *not* include the Areas to Address identification [i.e., "(+)a," "(–)b," "(++)c," etc., associated with each comment]. Further, the Item scores would *not* be provided as part of the feedback. The applicant would receive, in the feedback report summary, only the range in which the total score fell (e.g., 601–750). This information is only included here for the reader's benefit.

In an actual report, the (+) and (–) designations would be replaced with bullet (•) notations. Letters for areas would not be given. Scores would not be given (see sample, pages 334–345).

1994 Case Study: Feedback Report for Collins Technology

The following is an example of a feedback report. The report is developed as a consensus by the examining team members after each has completed an independent review of the full application. An actual feedback report would *not* include the Areas to Address identification [i.e., "(+)a," "(–)b," "(++)c," etc., associated with each comment].

Further, the Item scores would *not* be provided as part of the feedback. The applicant would receive in the feedback report summary only the range in which the total score fell (e.g., 601–750).

INTRODUCTION

The application report from your company has been evaluated in the competition for the Malcolm Baldrige National Quality Award. This feedback report,[1] which contains the findings of the Board of Examiners, is based upon the information contained in the written application report. It includes background information on the examination process, a summary of the scoring for your company, and a detailed listing of strengths and areas for improvement.

BACKGROUND

Application Review

The application review process began with the first-stage review, in which a team of six Examiners were assigned to each of the ## applications that met the requirements for evaluation.[2] Assignments were made based on the Examiners' areas of expertise and on their potential conflicts of interest. Each application was independently evaluated using a scoring system which was developed for the Award program and which was reviewed and put into practice using case studies in Examiner Preparation Courses. All Items were scored.

Based on the scores, the panel of Judges selected ## of the ## applications evaluated to go on for the second-stage review. If an application was not selected for the second-stage review, one Examiner was selected to review the comments of the other team members and prepare the feedback report.

---

1 The Collins feedback report is written as if the Collins application reached the consensus stage, but did not receive a site visit. At the scoring level achieved by Collins, a real application may have been selected for site visit. However, from a case-writing point of view, the creation of the kind of detail normally uncovered in a site visit was beyond the scope of a case study.

2 An actual feedback report would contain the numbers of applications reaching each stage of evaluation and tile distribution of written scores. These numbers may be obtained from tile Office of Quality Programs (National Insititute of Standards and Technology).

In the second-stage review, a team of six Examiners, including at least one Senior Examiner, conducted a telephone conference call to develop a consensus score for each Item and an aggregated list of comments. The Senior Examiner directed the consensus process to ensure the resolution of any scoring differences.

After the consensus process, the panel of Judges reviewed the scoring to verify that the evaluation process has been properly followed. Following this review, the Judges selected the applicants to receive a site visit. This decision was based not only upon the scores of an applicant but also upon the scores of *all* applicants, thus ensuring that site visits are made to all applicants who should have consideration in the final Award recommendation meeting. Because your application was not chosen for a site visit, the consensus team prepared this feedback report.

Of the ## applicants who went through the first-stage review, ## of those applicants continued on to the second-stage review. Of those companies, ## were selected to receive a site visit. Site visits were made to ## companies in the manufacturing category, ## in the service category, and ## in the small business category. The evaluation process is shown in Figure 1.

In the site visit process, the selected applicants are notified, and the site visit team prepares for the visit. Site visit issues are translated into a specific visit agenda, with each member of the team given a specific assignment. The site visit team meets prior to the visit to finalize all plans. While on the visit, team members meet periodically to review their findings and, if necessary, to modify the agenda. After the visit is complete, the team prepares a summary report of its findings. The stages of a site visit examination are shown in Figure 2.

Site visit reports, consensus scores, and Examiner comments are forwarded to the panel of Judges for final recommendations. The Judges separately consider the applicants in the manufacturing, service, and small business categories. They develop Award recommendations using a three-step process: (1) reviewing the reports of all site-visited applicants, (2) rank ordering or grouping the site-visited applicants, and (3) deciding whether or not to recommend one or both of the top two candidates in each category for the Award. The first step provides the basis for making recommendations. The second step identifies the highest rated applicants in each category based entirely upon a detailed comparison of the applications. The third step evaluates applicants against an "absolute" standard: the overall excellence and the appropriateness of the applicants as national role models. The Judges' evaluation process is shown in Figure 3.

During the Judges' recommendation meeting, strict rules involving conflict of interest are followed. Four major types of conflict are considered: (1) direct link such as current or recent employment or client relationship; (2) significant ownership; (3) business competitors of companies for which direct links or ownership exists; and (4) Award category conflict. Judges are allowed to vote only when they do not have any of these types of conflict.

Figure 1
Evaluation Process

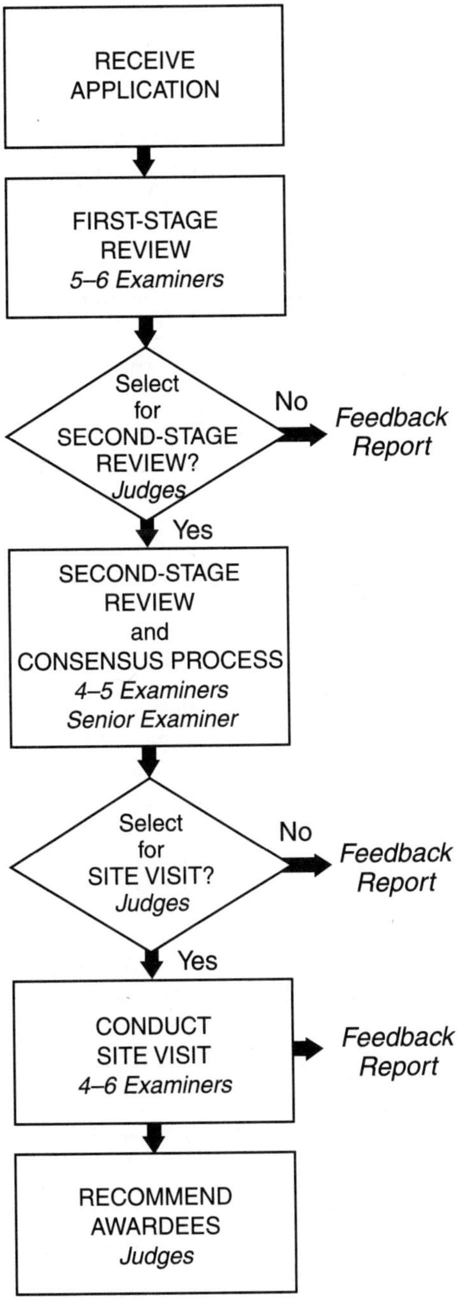

Figure 2
Site Visit
Examination

Step 1
Team Preparation:
- Review of Findings
- Review of Issues
- Site Visit Plan

Step 2
Site Visit:
- Presentations
- Interviews
- Observations
- Review of Records

Step 3
Site Visit Report:
- Issue Resolution
- Category Summaries
- Recommendations

Figure 3
Judges'
Evaluation
Process

Step 1
Panel of Judges Review:
- Scoring
- Comments
- Site Visit Reports

Step 2
Evaluation by Category
- Manufacturing
- Service
- Small Business

Step 3
Assessment of Top Companies:
- Overall Strength/Weakness
- Appropriateness as National Model

SCORING

The scoring system is designed to provide resolution among the very best companies and to facilitate feedback. The scoring Criteria are based on (1) evidence that a quality system is in place, (2) the depth of its deployment, (3) the results it is achieving, and (4) the length of time it has been in place.

Figure 4 provides an overview of the scoring for the ## applicants evaluated. The scores are divided into seven scoring ranges, and the percentage of applicants scoring in each range is given. There is a general description of the applicants scoring in each range.

It should be emphasized that Figure 4 represents the scores for the written applications only. Site visit teams investigate each of the seven Categories and find that some applicants are significantly stronger in some or all of the Categories than had been indicated by their original score; others are weaker. It can be concluded that some applicants would have moved up in range if they had been scored again, while others may have moved down.

Figure 4

Distribution of Written Scores

Range	Range Number	% Applicants in Range	Comments
0–125	1	##	Very early stages of developing approaches to addressing some basic Category requirements.
126–250	2	##	Early stages in the implementation of approaches. Important gaps exist in most Categories.
251–400	3	##	Beginning of a systematic approach, but major gaps exist in approach and deployment in some Categories. Early stages of obtaining results stemming from approaches.
401–600	4	##	Effective approaches and good results in most Categories, but deployment in some key areas is still too early to demonstrate results. Further deployment measures and results are needed to demonstrate integration, continuity, and maturity.
601–750	5	##	Refined approaches, including key measures, good deployment, and good results in most Categories. Some outstanding activities and results clearly demonstrated. Good evidence of continuity and maturity in key areas. Basis for further deployment and integration is in place. May be industry leaders.
751–875	6	##	Refined approaches, excellent deployment, and good to excellent improvement and levels demonstrated in all Categories. Good to excellent integration. Industry leaders.
876–1000	7	##	Outstanding approaches, full deployment, excellent and sustained results. Excellent integration and maturity. National and world leadership.

SUMMARY

Collins Technology scored between 601 and 750 points, indicating evidence of effective efforts in most Categories and outstanding efforts in Category 1.0, Leadership, and Category 4.0, Human Resource Development and Management.

The TQM Forum provides a systematic, visible, and integrated approach for the quality process. Senior executive involvement in this and other activities, such as customer calls, new employee orientation, and participating on various teams, is extensive.

A significant strength of Collins Technology is the communication which moves in all directions in the company. Monthly all-employee meetings are one of the many communication vehicles.

The annual employee survey provides feedback on multiple HR issues including overall employee satisfaction, recognition, and cooperation. Positive trends are shown for employee satisfaction, turnover, absenteeism, and number of accidents.

Collins Technology demonstrated at least effective efforts in all Categories but the Items needing the most improvement are in the area of supplier quality and support initiatives.

The Collins Technology application lacks sufficient detail regarding the Supplier Total Excellence procedures, supplier surveys, and ship-to-dock efforts to fully evaluate them. Deployment of these procedures seems limited at this time; only 9 suppliers out of 75 are designated as ship-to-dock.

In the support areas of the business there was no indication of quality activities in several departments such as computer information services, sales, marketing, and administration. Most striking is the absence of improvement activities for the LAN system. This seems to be an important data collection, analysis, and communication tool but no quality activities were addressed to make it more efficient and effective.

Overall, Collins Technology has many successful results from their efforts.

1.1 Senior Executive Leadership *(45 points)*

Strengths

- The TQM Forum provides an integrated and visible approach for leading the quality process.

  - The senior executives at Collins Technology are visible in the quality activities in a variety of ways. This includes permanent membership on the TQM Forum, calling on two customers per month, presentation of employee recognition awards, and conducting all-employee meetings.

  - The senior executives' involvement in communicating the quality message outside the company is extensive for a small company. It includes:

    - the CEO is on the Board of Directors of the Institute of Printed Circuits (IPC)

    - the company is a co-founder of the state Quality Expo

    - the company co-founded the local Quality Consortium and serves on the advisory board

    - over 20 presentations were given in 1993

    - actively supported the lobby efforts in Washington, D.C., for the continuation of National Quality Month.

  - The quality values are defined by a "Quality Policy" and a "Quality Promise."

  - A senior executive is present at all training classes and each executive is expected to teach at least one class per week.

  - All senior executives develop corrective action plans based on feedback from the employee survey.

Areas for Improvement

- The "Quality Promise" could be read by some as suggesting that people, not the system, are the problem.

> **Note:**
> This page is the same as the facing page except the comments are presented in the manner of a standard feedback report for an application that scored over 250 points. There are no references to the Areas to be Addressed, no indications of pluses or minuses, and no indication of the Item score as is shown on the facing page. The purpose of this is to focus on the intent of the Item, not the Areas or score.
>
> The rest of this feedback report will not follow these conventions, since the purpose of this report is to assist you in understanding the evaluation and feedback process, as well as enabling you to calibrate your quality "yardstick."

1.1 Senior Executive Leadership *(45 points)*

Strengths

(+) a The TQM Forum provides an integrated and visible approach for leading the quality process.

(++) a The senior executives at Collins Technology are visible in the quality activities in a variety of ways. This includes permanent membership on the TQM Forum, calling on two customers per month, presentation of employee recognition awards, and conducting all-employee meetings.

(+) a The senior executives' involvement in communicating the quality message outside the company is extensive for a small company. It includes:

- the CEO is on the Board of Directors of the Institute of Printed Circuits (IPC)
- the company is a co-founder of the state Quality Expo
- the company co-founded the local Quality Consortium and serves on the advisory board
- over 20 presentations were given in 1993
- actively supported the lobby efforts in Washington, D.C., for the continuation of National Quality Month.

(+) b The quality values are defined by a "Quality Policy" and a "Quality Promise."

(+) c A senior executive is present at all training classes and each executive is expected to teach at least one class per week.

(+) d All senior executives develop corrective action plans based on feedback from the employee survey.

Areas for Improvement

(–) b The "Quality Promise" could be read by some as suggesting that people, not the system, are the problem.

Score: 90%

1.2 Management for Quality *(25 points)*

Strengths

(+) a Quality values and customer focus are deployed from the TQM Forum to strategic business units and then to departments through the planning process.

(+) a Communication moves in all directions. Employees meet with internal customers each week to review progress and keep communication lines open.

(+) b The customer focus and quality values are reinforced at the monthly all-employee meetings where customer surveys are discussed and employees' questions are answered.

(+) c The TQM Forum reviews results from the department teams and three additional teams each month and provides resources for those needing help.

Areas for Improvement

(–) c The TQM Forum meets weekly and reviews three teams per meeting (50 teams total). Each team would be reviewed once every four months. This does not seem to be often enough.

(–) d It is not clear what information is obtained from the customer survey, employee survey, or *Red Book* that allows Collins Technology to evaluate and improve awareness and integration of quality values.

Score: 70%

1.3 Public Responsibility and Corporate Citizenship *(25 points)*

Strengths

(+) a For business ethics and public safety there are multiple sources of input.

(+) b The state has recognized the waste management system as a model.

(+) c The external promotion of quality awareness is varied and includes:

- corporate membership in professional societies
- availability of a corporate video
- presentations to state universities
- payment of employees' society membership dues.

Areas for Improvement

(–) a The principal goals for public health and safety are not clear.

(–) a The company claims no ethics violation. However, this assessment is made by the TQM Forum, hardly an independent appraiser of its own activities.

Score: 60%

2.1 Scope and Management of Quality and Performance Data and Information *(15 points)*

Strengths

(+) a Over 600 data/information points are used by Collins Technology including a broad base of customer, engineering, manufacturing, and accounting data.

(+) b The "Many-Point Data" checklist provides reliable data and standardized measurements. The LAN system allows for timely update and easy access to the data.

(+) c Each year the purpose, collection methods analysis, and documentation of the data are reviewed and evaluated.

Areas for Improvement

(–) a In the overview "...stability, predictable internal capacitance and tolerance for extreme environment..." were listed as key quality aspects but are not listed as key data.

(–) a The data identified for support areas seemed limited.

Score: 80%

2.2 Competitive Comparisons and Benchmarking *(20 points)*

Strengths

(+) a, b Key indices for product quality, internal operations, and customer satisfaction measures are benchmarked.

(+) a By getting information from their customers, Collins Technology compares themselves against their customers' best suppliers.

(+) c Teams brainstorm ways to close the gap identified through the benchmarking activities. These teams include customers and suppliers as appropriate.

(+) d Senior executives have attended training in benchmarking.

Areas for Improvement

(–) a The competitive and benchmark studies seem to be narrowly limited to senior executive participation.

(–) d It appears that only senior executives participate in the benchmark training.

Score: 70%

2.3 Analysis and Uses of Company-Level Data *(40 points)*

Strengths

(+) b A monthly operational review and weekly TQM Forum review is held to guide actions based on internal results.

(+) c Some comparisons are made between quality and financial results.

(+) d Analysis has led to some changes in the data that are collected.

Areas for Improvement

(–) a, b The description of the analysis did not provide enough specific detail to be able to evaluate it.

Score: 50%

3.1 Strategic Quality and Company Performance Planning Process *(35 points)*

Strengths

(+) a There is a defined process to annually develop a four-year strategic business plan, which covers quality, operations, financial, human relations, training, and capital goals and attainment plans (Figure 3.1).

(+) b The SBP process includes input from the gap analysis between "what should" and "what is the process doing" for key processes throughout the company.

(+) c The overall requirements are used by each department to develop their own attainment plan and goals. The goals are tracked in the *Red Book*.

(+) d A team has been formed to oversee improvements to the planning process.

Areas for Improvement

(–) c It is not clear how the overall requirements are deployed to the suppliers. (Reference is made to Item 5.4, but no direct answer was found there regarding this issue.)

Score: 70%

3.2 Quality and Performance Plans *(25 points)*

Strengths

(+) a Collins Technology addresses current situations and trends relative to customers and competitors. Aggressive goals are set for the current market niches.

(+) b The "Top 10" plans are deployed through the department plans.

(+) b, c There is a one- and four-year projection for the top three indicators.

(+) d There seems to be an understanding of future quality and performance requirements.

Areas for Improvement

(–) b, c Although the goals have been stated, the plans to achieve them have not been identified.

Score: 60%

4.1 Human Resource Planning and Management *(20 points)*

Strengths

(+) a, b There are HR goals for a variety of practices including training, recruitment, recognition, and mobility.

(++) c An annual employee survey is given to all employees. It includes questions regarding job satisfaction, cooperation, career development, and other related topics.

Areas for Improvement

(–) a The description of the plans is not specific enough to evaluate.

(–) a No indication of mobility and flexibility in job assignments above floor-level employees, e.g., career planning and job rotation for salary employees.

(–) a, b Plans for productivity improvement are not addressed other than implying that training will affect this in a positive sense.

Score: 60%

4.2 Employee Involvement *(40 points)*

Strengths

(+) a There are numerous opportunities for employee involvement (Figure 4.1) including three formal team structures.

(++) a Each cross-functional team (CFT) has a TQM Forum member assigned as a coach.

(+) a The ECI system allows employees to make suggestions. A response is given within 24 hours.

(+) b Operators have the ability to shut the line down when there is a problem. Assembly lines are equipped with green, yellow, and red lights to indicate status.

(+) c The extent and effectiveness of involvement is tracked through the number of teams, the percent of employees by category on teams, and the percent of suggestions implemented.

(+) d The trend in team participation and number of teams is positive for all categories of employees.

Areas for Improvement

(–) c Although data are tracked, it is not clear how results are linked to key quality requirements; i.e., are the teams working on the right things?

(–) c No data are presented on the extent of team member involvement (i.e., number of days, hours/meetings, etc.).

Score: 80%

4.3 Employee Education and Training *(40 points)*

Strengths

(+) a Core courses are provided to all employees and specific courses offered as identified through the *Training Action Team*.

(+) b The Station Qualification System program requires annual recertification.

(+) c, d The extent of training is tracked by a training team. The data show positive trends in extent of training.

Areas for Improvement

(–) c, d The quality and improvement of the education and training processes are not addressed.

Score: 70%

4.4 Employee Performance and Recognition *(25 points)*

Strengths

(+) a There are many different methods to recognize employees' accomplishments. Recognition is linked to achieving the SBP objectives.

(+) b Evaluation and improvement of the employee recognition process includes analyzing employee survey data on the topic.

(+) c The amount of employee recognition has increased over the last four years for all categories of employees.

(+) c Per the data in 4.1, employee satisfaction with recognition/rewards has been increasing since 1989 (measure of effectiveness).

Areas for Improvement

(–) c Employee satisfaction with recognition/reward is not reported by employee category so deployment cannot be checked.

Score: 80%

4.5 Employee Well-Being and Satisfaction *(25 points)*

Strengths

(+) a Safety teams have been formed to review health and safety factors. They review accidents each month and recommend corrective measures. A perfect score on an insurance company safety audit was received.

(+) a Collins' housekeeping program has been used as a model for industry in the area.

(+) b Special services include an employee assistance plan, day care center, and merit scholarship program for children of employees.

(+) c Employee satisfaction is determined through an annual survey.

(+) d The key indicators tracked over the last two to three years include:

- number of accidents
- absenteeism
- turnover
- employee satisfaction
- recruitment success.

Positive trends show in all 10 indicators.

Areas for Improvement

(–) a The activities of the safety teams sound like traditional "review, report, and recommend" safety reviews. It is not clear how they are preventing accidents although the accident rate is decreasing.

(–) d Except for absenteeism, no other benchmarks are given.

Score: 60%

5.1 Design and Introduction of Quality Products and Services *(40 points)*

Strengths

(+) a, b A cross-functional team uses the design guidebook during the design process. For half of the commercial products, a customer representative is on the team. When materials are critical, suppliers are on the team.

(+) a A preliminary design review is conducted using a checklist of topics. Design to cost techniques are used to control cost and FMEAs are used to prevent problems.

(+) b Design validation includes the use of computer simulation and designed experiments.

(+) c The design guide is reviewed and updated at the end of each design cycle. Cycle time is used as a measure of effectiveness.

Areas for Improvement

(–) a Although the design process sounds systematic, it is not clear how things get done. The actual translation of the customer requirements seems just as vague.

Score: 70%

5.2 Process Management: Product and Service Production and Delivery Processes *(35 points)*

Strengths

(+) a For each step in the total process, capability analysis is done, and the control plan (developed during the design stage) is updated as needed.

(+) a A formal problem-solving process, including corrective action, is used. Actions to correct out-of-control conditions are documented.

(+) b Performance information in the LAN and *Red Book* are reviewed by the TQM Forum.

(+) b Teams or individuals work on process simplification and alternative technologies.

Areas for Improvement

(–) b Although teams or individuals are trying to improve processes, it is not clear what process is used to accomplish this.

Score: 60%

5.3 Process Management: Business Processes and Support Services *(30 points)*

Strengths

(+) a Key indicators of quality are tracked in accounting, engineering, maintenance, and the lab.

(+) b Collins' seven quality tools are used for out-of-control occurrences.

(+) c Process flowcharts are used by the internal supplier/customer. Policy is to have employees spend 20 minutes per week with their internal customers.

Areas for Improvement

(–) a Deployment seems weak. There was no indication of quality activities in several departments such as computer information services, sales, marketing, and administration. The LAN system seems to be an important data collection, analysis, and communication tool but no quality activities were addressed to make it more efficient or effective.

(–) b It is not clear if the six-step problem-solving process (described in 5.2) is also used in the support areas.

(–) b Approaches for controlling processes appear to be reactive rather than proactive and prevention based.

Score: 40%

5.4 Supplier Quality *(20 points)*

Strengths

(+) a A close working relationship is maintained with the top 30 suppliers.

(+) b Supplier surveys are conducted by purchasing, engineering, and quality assurance.

(+) c A *Supplier Council* has been formed to improve feedback from the suppliers to Collins Technology.

Areas for Improvement

(–) a Limited proactive information was provided regarding the supplier *Total Excellence* procedures.

(–) a Expected performance levels of the principal requirements for the SRI was not provided.

(–) b The supplier survey description was not specific enough to determine what is assessed.

(–) d Limited deployment; only 9 suppliers (30% of the top 30) are designated as ship-to-dock/line suppliers.

Score: 50%

5.5 Quality Assessment *(15 points)*

Strengths

(+) a Many customer audits are conducted on the products. Systems, processes, and practices as reviewed through a four-level approach.

(+) b Repeat audit findings are used to track effectiveness.

Areas for Improvement

(–) a The assessments were not described in enough detail to determine if they are deployed across all practices in the company.

(–) b It is not clear how the assessment process is evaluated or improved.

Score: 60%

6.1 Product and Service Quality Results *(70 points)*

Strengths

(+) a There are favorable trends in the data reported.

Key factors from overview	Figure
• On-time delivery	6.8
• Competitive prices	6.1, 6.2, 6.3, 6.4, 6.5, 6.6, 6.7, & narrative
• State-of-the-art performance	6.17
• Precision and stability	5.4, 6.11
• Predictable internal capacitance	6.12 (?)
• Tolerance for extreme environment	"
• General cleanliness & workmanship	"
• Other (Eng. and mfr. lead times)	6.9, 6.10, 6.14

Note: Figure 6.13 belongs in Category 7.0.

(+) b Collins Technology is one of the leaders in those areas where comparisons were shown.

Areas for Improvement

(–) a More than half of the charts or reported numbers are cost related (belong in 6.2). There seems to be an imbalance between cost and other factors tracked.

(–) a Data on service standards (described in 7.2g as quotation response time, technical assistance response time, etc.) were not reported. The key characteristics were not all addressed.

(–) b The benchmarks seem low (yield, on-time delivery, and test yield). These numbers do not appear to be world-class levels.

Score: 60%

6.2 Company Operational Results *(50 points)*

Strengths

(+) a Material cost (Figure 6.2), productivity capacity gains (Figure 6.3), composite cost (Figure 6.5), cost of quality (Figure 6.17), and productivity (Figure 6.18) all show positive trends.

(+) b There are some benchmark levels shown and Collins is doing well in comparison.

Areas for Improvement

(–) a More charts were expected such as labor hours, inventory turns.

Score: 60%

6.3 Business Process and Support Service Results *(25 points)*

Strengths

(+) a The quality initiative includes a variety of areas including Accounting (Figures 6.21, 6.24a, 6.24b, and 6.24d), Engineering (Figure 6.22), Maintenance (Figure 6.24c), and Purchasing (Figure 6.24f).

(+) a The key measures in 10 trend charts show positive trends and have been tracked for at least two years.

Areas for Improvement

(–) b Except for "receivables over 30 days," there are no other comparisons for any other measures shown. The results presented are not representative of business and support service processes.

Score: 50%

6.4 Supplier Quality Results *(35 points)*

Strengths

(+) a A favorable trend was shown for the *supplier results indicator* (SRI). Dock-to-stock status was also reported.

(+) a Suppliers' ratings of Collins Technology shows improvement for the two years reported.

(+) b Benchmark data were shown for one key indicator.

Areas for Improvement

(–) a The data addresses the top 30 suppliers only.

(–) a Indicators described in 5.4a (e.g., on-time delivery and acceptance of delivered units) as key supplier requirements have no substantiating data reported.

Score: 40%

7.1 Customer Expectations: Current and Future *(35 points)*

Strengths

(+) a Market segments have been defined based on their willingness to use new technology.

(+) b Collins is able to participate as a key supplier in the customers' planning process.

(+) b The company has the ability to develop new technology and run prototypes in their process lab to assist the customer's R&D.

(+) c Through the SBP, improvements in determining customer requirements have been made, resulting in the customer support database and process lab.

Areas for Improvement

(–) b It appears that Collins Technology reacts to customers' needs but does not lead the way to show the customer what is possible.

(–) a, b The process to determine key features, short term and longer term, was not clear.

Score: 50%

7.2 Customer Relationship Management *(65 points)*

Strengths

(+) a There are business segment managers and sales coordinators for each business segment (Government, Industrial, and Commercial) to establish a better understanding of the customer needs.

(+) b Service standards are developed through the TQM Forum based on several sources of input from the customer.

(+) c EDI, extended hours, and the customer support database allow the customers easy access to employees that have current information.

(+) d A phone call follow-up occurs within one week of a shipment or quote.

(+) e Formal and informal complaints are documented and communicated across departments.

(+) f Customer-contact employees are empowered to make decisions up to $10,000.

Areas for Improvement

(–) a The process used to identify indicators of the state of the relationship with the customers is not clear, i.e., there are key measures of quality but not how they were selected.

Score: 80%

7.3 Commitment to Customers *(15 points)*

Strengths

(+) a A one-page document defines the contractual conditions and product warranty.

(+) b A four-year warranty is now offered compared to industry norm of one year.

(+) b When the business segment managers meet annually, one of the items they review is commitment to the customer.

Areas for Improvement

(–) a Although it is stated that "…a quality product, on-time, at a competitive price" is what the customer wants, it seems the warranty only covers the first part. There does not appear to be an on-time delivery guarantee.

Score: 60%

7.4 Customer Satisfaction Determination *(30 points)*

Strengths

(+) a An annual written survey is conducted covering all customers (used to be top 85% of sales). Multiple departments of the customer are surveyed including purchasing, quality, engineering, and manufacturing.

(+) b Collins Technology asks their customers for comparison to the best.

(+) c Survey questions and measurement scales are reviewed with the customers in order to improve the instrument.

(+) c There is a high (85%) response rate to the survey.

(+) c For the last 12 years a customer has not been lost for competitive reasons. (*Note:* The wording of their statement makes you wonder what were the reasons for losing customers.)

Areas for Improvement

(–) a The phone call follow-up data (7.2d) were not integrated or used in determining customer satisfaction as reported here.

(–) a Collins Technology asks for feedback from their customer but not from the end user. This could be done through the customer.

Score: 70%

7.5 Customer Satisfaction Results *(85 points)*

Strengths

(+) a Customer satisfaction data, representing 85% of volume, show a positive trend for the last three years (Figure 7.3). Customer rating for major customers (Figure 7.4) reached 100%.

(+) b Field quality (Figure 7.5) and customer complaints (Figure 7.6) show improvements.

Areas for Improvement

(–) a Data were not available for all customers and by business segment.

(–) a The customer satisfaction data (Figure 7.3) are a composite of many different indices. It is not clear if the indices were weighted by importance or just averaged together.

Score: 70%

7.6 Customer Satisfaction Comparison *(70 points)*

Strengths

(+) a Comparison to competitors show favorable results.

(+) b, c Collins Technology has increased their market share in all three business segments.

Areas for Improvement

(–) a Some data seem confusing or contradictory. Product quality is shown to be at 3.4 ppm in the table but warranty is at 2500 ppm (¼%). Customer complaints reported in the table is 0.3%, but Figure 7.5 shows only ~ 99% product with no problems.

Score: 70%

11 | Site Visit

INTRODUCTION

Some of the Baldrige applicants receive a site visit. Generally the highest scoring 15 to 20 applicants are selected by the panel of Judges for a site visit. The purpose of the site visit is to verify and clarify information. It is the final basis for selecting recipients. Companies using the Baldrige Criteria internally typically follow the Baldrige process so they too include site visits conducted by their own employees. This chapter discusses the purpose and keys to effective visits.

The site visit issues identified for Collins Technology are presented for each Item at the end of this chapter. Often there are many more issues than time permits the Examiners to explore. Consequently the site visit issues should only contain "need to know" issues, not "nice to know" issues. Even the need to know issues have to be prioritized. The site visit issues are directly linked to a comment written in the consensus report.

 ## PURPOSE

A site visit allows Examiners the opportunity to get a better awareness of the applicant and to investigate those areas most difficult to understand from a description in the amount of space allowed by the Award process. Deployment is particularly hard to determine from an application. A site visit is not intended to replace the application as an information source; instead, it augments it. There are two primary purposes for a site visit: to clarify points of uncertainty and to verify the information contained in the application.

SITE VISIT PROCESS

The steps involved in the site visit process are notification, preplanning, final planning, conduct visit, prepare report, and send report. Although this discussion focuses on the national model, it also describes most state, business, and professional processes. Each step is explained briefly.

The panel of Judges decides who will receive a site visit. At this point in time, the Judges do not know the names of the candidates. To assure an unbiased selection, the Judges are given the scores and consensus comments about the applicant but not the applicants' names. Although NIST has nine Judges, a fewer number of judges seems to be typical for state and internal company award processes. Once the Judges have selected the applicants who will receive a site visit, the American Society for Quality Control (ASQC), on behalf of NIST, notifies the applicants and those Examiners who have reviewed their applications.

A selected applicant may decline the site visit—this option has been exercised. The reasons a company would decline a site visit are varied:

- The company may feel that it is only in the initial stages of its quality journey and a site visit may be interpreted by the employees as an end or "gold star." This could divert some of the interest or intensity in the change process.

- The company may feel that it needs another year to fully deploy its quality strategies and obtain additional trend and comparative data.

- The company has started a major product program and feels that a site visit, at this time, would be too disruptive.

- The company has encountered operational or other problems since the application submission and feels that they would be unable to effectively demonstrate their quality system at this time.

- The company has undergone a significant management change since the application submission and feels that they would be unable to effectively demonstrate their quality system at this time.

Of course, this is also a gracious way "out" for a company that has done an excellent job of describing its intentions, not its accomplishments, along with the careful use of analysis to support (future) conclusions in their application.

Prior to the site visit, the Examiner team does an extensive amount of planning by phone, fax, mail, or in person. Administrative arrangements associated with the site visits are handled by ASQC. The team leader is one of the Examiners who has been involved in assessing the candidate but who also has been designated as a Senior Examiner. The Senior Examiner acts as a facilitator but has no more influence over comments, score, and site visit issues than any of the other Examiners.

The team meets to complete the planning the day prior to the visit. At this meeting there is a review of site visit issues, the agenda is finalized, assignments are allocated among the Examiners, and forms are filled out. It is desirable to identify who in the company will be questioned to minimize the time impact on company personnel.

The team must maintain control over the flow of activities of the site visit. Regardless of the number and extent of presentations the company may have prepared, the Examiner team has the final decision over what to see or not to see. Because the purpose of the site visit is to "verify and clarify," many of the presentations on activities which the company is proud of, and has described adequately in the application, will be left "waiting in the wings."

From the authors' experience, a typical site visit goes as follows. The site visit begins with the team leader presenting the agenda and explaining the team's activities. The applicant provides a welcoming introduction containing material of their choice. These two activities might consume the first hour. Next, a tour of the facility is conducted. The team covers any site visit issues pertaining to Category 1.0 with the leadership immediately after the tour. Now the team breaks up into smaller groups. Throughout the site visit, the Examiner team needs to caucus to make midcourse adjustments. As issues are clarified and verified, other issues may arise. In addition to midday caucus meetings, each evening the team shares and reviews their findings with the whole team. The team is together to cover issues pertaining to customer satisfaction. This discussion may occur midway through the site visit. At the end of the site visit, a closing meeting is held with the host at the host location. Unlike an "end of audit" meeting, feedback is not provided regarding the team's findings. That will be contained in the feedback report. The purpose of the closing meeting is to provide an opportunity for the Examiner team to inform the organization of future events, to return all material received during the site visit, and to thank them for their cooperation and efforts.

Immediately following the site visit the team writes a report for the panel of Judges. This summarizes findings and states whether the score should be increased, decreased, or remain the same. The actual score, however, is not modified. The Examiner team also writes the feedback report comments. The Senior Examiner may write the scoring summary or may have a team member do it.

The feedback report, as described in Chapter 10, is then sent to the applicant.

The applicant will be contacted by phone to notify them whether or not they were selected as a recipient. For the national process, if the company is a recipient of the Award, the call is made by the Secretary of Commerce. If the company did not receive the Award, then the call is placed by the director of the Awards Office at NIST.

EXPECTATIONS: FROM THE APPLICANT

If an organization is selected for a site visit, it is natural that they would be excited and want to showcase their successes and supply the site visit team with any amenity to make their visit enjoyable. Organizations are disappointed when the team declines to see all the presentations and take all the tours that have been set up. The team is not there to "enjoy" themselves and to be "wined and dined." They are visiting the organization to verify and clarify specific points identified during the evaluation process—and they have only a short period to accomplish this. The best thing an organization can do to make this an efficient and effective process is to be open to all communications before the visit, understand what the team is expecting, and be flexible during the visit. Be prepared to share records and documents pertaining to the application. Provide a conference room for the Examiners to caucus among themselves. Do not try to obtain conclusions from the Examiners during the visit. The feedback report will be based on the comments and observations of all the Examiners. Knowing the reaction of one Examiner will not tell you what the final report will contain. In general, be open, courteous, and helpful. This will go a long way toward making the site visit pleasant for both sides. To assist in preparing for a site visit, an applicant will often make contact with an organization that had a site visit (via an internal, state, or national process) in a previous year. This allows the applicant to pick up some useful hints and confirm the details of the site visit process.

EXPECTATIONS: FROM THE EXAMINERS

The Examiners on the site visit team must understand that they are directing the visit—and they must communicate this effectively. This does not mean that there should be an adversarial relationship between the team and the applicant. On the contrary, the team is visiting the applicant in order to better understand the organization's systems and processes, and, if the organization warrants, become an advocate of the applicant to the Judges for an award. The best way the Examiner team can make things easier for both sides is to make sure everything is well planned and implemented (see Figure 11.1).

PLANNING

At the national level, the site visit planning involves the Examiner team, administrator of the Award (ASQC), the NIST observer, and the representative from the candidate company. At the initial meeting, via telephone, the general requirements of the site visit (to clarify and verify elements of the application) will be discussed so that all participants have a common understanding of the challenges facing them. Also discussed will be the expected conduct of the company and the

Figure 11.1
Keys for Examiners to an Effective Site Visit

- Preplanning

- Defined Examiner assignments

- Control of site visit agenda

- Thorough documentation of findings during visit

- Summation of findings each day and immediately following visit

- Clear recommendation to Judges

site visit team. The company must maintain an atmosphere of openness and cooperation, and resist the natural inclinations of exhibition, ceremony, or defensiveness. The Examiner team must avoid any actions that might compromise its ethics or impartiality.

The Examiner team, through a series of meetings facilitated by a Senior Examiner, will identify significant site visit issues and prioritize these issues in order to set assignments and identify key individuals for interviews and presentations. It is important for the site visit team to act and project as a team throughout the site visit. Without a well-detailed plan with clear assignments to Examiners, the site visit could rapidly become unfocused. To develop this plan, the team should develop a profile of the organization that includes a broad outline and key details, focus on factors that are important to the business and on outcomes that could make a significant difference, and anticipate what the Judges are likely to ask. The team should adhere to its plan in order to efficiently gather the information they need, as well as to prevent the organization from "skimming the cream." However, the plan should be flexible enough to allow the team to make midcourse corrections without appearing to be unorganized.

In this time frame the company's representative will provide the team with a list of key company contacts and relevant organization chart(s) and facility chart(s). The administrator will work with the team to establish the overall schedule as well as the schedule by category and by site.

 **IMPLEMENTATION**

The final planning occurs the day before the site visit begins, with the team meeting in the evening. At the off-site planning meeting, the team will continue with team building and coordination activities. The team should be in complete agreement on the issues, priorities, individual assignments, and site visit agenda. In anticipation of the intense schedule, all parties must allocate sufficient time for traveling, meetings, discussions, etc.

To cover all the site visit issues, many teams will review certain issues together as a team and cover other issues as individuals. The team can make use of its diversity to assign specific issues to the Examiner with the best background. A guideline for this is that the entire team should be present for discussions on Categories 1.0 and 7.0. The rest can be dealt with in teams of two: one Examiner discussing the issue with the organization's representative and the other Examiner taking notes. This way the team has at least two sets of "ears" for every issue considered. Two Examiners go to a specific department and resolve all relevant site visit issues pertaining to that department regardless of Category of origin.

During the site visit, the Examiner team will be following a schedule that is physically draining while immersed in an atmosphere that is psychologically draining. Companies will have done an enormous amount of preparation in anticipation of the visit, with many companies wanting to know too much in advance. This can lead to information overload or intimidation if the company is allowed to control the agenda. The general overwhelming reception of the Examiner team can cause feelings of guilt in the Examiners because they cannot visit all people who would like to be visited.

The competition between the original agenda and the need for midcourse corrections is often made more complicated by the company doing a real-time assessment of questions, issues, and Examiners, and targeting their responses accordingly. The result is fatigue and pressure—and long days.

Throughout the site visit, the Examiner team must conduct themselves in a professional and ethical manner. The Examiners are instructed not to discuss personal or team observations and findings; conclusions and decisions; detailed observations of applicant's quality program, regardless of whether they are complimentary or critical; quality practices or team observations about other applicants; and the names of other Award program applicants. Further, the Examiners may not accept any kind of gift or leave with any of the applicant's materials. However, the team is encouraged to compliment the company for the cooperation of any of its units and/or employees as well as for the thoroughness of its presentations.

SITE VISIT ISSUES

The following is an example of the site visit issues for the Collins Technology application, which were developed through the consensus process.

1994 Case Study:
Site Visit Issues for Collins Technology

Site visit issues are *not* provided as part of the feedback report. They are developed by the examining team as issues that may require verification or clarification, if a site visit is indicated. The issues are included here for the reader's benefit.

SITE VISIT ISSUES

Site Visit Issues—1.1

- Review the agenda and minutes from recent TQM Forum meetings. What actions were taken as a result of these meetings?
- Look at corrective action plans developed by senior executives to improve their leadership style.

Site Visit Issues—1.2

- Determine what was done by the TQM Forum when a department or unit did not meet its goals.
- Find out what information is specifically used to evaluate and improve how well company values are integrated into business operations.

Site Visit Issue—1.3

- Clarify the goals for public health and safety.

Site Visit Issue—2.1

- Look at the "Many-Point Data" checklist to see if owner, benchmarks, methods of tracking, frequency of reviews, and basis of calculations are specified.
- Clarify why "...stability, predictable internal capacitance and tolerance for extreme environment..." were not listed.

Site Visit Issues—2.2

- Determine who participates in benchmarking training and studies.
- Review information obtained from the customer regarding Collins compared to other suppliers. Is this information from multiple customers? How current is the information? How is it obtained?

Site Visit Issue—2.3

- Clarify what analysis is done on the data described in Item 2.1 and what correlation studies are done.

Site Visit Issue—3.1

- Walk through the strategic planning process used to develop the most recent plan. How were suppliers included? How were the corporate plans cascaded to departments? How was the process improved?

Site Visit Issue—3.2

- Review the plans that have been developed to achieve the short- and long-term goals. Goals were included in the application but not plans.

Site Visit Issues—4.1

- Review the HR plan. Does the plan contain specific strategies that address training, recognition, and mobility?

- Determine what opportunities are offered to salary employees for mobility.

Site Visit Issues—4.2

- Talk to operators. Have they shut the line down? Do they feel there is any penalty from management if they do stop the line?

- Cost of quality (CoQ) has decreased per Figure 6.19. Determine to what extent employee suggestions have contributed to this.

Site Visit Issue—4.3

- Talk to employees (salary and floor). What training have they had? Do they feel it is sufficient to do their jobs? Do they feel management supports the use of the newly acquired skill or education?

Site Visit Issue—4.4

- Review the "employee satisfaction with recognition" data sorted by Category to determine the deployment. What percent of employees does this satisfaction data represent?

Site Visit Issue—4.5

- Review the employee satisfaction survey. How is the survey distributed? What percent of employees respond (by Category)? What analysis is done? What feedback is provided to employees?

Site Visit Issues—5.1

- Clarify how the voice of the customer is translated. Walk through a recent new product development to see what tools were used and which departments were involved.

- Review the *Design Guide*. What changes have been made to it in the last few years?

Site Visit Issues—5.2

- Walk through the process that led to a resolved problem. Were the symptoms documented and quantified? Who got involved? What tools were used? What corrective action was taken? Were similar processes improved also?

- Talk to team members working on process simplification. What do they do, how often do they meet, what has happened to date?

Site Visit Issues—5.3

- Clarify if other support areas have an ongoing improvement effort under way. What have the efforts been to improve the LAN system?

- Talk to people in the support areas. What efforts have there been to increase quality? What training have they had? Are they familiar with the Collins seven tools? Do they use the six-step problem-solving process?

Site Visit Issue—5.4

- Review the *Total Excellence* procedure. What information does it contain?

- Determine what is covered in a supplier survey. Look at some recent surveys.

Site Visit Issue—5.5

- Establish what nonproduct assessments are done, by whom, how often, and covering what topics.

Site Visit Issues—6.1

- Investigate the benchmark levels. How were these numbers set? What are industry averages? The benchmark numbers reported seem low.

- Review data related to key factors other than cost.

Site Visit Issue—6.2

- Determine if other operations-related data are available. This might be for labor hours or inventory turns.

Site Visit Issues—6.3

- Discuss how the data in Figure 6.25 (five charts) are calculated. Does it represent 100% of the repairs/ bills/closures? Is it a sample?
- Are benchmarks available for some of the data?

Site Visit Issues—6.4

- Determine how the reported SRI is calculated. Is equal weighting given to all measures that comprise this composite?
- What data are tracked on the suppliers not in the top 30?

Site Visit Issues—7.1

- How are key features determined?
- Verify that the customer support database has the described information, is accessible, and up to date.
- Review R&D activities. Their application does not address their R&D activities. Is Collins Technology leading the customer or just reacting to customer needs?

Site Visit Issues—7.2

- Determine how the service standards were established. How does Collins know they are working on the right things regarding the customer relationship?
- Establish what information is collected in the one week follow-up phone calls.

Site Visit Issues—7.3

- Review the one-page warranty. Is it clear and easy to interpret?
- Is there an on-time delivery guarantee?

Site Visit Issues—7.4

- Review the current customer satisfaction survey. How does Collins Technology select the people who are to receive the survey (for example, who in manufacturing will receive it)? What analysis is done on the survey? How are the results correlated to internal measures?

- Does Collins have data from the end user regarding satisfaction?

Site Visit Issue—7.5

- Walk through the process of converting the raw survey data into a customer satisfaction index.

Site Visit Issues—7.6

- Clarify the various data that seem confusing. Product quality is reported to be 3.4 ppm but warranty is at 2500 ppm (¼%). The customer complaint percentage reported in the table is .3% but Figure 7.5 shows less than 99% of the product with no problem.

- Discuss the source of the comparison data. How current is it? How many customers provided input into it?

1994 Case Study:
Score Worksheet for Collins Technology

The individual and total scores would *not* be provided as part of the feedback. The applicant would receive in the feedback report summary only the range in which the total score fell (e.g., 601–750).

Examiner Name _____ *Sample Scores* _____

Applicant Name _____ *Collins '94* _____ Applicant Number _____

SUMMARY OF EXAMINATION ITEMS	Total Points Possible **A**	Percent Score 0–100% (10% units) **B**	Score (A × B) **C**
1.0 LEADERSHIP *95 POSSIBLE POINTS*			
1.1 Senior Executive Leadership	45	90 %	40.5
1.2 Management for Quality	25	70 %	17.5
1.3 Public Responsibility and Corporate Citizenship	25	60 %	15.0
Category Total	**95**		73.0
	SUM A		SUM C
2.0 INFORMATION AND ANALYSIS *75 POSSIBLE POINTS*			
2.1 Scope and Management of Quality and Performance Data and Information	15	80 %	12.0
2.2 Competitive Comparisons and Benchmarking	20	70 %	14.0
2.3 Analysis and Uses of Company-Level Data	40	50 %	20.0
Category Total	**75**		46.0
	SUM A		SUM C
3.0 STRATEGIC QUALITY PLANNING *60 POSSIBLE POINTS*			
3.1 Strategic Quality and Company Performance Planning Process	35	70 %	24.5
3.2 Quality and Performance Plans	25	60 %	15.0
Category Total	**95**		39.5
	SUM A		SUM C
4.0 HUMAN RESOURCE DEVELOPMENT AND MANAGEMENT *150 POSSIBLE POINTS*			
4.1 Human Resource Planning and Management	20	60 %	12.0
4.2 Employee Involvement	40	80 %	32.0
4.3 Employee Education and Training	40	70 %	28.0
4.4 Employee Performance and Recognition	25	80 %	20.0
4.5 Employee Well-Being and Satisfaction	25	60 %	15.0
Category Total	**60**		107.0
	SUM A		SUM C

Examiner Name ___*Sample Scores*___

Applicant Name ___*Collins '94*___   Applicant Number _____

SUMMARY OF EXAMINATION ITEMS	Total Points Possible **A**	Percent Score 0–100% (10% units) **B**	Score (A × B) **C**
5.0 MANAGEMENT OF PROCESS QUALITY	*140 POSSIBLE POINTS*		
5.1 Design and Introduction of Quality Products and Services	40	70 %	28.0
5.2 Process Management: Product and Service Production and Delivery Processes	35	60 %	21.0
5.3 Process Management: Business and Support Service Processes	30	40 %	12.0
5.4 Supplier Quality	20	50 %	10.0
5.5 Quality Assessment	15	60 %	9.0
Category Total	**140**		80.0
	SUM A		SUM C
6.0 QUALITY AND OPERATIONAL RESULTS	*180 POSSIBLE POINTS*		
6.1 Product and Service Quality Results	70	60 %	42.0
6.2 Company Operational Results	50	60 %	30.0
6.3 Business and Support Service Results	25	50 %	12.5
6.4 Supplier Quality Results	35	40 %	14.0
Category Total	**180**		98.5
	SUM A		SUM C
7.0 CUSTOMER FOCUS AND SATISFACTION	*300 POSSIBLE POINTS*		
7.1 Customer Expectations: Current and Future	35	50 %	17.5
7.2 Customer Relationship Management	65	80 %	52.0
7.3 Commitment to Customers	15	60 %	9.0
7.4 Customer Satisfaction Determination	30	70 %	21.0
7.5 Customer Satisfaction Results	85	70 %	59.5
7.6 Customer Satisfaction Comparison	70	70 %	49.0
Category Total	**300**		208
	SUM A		SUM C
GRAND TOTAL	**1000**		**638**
			D

For Supplemental Reports Only

	Total Score	×	Percent Sales	=	
Basic Report	_____	×	_____	=	_____
Supplemental Report 1	_____	×	_____	=	_____
Supplemental Report 2	_____	×	_____	=	_____
Supplemental Report 3	_____	×	_____	=	_____
Supplemental Report 4	_____	×	_____	=	_____
Supplemental Report 5	_____	×	_____	=	_____

12 | Getting Started: Organization and Follow-Through

INTRODUCTION

Many topics have been discussed so far, dealing with the philosophy, elements, and structure of the MBNQA Criteria and Award process. The emphasis has been on understanding and using the Criteria for self-assessment and improvement. Ideas were presented for successfully using the Criteria (Chapter 3). This chapter addresses the organizational issues of getting started. If a company wants to use the Criteria for self-assessment and improvement but needs some advice on organization and follow-through, this chapter will help.

 ## ATTITUDE

Use of the Baldrige Criteria to guide a quality initiative has many benefits. The Criteria provide a framework for the organization's total quality leadership system. It helps the organization achieve consensus on what needs to be done. By doing a self-assessment, energy can then be focused where it is most needed. Over time, the Criteria act as a stabilizing force to maintain the positive direction.

But is the organizational climate or culture receptive to doing an honest self-assessment? This is an important issue. If there is fear in the organization, the assessment may yield only what the boss wants it to say or what the employees think the boss wants it to say. The self-assessment should be viewed as an improvement tool—not as an audit.

Peter Senge, in describing attitudes toward a vision,[1] discusses various aspects of commitment. If an individual is *committed* to a vision or plan, they want it and will make it happen. If someone is at the stage of *enrollment*, they want the

[1] Peter Senge, *The Fifth Discipline*, Doubleday, New York, 1990.

vision and will do whatever can be done within the spirit of the law. The next level is called *genuine compliance.* This is where a person sees the benefits of the vision and is a good soldier, doing everything that is expected and more. *Formal compliance* involves seeing the benefits of the vision but only doing what's expected and no more. With *grudging compliance* the benefits are not recognized but the individuals do enough of what's expected to stay out of disfavor with their boss.

If the plan is to do an honest self-assessment and use the results for improvement, then an organization needs to have at least *genuine compliance* by all individuals involved. The commitment level is preferred but success can be achieved if the genuine compliance or enrollment level is obtained.

⊙━ Model for Change

A model for change that has been used to get people to the enrollment level is shown in Figure 12.1. It is not a mathematical formula; rather, it is a concept. There needs to be a dissatisfaction with the present system. If people are happy with the present system then the attitude is "why change?" A vision of the new state, of what can be achieved, is critical. If there is dissatisfaction but no vision of a better state, all we have created is frustration. The third piece of the model is the implementation steps. The steps necessary to start the journey need to be defined. With these three elements—dissatisfaction, vision, and implementation steps—the natural resistance to change can be overcome.

To create the dissatisfaction with the current ways or create the desire for something better, sharing facts and data help. Customer loyalty and satisfaction are not growing—this could be a reason for change. A process is taking seven weeks but only involves two hours of actual work—this could be a reason for change. The natural owners of a process are probably already aware of the need to do things differently. If not, looking outside your organization to see how outstanding companies perform may provide ideas and information that lead to change.

Figure 12.1
Model for Change
$$D * V * I > R$$

D = Dissatisfaction with the existing process; desire for something better

V = Vision of what can be accomplished

I = Implementation steps

R = Resistance to change

MODEL FOR GETTING STARTED

A macro model for getting started has been developed based on the premise that the self-assessment is to be used as an improvement tool, not an audit. The model is described here in narrative form and is represented in Figure 12.2 as a process flow. It is based on the assumption that a pilot (or multiple pilots) will be done before the assessment is carried out on the entire organization. For smaller companies this may not be necessary and could be modified accordingly.

This model is written *from the perspective of the facilitator* of the assessment process.

1. Conduct high-level training for the leadership of the organization to assure their understanding, buy-in, and participation. This training may be offered in four- to eight-hour sessions that include a workshop, allowing the executives an opportunity to read and critique the leadership portion of an outstanding company. The application used could be a case study such as Collins Technology or one of the studies used for Examiner training by NIST and sold by ASQC.

2. Identify an organization (business unit or division) that is willing to volunteer to be the pilot. Advise the volunteer organization—the applicant—of the magnitude of the commitment. It is essential that the managers of the volunteer organization be able to dedicate sufficient time to the self-assessment effort.

3. Develop a letter of understanding. The facilitator of the assessment process and the senior executive of the volunteer organization should write a short document identifying what will be done, the commitment required by the applicant, what the applicant should expect to gain from this activity, and how results will be measured. A timetable should also be developed.

4. Select the Examiners to participate on this particular assessment. The Examiners should be individuals who will not be part of the application development teams. The Examiners are to provide an unbiased review of the written response to the Criteria. This would be difficult if they are part of the information gathering and writing activities. Smaller companies may decide to bring in outside resources as Examiners.

   The Examiners should have been trained shortly before this event. If they have not had recent training, then conduct the necessary training to review these items:

   - Criteria

   - Examination process

   - How to write comments and site visit issues

- How to reach consensus
- How to write a feedback report
- How to conduct a site visit.

This training could be similar to what ASQC provides in their public offering, which is a three-day Baldrige assessment class.

5. Educate the applicant on the Baldrige Criteria. At a minimum, the people that are expected to be involved in writing the response to the Criteria should be trained. This training should cover the Baldrige Criteria and the scoring system and should include learning how to collect data and information already in existence in the organization, and how to keep the focus on self-improvement, not on an award should also be addressed.

It is recommended that others in the organization also be trained in the Criteria so they understand the road map that will be guiding improvement efforts. This training need not be as deep as the training for the key players.

6. Assist the applicant in its interpretation of the Criteria or other aspects where they seem to need guidance. Although it is primarily the applicant's responsibility to gather the information and write the application, the facilitator of the assessment process may provide clarification when it is requested by the applicant.

7. The applicant writes an application by providing responses to the Areas to Address from the Criteria.

8. Once the application is complete, each member of the Examiner team does an independent review. They each write comments and develop a score for every Item without discussing their opinion with the other Examiners (or others in the organization). The application should be treated in a confidential manner. The comments should typically be limited to three to five key comments per Item with more detailed Items resulting in five to eight comments.

9. The Examiners hold a consensus meeting attended by their team members only. Comments and scores are discussed and consensus reached. Site visit issues are developed and agreed on.

10. A site visit is held to clarify and verify the information from the application. Given the limitation in pages, the applicant may not have described a process or results in sufficient detail for an Examiner to draw a firm conclusion, so the site visit allows an opportunity to clarify information. The second aspect of the site visit is to verify information. Key strategies, activities, and results need to be verified.

11. The Examiner team writes a feedback report and identifies in which range (e.g., 401–600) the consensus score lies. The comments in the feedback report are of great significance to the applicant (or at least they should be), so the Examiners need to take care that the comments are accurate and reflect the key points.

12. After the applicant has had time to review the feedback report, a debriefing meeting should be held. During this meeting, the leadership, applicant, and the Examiner team should capture lessons learned. This should include discussion on what could be done differently to improve the assessment process.

Figure 12.2
Model for Getting Started

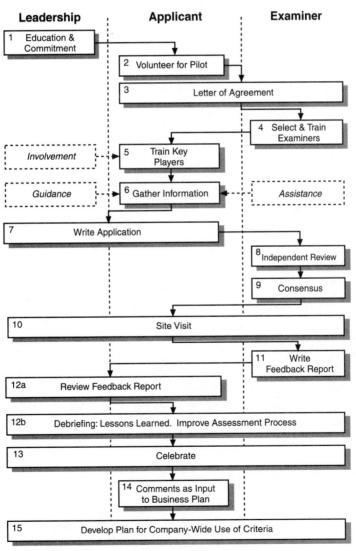

13. Take time to celebrate. The leadership, applicant, and Examiner team should celebrate together as a group.

14. The applicant should use the comments from the feedback report as input to its business plan. Areas identified as needing improvement should be addressed. Also, areas identified as strengths should be built on.

15. With assistance from the applicant and the Examiner team, the company should develop a plan for all units to use the Criteria for self-assessment and improvement.

PERCEIVED BARRIERS TO USING THE CRITERIA

Some people are quick to tell you why a plan won't work. To be prepared for that, consider the following common excuses and some ideas to overcome the obstacles of using the Criteria for self-assessment and improvement. Four common barriers are:

- Time required to prepare the written responses to the Criteria (the application)

- Fear of assessment findings

- No perceived need ("We're doing OK.")

- Lack of trained Examiners to do assessment.

The time constraint is the most common reason given for why a self-assessment using the Baldrige Criteria cannot be done. The plate is already full. We are too busy fighting fires to talk about fire prevention. In terms of the *model* for *change* (mentioned earlier in this chapter) these people either are not dissatisfied with the present system, do not share the vision, or do not understand the implementation steps. If time is not made for prevention, the fire fighting mode will continue. To reduce the amount of time spent and still get the organization to take the first small step, a pilot application based on 10 to 15 Items could be done. In the authors' opinion, based on the 1995 Criteria, the key Items to address would be Items 1.1, 2.3, 3.1, 4.1, 4.4, 5.1, 5.2, 6.1, 6.2, 6.3, 7.1, 7.3, 7.4, and 7.5.

Another idea to address the concern about limited time is to look for business units that are most interested—those that share or at least understand the vision. This does not shorten the time needed, but these organizations would be more receptive to doing a self-assessment. As they begin to have success, other units may become interested.

The response about time, or lack of it, may be a reflection that the assessment is seen as an exercise. This can be countered by stressing the benefits of a self-evaluation and improvement plan.

Fear is the second most common reason people may oppose a self-assessment. They often do not label it as such, but when their resistance is explored, that is the basic concern—fear. There is fear that their "dirty laundry" will be exposed, and fear that change from the status quo will be needed. One way to minimize this is to keep the results internal to the unit. Scores should not be posted for individual units. Keep the focus on using results for improvement not comparison. Minimize the adverse consequences of low scores in the early years of using the Criteria.

The third reason someone may be opposed to doing a self-assessment is the perception that there is no need—i.e., there is no dissatisfaction with the present system. Responses like "Everything is great" or "If it ain't broke, why fix it?" are indications of this barrier. One way to change that perception is to read, circulate, and discuss articles by companies who are Award recipients. Attend a conference that includes participation by Award recipients such as the annual Quest for Excellence conference held every February. Learn what leading companies are doing so people have a sense of the "height of the bar." They need to know what the new levels of expectation are. Benchmarking your company against others considered to be outstanding could also open people to new ideas. However, as described later in this chapter, you should study your own process before benchmarking others. Benchmarking should not be the first step in responding to this barrier.

The last excuse—the lack of trained Examiners—is an easy one to address. The easiest way to overcome this is for the company to support an employee to become an Examiner for the national Award. This will get at most three people trained in any given year at the national level.[2] Alternatively, send people to training focused on developing internal Examiners such as the ASQC training.[3] Another easy solution is to contract a national Board of Examiner member to support your effort of developing internal experts. In the state of Michigan when a state award was developed and patterned after the MBNQA, 37 people were identified in the state that are or were on the Board of Examiners for the MBNQA. Michigan trained 60 Examiners. Many other states have similar numbers to report.

---

[2] NIST allows a maximum of three people from one company to serve on the Board of Examiners in any year. The limit is one per company for consulting firms.

[3] The American Society for Quality Control (ASQC) is an international professional society consisting of more than 120,000 members. Services provided include training. Their headquarters is in Milwaukee, WI, 800/248-1946.

BENCHMARKING

Benchmarking information is requested in several Categories. It is being presented here not only as a method to get new ideas in general for use in the company but also as a way to get people to see the benefits of using the Baldrige Criteria for self-improvement.

Benchmarking is a learning process. It provides an opportunity to learn how others have accomplished a task in an efficient and effective manner. Two major issues are often overlooked with regard to benchmarking. First, before ideas from other companies are sought, it is critical that you understand the methods and performance of your own system. Second, in learning from others, seek an understanding of how and why. From this you can incorporate what will optimize your system. Copying is not the answer. What worked for one company may not work for another.

Some of the benefits of benchmarking include increased customer satisfaction through maximum value products and services. Because the process begins by studying your own activities, it becomes a self-evaluation, which can lead to new insights. In studying an activity, for example, order entry, and discovering there are six hand-offs of information, it is no surprise to learn that time is lost and errors are made.

Benchmarking can assist the company's efforts to change. Suppose an organization is getting grudging compliance (an attitude of doing something only because there is no choice as described earlier in this chapter) when trying to use the Baldrige Criteria for self-improvement. The organization may want to benchmark other companies so the people resisting the effort see firsthand how the Baldrige assessment effort has benefited others.

Benchmarking outside of the company provides neutral examples. Through benchmarking, technological breakthroughs, cultural issues, and innovative applications of tools can be identified in other organizations.

Benchmarking Pitfalls

The late Dr. W. Edwards Deming called the Baldrige Award nonsense.[4] One might get the impression that he saw no redeeming value in it. Not true. He did get upset with some issues. Benchmarking was one of those issues. When one of the authors of this book was a Baldrige Examiner, she and a Judge who was also a student of

---

[4] Conversations with Dr. Deming resulting in a letter dated September 20, 1991, of recommended changes to NIST.

Dr. Deming took the opportunity to discuss the Baldrige Criteria with him. There were several major issues that did cause Dr. Deming concern. The misuse regarding benchmarking was one of them. Some concerns about benchmarking were expressed in Chapter 4 of this book but others are explained here.

To copy an idea, method, practice, or system does not assure success. In the 1970s, many people were going to Japan to "learn the secrets." A common conclusion brought back was the notion of quality circles. The concept was "If we could just somehow get our employees to submit more ideas, we would be a better company." The people drawing this conclusion did not realize there was a business climate, a business (not ethnic) culture, that needs to be in place in order for the employee ideas to even be heard. It makes no sense to put employees together to work on issues if they are not trained to use data and draw conclusions. Typically, the data were not even available to them. It makes no sense to ask employees for input on how to make processes better if the orders from the top require the business to take a short-term focus and only put out fires. Mixed messages are sent when employees are asked to work together in teams but the recognition system is still promoting the lone ranger, the superstar as the hero. It makes no sense to ask employees how to improve a process when the process may not need to exist. For example, employees are asked "How can we shorten the cycle time on expediting parts?" It would be better to ask why do we need to expedite—is the system so slow that expediting is becoming the norm? Ask an automotive supplier what its air freight bill is for expediting special orders. You will wonder how they stay in business with so much being spent on express delivery of parts. Expediting seems to be the norm in some companies; i.e., getting the work done in spite of the system rather than because of the system.

At one company, accounting had to post "walk through" hours; specific hours when people could walk through their purchase order. It had become the norm. No one followed the regular process because it took weeks or months to get the purchase order. If you walked it through by hand carrying the paperwork to the eight people that needed to process it, then it only took a week. "Why do eight people need to touch this paperwork?" or "Is paper even required?" are better questions than "How can we get the paper through these eight people faster?"

It is a good idea to look outside your company and outside your industry to get new ideas but don't expect to find the "one size fits all" answers that you can copy.

A second pitfall to benchmarking is the potential to become a follower not a leader. If a business waits for someone else to try out an idea and then wants to copy it, they will find themselves always trying to catch up.

A third concern can be compared to the "forest and trees" view. In benchmarking other companies, people may observe the trees but not see the forest, the bigger picture. A systems view is important. Observing and understanding the tools being used is not enough. The bigger picture needs to be understood.

Some people think benchmarking should be restricted to their industry, otherwise there is an apples to oranges comparison. If the study is restricted to their industry, the big fish in the little pond feeling may be promoted. A company may observe they are the leader, the big fish, but the pond where they looked was small. If a company wants to improve its process for tracking parts, and material then UPS or Federal Express may be a useful company to study. They track packages as small as $8\frac{1}{2}$ by 11 inches and know where the package is at all times. Some companies have trouble tracking the location of a product that is many times larger than that.

Another pitfall is to think that benchmarking is just a buzzword, a trend that will soon pass. Employees who think this may go through the motion as if they are attempting to gather some information but they are really just passing time. This situation can occur when the vision of continual improvement is not fully understood. In this case benchmarking is considered as a competing task rather than an activity which helps and interfaces with the total self-assessment and improvement processes. This view of benchmarking may be reinforced if management focuses on getting and using only the measures, rather than understanding how and why the measurable results came to be.

If a company has a benchmark team that gathers the information for everyone else in the company, this should be a red flag of trouble. The natural owners of the process should be involved in the study. It is all right to have a specialist as a resource; someone designated as the benchmark champion to be actively seeking information through professional organizations and journals. But the involvement by those who will be using the information is critical. The pitfall of limited participation becomes obvious when the good ideas learned are not embraced by the organization. There is no buy-in to the ideas without understanding.

In desperation managers look for quick results to solve a long-term existing problem only to find there is no quick fix. Whenever there is an unreasonable urgency, or a panic, to find a solution to a problem via benchmarking, there is a good chance the benchmarking effort will fall short of its potential and expectations. In this "just don't stand there, do something" mode, the benchmark team can forget that the major part of their activity is to plan (i.e., Plan-Do-Study-Act—see next section) and understand the processes.

О⊞   Focusing on the metrics (the numbers), not the understanding of the practices (the "why") will also result in limited success. Getting a new goal and holding it out for the old system to meet is going to cause frustration. Learning that the accounting departments at some companies can close their books in one day and making this your goal without changing the current accounting system will cause frustration. If your accounting department could have closed the books in one day under the old system, they would have. Goals, without the means to achieve them, will also lead to distortion of the facts. If people are recognized and rewarded based on meeting a goal, data will often be forced to show the goal was met.

О⊞ Benchmarking Steps

There certainly are a lot of potential pitfalls. What are the steps to consider to be successful in obtaining new ideas and information from a benchmarking activity? The steps can be presented in the cycle of Plan-Do-Study-Act (PDSA).[5]

Plan

- Identify key processes

- Establish "why"—what will be done with benchmark information

- Identify "who" (generic) will use information

- Meet, discuss, educate users

- Determine commitment and sense of urgency

Do

- Select business function to be benchmarked

- Develop operational definition of business function including all requirements, e.g., timing, productivity

- Select people for benchmark study

- Analyze your process

Study

- Study your process findings

Act

- Meet with benchmark study group to establish second PDSA

---

[5] The PDSA cycle is discussed in Chapter 10.

With the completion of the first cycle of PDSA, the team may have identified several areas for improvement. This information should be passed on to a local process improvement team as part of its ongoing continual improvement activities. The second PDSA cycle is comprised of these steps:

Plan

- Identify companies to study

- Determine data to be collected

- Develop operational definitions, including all requirements

Do

- Collect the data utilizing research, phone calls, site visits

Study

- Study best practices

- Determine current and predicted gap

- Determine root causes of their success

- Identify "enablers" of the process

- Identify "inhibitors" in your process

Act

- Communicate findings

- Integrate findings into your strategies for improvement

- Establish timetables and activities

With the completion of the second cycle of PDSA, the team will have identified stretch goals as well as the means to achieve them. Implementation, with reviews and feedback, is now the focus of the PDSA.

Plan

- Establish timetable for ongoing reviews of progress

- Determine timetable for updating benchmark information

Continue on with the Plan-Do-Study-Act cycle.

Benchmarking Sources of Information

Benchmarking involves getting information and new ideas about best practices. This can be done through the use of library literature searches, material from the source of interest, electronic media searches such as the Internet, talking to industry experts, reviewing material from professional societies, partnerships with others brought together due to a common interest, or on-site visits. Oftentimes people think of the last point as being the essential aspect of benchmarking. Actually, if the other sources are used extensively, the on-site visit may not add anything substantially.

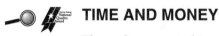 ## TIME AND MONEY

Through our consulting practice we hear excuses about why something is or is not going to work. With the Baldrige process, the story we have heard too many times is that it takes too much time and costs too much money to do such a thorough assessment. If someone begins to tell us it costs tens of thousands of dollars to do the assessment, we ask them to clarify this amount. Often, the total amount includes not only the assessment activities, but also the improvement activities. Some people are of the opinion that "If we were not doing the MBNQA assessment, we would not be making these improvements. Therefore, the cost of the improvements should be part of the MBNQA budget." The key point is not where the improvement budget should reside, but that all persons involved realize that this is a long-term investment with most of the costs (and all of the returns) coming from the improvements being implemented.

It may cost a company a significant amount of time to make the improvements that are identified. The amount of effort required depends on how far the company is on their journey. A company that has been successful operating "from the gut" and now wants to collect and use specific data is going to invest time and money in establishing such a system. To give the reader a sense of the time and money invested, the following information is offered.[6]

Different companies track their investments in different ways. Some companies may have included some of the effort that went into making improvements while others reported just the assessment effort and others considered the assessment as part of their business process and did not track it separately. The amount of effort expended is not a function of company size. The following is a list of the Award recipients indicating the number of full-time equivalent employees and the time and money investment when reported by the recipient. Note also that only 4 out of the 22 Award recipients have 40,000 or more employees while 10 organizations have 2500 or fewer employees, indicating that the Baldrige process is not just for the "big boys."

[6] This information was obtained through public material or public forum.

Globe Metallurgical *(Recipient—1988)*

- 210 employees

- Application developed by one vice president working over one weekend

Westinghouse Commercial Nuclear Fuel Division *(Recipient—1988)*

- 2000 employees

Motorola *(Recipient—1988)*

- 99,000 employees

Xerox Business Products *(Recipient—1989)*

- 50,200 employees

- 26 people working six months full time

Milliken & Company *(Recipient—1989)*

- 14,300 employees

IBM Rochester *(Recipient—1990)*

- 8100 employees

- 12 people off site 21 days to write the self-assessment

Cadillac *(Recipient—1990)*

- 10,000 employees

- 24 people "part time" for a year (although the people were told this was a part-time activity, for many it became full time—in addition to their regular assignments)

Federal Express *(Recipient—1990)*

- 90,000 employees

Wallace Company, Inc. *(Recipient—1990)*

- 280 employees

Solectron *(Recipient—1991)*

- 2100 employees
- Total of 800 hours for application development

Zytec (Recipient—1991)

- 748 employees
- Total of 767 hours for application development in 1990
- $9000 including application fee and trip to Motorola and Xerox for assistance in the Baldrige process
- $8300 in 1991 prior to site visit

Marlow Industries (Recipient—1991)

- 160 employees

AT&T Transmission (Recipient—1992)

- 12,000 employees
- Applied 1991 (received site visit) and 1992
- 20 people working part time for a year

TI Defense and Electronics (Recipient—1992)

- 15,000 employees
- Applied 1990, 1991 (received site visit) and 1992
- Senior management plus 50 employees

AT&T Universal Card Services (Recipient—1992)

- 2500 employees

Ritz-Carlton Hotel (Recipient—1992)

- 11,500 employees

Granite Rock (Recipient—1992)

- 400 employees

Eastman Chemical *(Recipient—1993)*

- 17,750 employees

- Applied 1988 (received site visit)

Ames *(Recipient—1993)*

- 445 employees

AT&T Consumer Communications Services *(Recipient—1994)*

- 44,000 employees

GTE Directories Corp. *(Recipient—1994)*

- 5150 employees

Wainwright Industries, Inc. *(Recipient—1994)*

- 275 employees

SUMMARY

This chapter has pulled together many of the topics discussed so far in the book by addressing how to get started. The importance of the attitude and culture were discussed including the notion of commitment and compliance. A model for change was presented. Advice to candidates and a process flow were included.

Common barriers were identified and ways to overcome them were presented. Details pertaining to benchmarking were discussed. Some facts and figures for time and money spent by recipients were shown. For a company that wants to use the Criteria but needs some advice on first steps, this chapter should have been helpful.

13 | Comparison Between the MBNQA Criteria and the ISO 9001 Standard

INTRODUCTION

A comparison between the 1995 Malcolm Baldrige National Quality Award (MBNQA) Criteria and the International Organization for Standardization (ISO) 9001—1994 standard[1] is discussed in this chapter and illustrated in figures. Any comparison will be subjective due to the descriptive nature of MBNQA and the prescriptive focus of ISO 9001. In the figures, the interrelationships are indicated by listing the Items across from each other. These relationships are shown if the Criteria and ISO standard had something in common. This correspondence is not meant to indicate equality of requirements but that the specific element is addressed by both documents. Typically the ISO 9001 standard is a subset of the MBNQA Criteria and is only one example of how a portion of the Criteria could be met.

The main overlap between the ISO 9001 standard with the MBNQA Criteria is in the Criteria's Category 5.0, Process Management, which addresses the methods and processes for design, production, delivery, support areas, and the management of supplier performance.

---

[1] While the standard is referred to as ISO 9000 in the general literature, the ISO standard 9001, 9002, or 9003 are the requirements standard. Conformity to ISO 9000 standard refers to the use of one of these three standards in third-party registration or other assessment activities. This chapter focuses on ISO 9001 because it has the broadest scope.

BACKGROUND

The ISO[2] is a worldwide federation of the national standards bodies of 91 countries. The American National Standards Institute (ANSI) is the member body representing the United States. ISO was founded in 1946 to promote the development of standards and related activities to facilitate the exchange of goods and services worldwide.

The ISO is made up of 180 technical committees, each responsible for an area of specialization. In 1979, ISO formed Technical Committee 176 on Quality Management and Quality Assurance to harmonize the increasing international activity in quality system requirements. On the basis of the work of this committee, the ISO 9000 Standard Series on quality management and assurance was published in 1987. In that same year, the United States adopted the ISO 9000 series as the ANSI/ASQC Q90 series.

Genealogy[3]

In 1959, the U.S. Department of Defense (DoD) established a Quality Management Program designated MIL-Q-9858. It was revised four years later to MIL-Q-9858A, which is still current today. In 1968, the North Atlantic Treaty Organization (NATO) essentially adopted the provisions of MIL-Q-9858A in the form of the Allied Quality Assurance Publication-1 (AQAP-1). In 1970, the Ministry of Defence of the United Kingdom adopted the provisions of AQAP-1 as its Management Program Defence Standard DEF/STAN 05-8. In 1979, the British Standards Institute (BSI) developed the first commercial quality management system standard—BS 5750. From these predecessors, ISO created the ISO 9000 Standard Series, adopting most of the elements of BS 5750.

ISO 9000 STANDARD SERIES

Although it is not the intent of this book to give a detailed description of ISO 9000, a brief synopsis is provided. The ISO 9000 series of international standards (see Figure 13.1) embodies a rationalization of the many and various national approaches to the development of quality systems. These standards and guidelines were developed to complement relevant product or service requirements given in the technical specifications. They are generic, not specific to any particular product or industry.

Third-party auditing—registration by a body providing recognition of conformity to the ISO 9000 standard—is applied in at least 32 countries. In the United States, the American Society for Quality Control (ASQC) maintains a Registrar

---

2 ISO is not an abbreviation of the name of the organization The International Organization for Standardization; it is from the Greek prefix *iso* derived from the Greek *isos*, meaning equal (from "Why the Term 'ISO?'", ISO 9000 Journal, Vol. J-3,1994).

3 Based on NIST reports NISTIR 4721 and NISTIR 5122 by Maureen Breitenberg, April 1993.

Figure 13.1
ISO 9000 Standard

ISO Standard	Description
9000	Quality Management and Quality Assurance Standard
Part 1	Guidelines for selection and use (1987, revision 1994)
	The purpose of these documents is to
	(a) Clarify the distinctions and interrelationships among the principal quality concepts, and
	(b) Provide guidelines for the selection and use of a series of standards on quality systems that can be used for internal quality management (ISO 9004) and for external quality assurance (ISO 9001, 9002, 9003).
Part 2	Generic guidelines for the application of ISO 9001, 9002, and 9003 (1993)
Part 3	Guidelines for the application of ISO 9001 to the development, supply, and maintenance of software (1991, reissue 1993)
Part 4	Application for dependability management (1993)
9001	Quality Systems—Model for Quality Assurance in Design/Development, Production, Installation, and Service (1987, revision 1994)
9002	Quality Systems—Model for Quality Assurance in Production and Installation (1987, revision 1994)
9003	Quality Systems—Model for Quality Assurance in Final Inspection and Test (1987, revision 1994)
9004	Quality Management and Quality Systems Elements
Part 1	Guidelines (1987, revision 1994)
Part 2	Guidelines for services (1991, reissue 1993)
Part 3	Guidelines for processed material (1993)
Part 4	Guidelines for quality improvement (1993)
Part 5	Guidelines for quality assurance plans (draft international standard)
Part 6	Guide to quality assurance for project management (working draft)
Part 7	Guidelines for configuration management (draft international standard)

Accreditation Board (RAB) to accredit registrar companies and certify auditors. An organization can have its quality system audited to ISO 9001, 9002, or 9003 by an accredited registrar. If its documented quality system is in conformance with the appropriate standard, and operating practices conform to the requirements, the organization will receive a certificate of registration (i.e., will be "certified" by the registrar). This registration can be used as evidence of compliance to the ISO standards to satisfy specific customer requirements.

ISO 9000 registration schemes are intended to provide a common basis for assuring buyers that specific practices, including documentation, are in conformance with the supplier's stated quality systems. In addition to registration to meet customer requirements, some organizations use the ISO 9000 standard to bring basic process discipline to their operations.[4]

Is ISO 9000 Required?

One of the most common misconceptions about ISO 9000 is that registration is required to do business within the European Community (EC). Registration is not a legal requirement—even for regulated product groups (see Figure 13.2). Among the standards specified by the EC directives for these groups is a quality assurance standard, but not specifically ISO 9000 registration. ISO 9000 registration will satisfy the standard, but alternatives are also outlined.

The globalization of ISO 9000 is being driven by commercial requirements, not legal fiat. International companies, such as Siemens, Motorola, Xerox, and Du Pont, are encouraging their suppliers to seek registration. The American automotive companies, Chrysler, Ford, and General Motors, are using ISO 9001—1994 as the foundation of their joint *Quality System Requirements—QS-9000*. This document defines the companies' quality system expectations for internal and external production and service parts suppliers. Further, the companies have harmonized

Figure 13.2
Regulated Product Groups

Regulated Product Groups

- Active implantable medical devices
- Construction equipment
- Electromagnetic compatibility
- Gas appliances
- Low-voltage equipment
- Personal protective equipment
- Simple pressure vessels
- Telecommunications terminal equipment
- Toys
- Weighing machines

Product Groups Proposed for Regulation

- Amusement park and playground equipment
- Cable ways
- Equipment used in explosive atmospheres
- Furniture flammability
- Lifting appliances
- Measuring and testing instruments
- Medical devices
- Recreational craft
- Used machinery

[4] C. Reimann and H. Hertz, "The Malcolm Baldrige National Quality Award and ISO 9000 Registration: Understanding Their Many Important Differences," NIST, Gaithersburg, MD, 1993.

supplemental and unique quality systems requirements in order to reduce the administrative and paperwork burdens on their suppliers.

In addition to industry, governmental agencies are also adopting the ISO 9000 family. In the United Kingdom, the Ministry of Defence has revised DEF/STAN 05-8 to reflect the provisions of ISO 9001–9004 (and renumbered to DEF/STAN 05-21 through 24). Within the United States, government agencies, such as DoD, Food and Drug Administration, and the Department of Energy, have either adopted or are considering adopting the ISO 9000 series. Other agencies, such as the Federal Aviation Administration, General Services Administration, NASA, and the Nuclear Regulatory Commission, are looking into the usefulness of the ISO 9000 standards within the context of their regulatory and procurement programs.

COMPARISONS

With the global emphasis on the ISO 9000 series, a natural question is how does the ISO 9000 standard compare to the MBNQA Criteria? First of all, any comparison will be subjective due to the descriptive nature of MBNQA versus the prescriptive focus of ISO 9000 series.[5] The ISO 9001 standard is used in the following comparisons since it is the most comprehensive of the three standards to which a company can be registered.

Curt Reimann and Harry Hertz state that:[6]

the ISO 9001 requirements address less than 10% of the scope of the Baldrige Criteria, and do not fully address any of the 24 Criteria Items [in the 1993 and 1994 Criteria]. All the ISO 9001 requirements are within the scope of the Baldrige Award. Due to the broader nature of the Baldrige Criteria and assessment, a rigorous audit of a printed quality manual and compliance with its procedures does not occur during a Baldrige assessment.

As a measure of overlap, the ISO 9001 requirements could result in approximately 250 out of 1000 points.[7]

---

[5] Some people feel that the word "prescriptive" has negative connotations, especially when used in reference to the ISO 9000 series. This is not our intent. Rather, we are attempting to identify the focus of the standards. One of our reviewers, Edwin Shecter thinks that control, not prescription, better describes the focus of the ISO series. He states that "ISO [standards] seek control while Baldrige seeks improvement."

[6] *Ibid.*

[7] R. Majerczyk and D. DeRosa, "ISO 9000 Standards: The Building Blocks for TQM," *48th AQC Proceedings*, ASQC, Milwaukee, WI, 1994.

On the other hand, an ISO audit reviews an organization's quality system with a depth and detail that far exceeds the MBNQA review process. It would be possible for an organization to fail an ISO audit while doing well on a Baldrige assessment. The difference would be the completenesss of the deployment and maintenance of documentation, or lack therof, in the areas common to the MBNQA and ISO 9000 series.

 ## Major Concepts in MBNQA Not Addressed in ISO 9001—1994

The major differences between the MBNQA Criteria and the ISO 9001 standard lie not in their individual elements but in the core values and concepts (see Chapter 1 for details of Baldrige key concepts).

Customer-Driven Quality, Leadership, Long-Range Planning

The primary focus of the MBNQA Criteria is customer-driven quality. Customer-driven quality is a strategic concept and must be linked to long-range planning. The MBNQA Criteria consider quality management as part of the total business system whose goals include achieving customer delight. It emphasizes quality system alignment and proactive continual improvement cycles at all levels and in all parts of the company. Extensive leadership involvement in the quality effort is fundamental. The senior leaders of a company are expected to create and constantly and consistently reinforce a customer orientation with clear and visible quality values and expectations.

The scope of ISO 9001 includes the concept of customer satisfaction:

> The requirements specified in this International Standard are aimed primarily at achieving customer satisfaction by preventing nonconformity at all stages from design through servicing.

However, the rest of the standard reflects its role as minimum requirements for a quality assurance system. Further, the standard does not require a long-range outlook, relying instead on the periodic audits of the registration process.[8] Although ISO 9000 registration is frequently becoming a customer requirement, the ISO 9000 registration system further limits the relationship of quality and the customer:[9]

---

[8] Most registrations are valid for a three-year period and require a complete reassessment for continued registration. Some registrars also offer surveillance auditing, usually on a six-month schedule.

[9] C. Reimann and H. Hertz, "The Malcolm Baldrige National Quality Award and ISO 9000 Registration: Understanding Their Many Important Differences," NIST, Gaithersburg, MD, 1993.

The ISO 9000 registration focus on conformity means that it does not fully address either customer-driven quality or the meaning of quality given in ISO 9000 guidance documents. ISO 9000 registration addresses a narrower meaning of quality, a meaning that includes only one aspect of quality—consistency in the production of a product or service. This distinction is important, particularly for service organizations or for manufacturers which seek service and/or relationship differentiation from their competitors based upon overall *value* delivered.

Although "management with executive responsibility for quality" is required to define and document the quality policy and ensure that the quality system "requirements are established, implemented, and maintained," personal involvement by the senior executives is not part of the standard. That is, the standard requires management's authorization—not their commitment and involvement.

Continual Improvement

The Criteria include interrelated learning cycles. Methods for selecting data, benchmarks, and data analysis are interrelated and support the total business planning process and plans. This reflects the emphasis of continual improvement in the Criteria. This includes both incremental and breakthrough improvement in the overall operations and in every process.

ISO 9001 does not contain a requirement for a plan for continual quality improvement.[10] There are elements of improvement and prevention contained in the standard, but primarily as reaction to nonconformity. In fact, if an organization uses traditional approaches to quality procedures, continual improvement could be impeded by the system. With the traditional model, all activities are defined in the procedures; even to the level of specifying operating instructions and sample sizes and frequencies for monitoring activities. Any change in the process would require formal procedural changes. With continual improvement we want the process to change—to improve. Under the traditional scenario, either the process would not be changed to avoid the paperwork, or the change may be implemented and the procedural paperwork neglected until the next audit.

Yet the standard can positively impact the continual improvement process. AT&T uses the MBNQA Criteria for its internal assessments for process improvement but also uses the ISO 9000 standard. They feel that "ISO 9000 compliance can be an excellent start for that journey to a world-class quality system. ISO 9000 requires that an organization define and document the way it does business. This is essential since you cannot continually improve your processes if you do not

---

10 There is such a clause on quality improvement in the ISO 9004-1—1994 guidelines.

know what they are to begin with."[11] Robert Peach notes that "while the capability of ISO 9001 to provide a *foundation* for quality improvement is recognized, a review of the content of the ISO 9004-4 "Guidelines for Quality Improvement" demonstrates that much more is needed to accomplish continuous improvement than the minimum requirements of ISO 9001."[12]

Employee Participation and Development and Partnership Development

Also reflected in the dynamic linkages among the Criteria requirements are the core values of employee participation and development—to enable employees to develop their full potential; and partnership development—to create win–win situations among all groups involved. This goes beyond work skills training, as required in the ISO 9001, and includes employee satisfaction, involvement, and recognition as well as partnerships between labor and management between the supplier and the company, and between the customer and the company. ISO 9001 addresses human resource issues solely through the training requirement— procedures for identifying training needs, qualification of personnel performing specific tasks, and maintenance of appropriate records of training.[13]

Fast Response

ISO 9001 does not address the core value of fast response–cycle time reduction, nor the relationships of cycle time to productivity and quality.

Corporate Responsibility and Citizenship

ISO 9001 does not address public responsibility or corporate citizenship.

Management by Fact/Results

Finally, the Criteria are directed toward results as well as approach and deployment. They ask for data showing positive trends and favorable comparisons to industry leaders and benchmarks throughout the organization and customer satisfaction results showing positive trends and favorable comparisons to industry leaders and benchmarks.

> ISO 9000 registration does not use outcome-oriented results or improvement trends in the assessment process. Thus, registration does not require demonstration of high quality, improving quality, efficient operations, or similar levels of

[11] R. Majerczyk and D. DeRosa, "ISO Standards: The Building Blocks for TQM," *48th AQC Proceedings*, ASQC, Milwaukee, WI, 1994.

[12] R. Peach, "Planning the Journey from ISO 9000 to TQM," *48th AQC Proceedings*, ASQC, Milwaukee, WI, 1994.

[13] C. Reimann and H. Hertz, "Malcolm Baldrige National Quality Award and ISO 9000 Registration: Understanding Their Many Important Differences," NIST, Gaithersburg, MD, 1993.

Figure 13.3

Major Concepts in MBNQA Not Present in ISO 9000—1994

- Proactive continual improvement
- Extensive leadership involvement in quality effort
- Methods for selecting benchmarks and data analysis
- Business planning process and plans
- Employee satisfaction, involvement, and recognition
- Data showing positive trends and favorable comparisons to industry leaders and benchmarks throughout the organization
- Customer focus through service standards, customer satisfaction, and proactive systems
- Customer satisfaction results showing positive trends and favorable comparisons to industry leaders and benchmarks

quality among registered companies… The ISO 9000 assessment involves an evaluation of the organization's quality manual and working documents, a detailed site audit to ensure conformance to stated practices and periodic re-audits after registration. The ISO 9000 focus is on the documentation of a quality system and on the conformity to that documentation.[14]

Major Concepts in ISO 9001—1994 Not Explicitly Addressed in MBNQA

The most visible characteristic in ISO 9001—1994 is the requirement to "establish and maintain documented quality systems manual and procedures." Because of this, document control is directly or indirectly interrelated to most of the standard's requirements. Although the Criteria do not explicitly address this area, some might say that good management processes would yield the same results. However, a company could have quality excellence and be highly innovative (e.g., be a MBNQA Award recipient) and still have difficulty passing an ISO 9001 audit.[15]

Figure 13.4

Major Concepts in ISO 9000—1994 Not Explicitly Addressed in MBNQA

- Document control
- Product identification and traceability
- Receiving and final inspection and test requirements
- Control of nonconforming material

---

14 *Ibid.*

15 For example, see J. L. Lamprecht, *Implementing the ISO 9000 Series*, Marcel Dekker, New York, 1993.

Some concepts addressed by the standard show its assurance focus and prescriptive nature. These include product identification and traceability, receiving and final inspection and test requirements, and control of nonconforming material. These activities are "safety nets" and reaction plans and reflect our lack of confidence in the established process—and our belief in Murphy's law. Since the standard does not address the total business system, it is logical to include some sort of safeguard. We should remember that many of the traditional safeguard activities (e.g., receiving inspection, 100% in-process inspection, final inspection, compliance audits) are generally considered to be non-value-added in that these activities do not increase the value of the product or service. However, these activities are initially not valueless because they do find and remove nonconformances and, thus, assure the customer of a delivery with improved (reduced) defect levels. The MBNQA approach is to improve the process continually until there is no need for this type of activity because all the customers—internal and external—have confidence in the process.

SUMMARY

"ISO 9000 is foundational, yet not state-of-the-art like Baldrige," says Robert Peach, chairman of the Registration Accreditation Board (RAB).[16] Joseph Klock, director for AT&T's Quality Registrar, notes that "ISO is definitely moving toward TQM. But, for now, it is a basic standard, a minimal requirement. If companies are currently doing ISO 9000, it can be worth 200 to 300 points in Baldrige."[17] Likewise, if companies are using the MBNQA Criteria for self-improvement, they are in a much better position to implement an ISO series standard. Wallace Co., Inc., Vice President Dave Coulter says his company's 1990 MBNQA Award made achieving ISO 9000 registration a less intimidating process. ". . . Baldrige instilled an ability to gather as a team, brainstorm the process, [and] see what we needed to do. . . ." He also noted that even a Baldrige Award recipient can't relax when it comes to ISO registration.[18]

So, how should you proceed: ISO 9000 or Baldrige? Steven Bergeron of Unitrode Integrated Circuits Corporation feels that:[19]

---

[16] G. Spizien, "The ISO 9000 Standards: Creating a Level Playing Field for International Quality," *National Productivity Review,* Summer 1992.

[17] *Ibid.*

[18] "First RAB Registered Company Also Baldrige Winner," *Quality Systems Update,* July 1991.

[19] Steven Bergeron, private communcation, 1994.

Given market demand, ISO will win in the short run (3–5 years). One of the misconceptions of ISO is the coverage issue with the MBNQA Criteria. ISO can cover everything in Baldrige *if* the company chooses it. Explanation: Companies typically achieve registration by introducing business practices compliant to the Standard. Most companies and many consultants stop their effort at minimum compliance. However, there are no maximum requirements. Indeed the intent of ISO is not to stop at the minimum. All companies have to do is declare a business practice(s) as part of their Quality Manual, document and use it. Auditors will then audit for effectiveness the newly included business practice(s). Here lies the bridge with Baldrige. A company's Quality Manual could in fact include everything in Baldrige.

The decision is yours. If you are looking at the ISO 9000 series simply to satisfy a customer's requirement for registration, then you have a long way to go on your quality journey. If you are looking for a structured approach to get started to understand your processes with the intent of continual improvement and customer delight, then the ISO 9000 series is an excellent candidate.

Comparison Figures

If you do decide to employ both the MBNQA Criteria and the ISO 9000 series standards in your quality journey, Figures 13.5, 13.6, and 13.7 may be of use. In these figures, the interrelationships between the ISO 9001 and the MBNQA Criteria are indicated by listing the Items across from each other. Entries are shown if the Criteria and standard have something in common. This correspondence is not meant to indicate equality of requirements but that the specific element is addressed by both documents.

If you are already involved with ISO 9000, these figures indicate where you should have activities and information for responses to MBNQA Items. Conversely, if you start with the MBNQA Criteria first, the figures point out those areas that should be considered for a customer (ISO 9000)-focused review and continual improvement.

In conclusion, in the authors' opinion, the comparisons given here show ISO 9001 to be a viable building block for companies beginning their quality journey. But ISO 9001 registration does not take the company as far on the journey as the MBNQA Criteria.

Figure 13.5

MBNQA Criteria Compared to ISO 9001 Standard*

MBNQA Criteria	**ISO 9001 Standard**
1.0 Leadership	
1.1 Senior Executive Leadership	4.1.1 Quality Policy
	4.1.1 Quality Policy
	4.1.2 Organization
	4.1.3 Management Review
1.2 Leadership System and Organization	4.17 Internal Quality Audits
1.3 Public Responsibility & Citizenship	
2.0 Information & Analysis	
	4.5 Document and Data Control
2.1 Management of Information and Data	4.11 Inspection, Measuring, & Test Equipment
2.2 Comparisons & Benchmarks	
2.3 Analysis and Use	
3.0 Strategic Planning	
3.1 Strategy Development	4.2 Quality Systems
3.2 Strategy Deployment	
4.0 Human Resource Dev & Mgt	
4.1 HR Planning and Evaluation	
4.2 High Performance Work Systems	
4.3 Employee Educ, Training, & Dvlp	4.18 Training
4.4 Employee Satisfaction	

* Based on ISO 9001—1994 and the 1995 MBNQA Criteria.

Figure 13.5
MBNQA Criteria Compared to ISO 9001 Standard (*continued*)

MBNQA Criteria	ISO 9001 Standard

5.0 Process Management

	4.3 Contract Reviews
5.1 Design Process	4.4 Design Control
5.2 Process Management: Product	4.2 Quality Systems
and Service Production & Delivery	4.9 Process Control
	4.10.2 In-Process Inspect & Test
	4.11 Inspection, Measuring, & Test
Equipment	
	4.14 Corrective and Preventive Action
	4.15 Packaging & Delivery
	4.19 Servicing
	4.20 Statistical Techniques
5.3 Support Services	
5.4 Mgt of Supplier Performance	4.3 Contract Reviews
	4.6 Purchasing

6.0 Business Results

	4.10.2 Receiving Inspection & Testing
	4.10.5 Inspection and Test Records
6.1 Product & Service Quality Results	4.16 Control of Quality Records
6.2 Operational & Financial Results	
6.3 Supplier Performance Results	4.16 Control of Quality Records

7.0 Customer Focus & Satisfaction

7.1 Customer & Market Knowledge	4.4.4 Design Input
7.2 Customer Relationship Mgt	
7.3 Satisfaction Determination	
7.4 Customer Satisfaction Results	
7.5 Customer Satisfaction Comparison	

Figure 13.6
ISO 9001 Standard Compared to MBNQA Criteria*

ISO 9001 Standard	MBNQA Criteria
4.1.1 Quality Policy	1.1 Senior Executive Leadership
	1.2 Leadership System & Organization
4.1.2 Organization	1.2 Leadership System & Organization
4.1.3 Management Review	1.2 Leadership System & Organization
4.2 Quality System	1.2 Leadership System & Organization
	3.1 Strategy Development
	5.2 Production & Delivery
4.3 Contract Review	5.4 Mgt of Supplier Performance
4.4 Design Control	5.1 Design Process
	7.1 Customer & Market Knowledge
4.5 Document and Data Control	2.1 Mgt of Information and Data
4.6 Purchasing	5.4 Mgt of Supplier Performance
4.7 Customer Supplied Product	
4.8 Product ID & Traceability	
4.9 Process Control	5.2 Production & Delivery
4.10 Inspection & Testing	5.2 Production & Delivery
	5.4 Mgt of Supplier Performance
	6.1 Product & Service Quality Results
4.11 Inspection, Measuring, and Testing Equipment	2.1 Mgt of Information and Data
	5.2 Production & Delivery
4.12 Inspection & Test Status	
4.13 Control of Nonconforming Product	
4.14 Corrective and Preventive Action	5.2 Production & Delivery
4.15 Packaging & Delivery	5.2 Production & Delivery
4.16 Control of Quality Records	6.1 Product & Service Quality Results
	6.3 Supplier Performance Results
4.17 Internal Quality Audits	1.2 Leadership System & Organization (1.2c Review of Company's Performance)
4.18 Training	4.3 Employee Educ, Training, & Dvlpmnt
4.19 Servicing	5.2 Production & Delivery
4.20 Statistical Techniques	5.2 Production & Delivery

* Based on ISO 9001—1994 and the 1995 MBNQA Criteria.

Figure 13.7

Cross-Reference List for ISO 9000 Series Quality System Elements*

Item Number in ISO 9004	Title	ISO 9001	ISO 9002	ISO 9003
4	Management responsibility	4.1	⊙	○
5	Quality system principles	4.2	⊃	⊙
5.4	Auditing the quality system (internal)	4.17	⊙	□
6	Economics — Quality-related cost considerations	—	—	—
7	Quality in marketing (Contract review)	4.3	⊃	□
8	Quality in specification and design (Design control)	4.4	□	□
9	Quality in procurement (Purchasing)	4.6	⊃	□
10	Quality in production (Process control)	4.9	⊃	□
11	Control of production	4.9	⊃	□
11.2	Material control and traceability (Product identification and traceability)	4.8	⊃	⊙
11.7	Control of verification status (Inspection and test status)	4.12	⊃	⊙
12	Product verification (Inspection and testing)	4.10	⊃	⊙
13	Control of measuring and test equipment (Inspection, measuring, and test equipment)	4.11	⊃	⊙
14	Nonconformity (Control of nonconforming product)	4.13	⊃	⊙
15	Corrective action	4.14	⊃	□
16	Handling and postproduction functions (Handling, storage, packaging, and delivery)	4.15	⊃	⊙
16.2	After-sales servicing	4.19	□	□
	Quality documentation and records (Document control)	4.5	⊃	⊙
17.3	Quality records	4.16	⊃	⊙ ○
18	Personnel (Training)	4.18	⊙	
19	Product safety and liability	—	—	—
20	Use of statistical methods (Statistical techniques)	4.20	⊃	⊙
	Purchaser-supplied product	4.7	⊃	□

Key (with ISO 9001 — 1994 as basis):
- ⊃ Full requirement
- ⊙ Less stringent than ISO 9001
- ○ Less stringent than ISO 9002
- □ Placeholder – no requirement
- — Requirement not present

Note: The quality system requirements in ISO 9001, 9002, and 9003 are identical in many cases, but not in every case.

* Adapted from ISO 9000—1987 reflecting the 1994 revisions

14 | The Future of MBNQA

INTRODUCTION

The goals of the MBNQA process are to help organizations (1) deliver ever-improving value to customers and (2) improve overall operational performance. True to its philosophy, the MBNQA process is itself subject to continual improvement. Since its initial release in 1988, the Criteria have undergone several changes based on the feedback of the users. With the exception of the seven-category framework, all other supporting materials are either new or incorporate major lessons learned.[1] Although the most visible changes are in the Criteria and supporting documentation, the greatest change is the evolution of the Criteria from a means of evaluating a total quality system to road map and yardstick to help organizations on their journey toward total system (performance) excellence.

In the future it is expected that the major changes will not be in the Criteria itself but in the process. Already bills are in the legislature to eliminate the arbitrary two awards per eligibility category. Also being developed is the expansion of the Award process to include health care and educational organizations. Non-award versions of both of these Criteria have been released for 1995. With this evolution of the MBNQA, the use of the Criteria is expected to grow, with most organizations focusing on self- improvement.

 ## THE EVOLUTION OF THE MBNQA

To understand where the MBNQA process is headed, it is useful to look at its own continual improvement journey. From the very beginning, it was recognized that the MBNQA program itself must be a "rapid-cycle-time, learning organization"[2] in order to fulfill the roles necessary to support the intent of Public Law 100-107

[1] J. S. Neves, and B. Nakhai, "The Evolution of the Baldrige Award," *Quality Progress,* June 1994.

[2] C. Reimann and H. Hertz, "The Malcolm Baldrige National Quality Award and ISO 9000 Registration: Understanding Their Many Important Differences," NIST, Gaithersburg, MD, 1993.

(see Chapter 5). These changes incorporate lessons learned from many sources and listening posts—from the experiences of the Board of Examiners and the applicants with the Criteria and the evaluation and feedback process to published successes as well as deficiencies in "TQM" and other management approaches. In addition to this, NIST annually sponsors an "MBNQA Improvement Meeting." Current and former members of the Board of Examiners, current and past applicants, friends of the MBNQA program and staff meet in a workshop to discuss how the Criteria and process can be improved. Written comments are welcomed from those interested parties unable to attend the meeting.

Figure 14.1

Evolution of the MBNQA Major Focus

Year	Major Focus
1988	"to test all elements of a total quality system"[F1]
1989	"to permit evaluation of the strengths and areas for improvement in the . . . quality systems and quality results."[F2]
1990	"to make possible a *diagnosis* which is the basis for *feedback reports*"[F3]
1991	"to be a quality excellence standard for organizations seeking the highest levels of overall quality performance and competitiveness... by focusing not only on results but also upon the conditions and processes that lead to results. ... offers a framework that can be used by organizations to tailor their systems and processes toward ever-improving quality performance"[F4]
1992	"to project key requirements for delivering ever-improving value to customers while at the same time maximizing the overall productivity and effectiveness of the delivering organization. ... the Criteria need to be built upon a set of [core values and concepts] that together address and integrate the overall customer and company performance requirements."[F5]
1993	"to support dual, results oriented goals: • delivery of ever-improving value to customers; and • improvement of overall company operational performance."[F6]
1994	"to help companies enhance their competitiveness through focus on dual, results oriented goals: • delivery of ever-improving value to customers, resulting in improved marketplace performance; and • improvement of overall company operational performance."[F7]
1995	"The Criteria continue to evolve toward better definition of performance management and better integration of performance management with overall business management."[F8]

[F1] NIST, *MBNQA Applications Guidelines 1988*, NIST, Gaithersburg, MD, 1988, Preface.

[F2] NIST, *MBNQA Applications Guidelines 1989*, NIST, Gaithersburg, MD, 1989, p. 14.

[F3] NIST, *1990 MBNQA Applications Guidelines*, NIST, Gaithersburg, MD, 1990, p. 18.

[F4] NIST, *1991 MBNQA Applications Guidelines*, NIST, Gaithersburg, MD, 1991, p. 2.

[F5] NIST, *1992 MBNQA Award Criteria*, NIST, Gaithersburg, MD, 1992, p. 2.

[F6] NIST, *1993 MBNQA Award Criteria*, NIST, Gaithersburg, MD, 1993, p. 2.

[F7] NIST, *1994 MBNQA Award Criteria*, NIST, Gaithersburg, MD, 1994, p. 2.

[F8] NIST, *Draft 1995 MBNQA Award Criteria*, NIST, Gaithersburg, MD, 1994, p. 1.

The most visible changes in the MBNQA process have been in the Criteria and the Criteria booklet. Many of the changes in the booklet have been made in support of the educational role of MBNQA. However, the greatest change has been the evolution of the focus of the Criteria. As can be seen in Figure 14.1, the focus of the Criteria has been evolving from quality excellence to a total system's performance excellence and integration with overall business leadership and from an evaluation tool to a road map and yardstick.

This evolution is also seen in the changes in the Category names:

- *Category 3:* 1988, Strategic Quality Planning; 1995, Strategic Planning

- *Category 5:* 1988, Quality Assurance of Products and Services; 1992, Management of Process Quality; 1995, Process Management

- *Category 6:* Results from Quality Assurance Products and Services; 1989, Quality Results; 1992, Quality and Operational Results; 1995, Business Results.

The evolution of the Criteria toward a total systems approach is a positive move. It is consistent with Dr. Deming's "A System of Profound Knowledge,"[3] which calls on the organization to have

- Appreciation for a System

- Knowledge of Variation

- Understanding of the Theory of Knowledge

- Understanding of Psychology.

However, in the authors' opinion, there also seems to be a "softening" of the evaluation scale. For example, the 1993 scoring guidelines called for "more emphasis on problem prevention than on reaction to problems" while the 1994/5 scoring guidelines put "more emphasis on improvement than on reaction to problems." Some may say that "improvement" is better than "prevention." Unfortunately, if the focus is improving the results of the process rather than understanding the process and eliminating sources of variation (i.e., prevention), the result can be a suboptimized process.

Another example is Area to Address 5.1b, which, in 1994, asks for how "designs are reviewed and validated, taking into account key factors . . .," whereas in 1995 it asks for how "designs are reviewed and/or tested in detail to ensure a trouble-free launch." Traditional engineering testing, especially "bogey" testing, is after-the-fact inspection rather than a proactive tool to understand the design

---

3 W. E. Deming, *The New Economics*, MIT Press, Cambridge, MA, 1993.

more fully. In this type of testing, the goal is to get the design to "pass" instead of developing the highest performance design. Although planned experimentation and testing is a vital and necessary part of the engineering design, validation, and verification process, this relaxation of requirements does allow as a valid response the engineering equivalent of 100% inspection.

Although this softening of the Criteria and guidelines should result in increased scores, it, fortunately, should not have an adverse effect on organizations using the MBNQA process for self-improvement.

 ## THE FUTURE

With the continual improvement of the MBNQA process, the theme and the focus of the Criteria will continue to evolve toward a more comprehensive road map and yardstick for total system performance management. In 1995 the Criteria and guidelines have increased emphasis on

- Key business drivers derived from business planning

- Economic factors in setting priorities in process management

- Building workforce capabilities as a key part of building company capabilities

- High-performance, more flexible, more responsive work organizations and the critical mechanisms that build and sustain them

- Continual learning as well as continual improvement.[4]

In the future, the major changes will be seen not in the Criteria but in the process. With the activity in the legislature to eliminate the arbitrary two Awards per eligibility Category, the contest aspect (i.e., the win–lose character) of the Award process will be eliminated. Instead, the process will be able to identify and recognize *all* excellent organizations among the applicants. Also being developed is the expansion of the Award process to include organizations not currently eligible such as health care and educational organizations. Many of these organizations are currently eligible for many state and professional organization award processes.

In 1995 NIST is conducting pilot programs with education and health care as eligibility categories in parallel with the Baldrige process for businesses. The Criteria and evaluation process to be used in these pilot programs are derived and comparable to a mix of the 1994 and 1995 MBNQA Criteria. However, education and health care organizations participating in this pilot program will not be

---

[4] See also *1995 MBNQA Award Criteria*, NIST, Gaithersburg, MD, 1994, p. 18.

eligible for awards. The results and experiences gained in this program will be used to determine if there is an interest and readiness on the part of education and health care organizations to participate in such a national-level program and, if so, what improvements are necessary to provide the maximum value to the participants and to other U.S. education and health care organizations.

And what about the "competition" between the MBNQA and ISO 9000 standards? Brad Stratton, editor of *Quality Progress*, believes that the Baldrige Award is on the verge of a renaissance.[5] This belief is influenced by the recent actions by senior policy officials in the EC's Directorate–General III for Industry, a high-level EC standards regulatory group. Members of this group have questioned the effectiveness of ISO 9000 as a tool to drive quality. They are calling for creation of a European-wide quality program that could include the European Quality Award in order to help European firms infuse quality into their organizations. That is, the group wants to deemphasize certification and reemphasize quality practices. On another front, John King, director general of the European Foundation for Quality Management, has hinted at a possible collaboration between his group and the MBNQA sponsors.[6]

Even with the continual improvement of the Criteria and MBNQA process and the growing acceptance of the Criteria worldwide as an effective road map for quality and performance excellence, the number of applicants for the Award is not expected to grow dramatically. Many organizations feel that focusing on an "award" sends the wrong message to their employees or that they would rather receive the recognition of the marketplace and their customers than spend the time "competing for an award." These same organizations also recognize the need for a well-defined and consistent road map and yardstick of quality and performance excellence. Consequently, the use of the Criteria is expected to grow, with most organizations focusing on self-improvement with internal feedback processes or using the increasing number of state award processes for evaluation and feedback.

[5] Editorial in *Quality Progress*, August, 1994.
[6] A. Zuckerman, "EC Drops Ticking Time Bomb," *Industry Week*, May 16, 1994.

15 | Summary and Conclusions

The focus of this book has been to help organizations on their journey toward quality and performance excellence. The exact path taken by an individual company will be unique to that organization, since every company has its own distinctions. However, among the organizations doing well on their quality journey, there are some common traits. Foremost is a *customer focus* throughout the organization. To employees in these organizations, a customer is not only the final recipient of the product or service, but anyone who could be impacted by their activities—from the next operation, to shipping and servicing, to the traditional "customer." But a philosophical orientation is not enough. These companies have *aggressive quality goals and strategies* to achieve their quality excellence. Although these strategies recognize the impact of every element in the total system—*quality in all operations*—it is the leadership that creates the quality culture. The *personal involvement of the leadership* enables and provides the constant reinforcement necessary for the cultural change to *continual improvement*.

With a total systems approach, it is recognized that all knowledge does not reside within the company. Excellent companies encourage the *involvement of customers and suppliers* in the product and process development and implementation cycles. Further they seek new ideas and *innovation* within the organization as well as outside the company (via benchmarking). In support of their strategy implementation, the companies have established *strong information systems* and *comprehensive training*.

There are also some common characteristics in organizations needing improvement. Management in these companies always seem to know, and use conversationally, the latest buzzword(s) and trends. However, they are always too busy with the "real" activity of the company—making a profit—to become personally involved with the quality system. This *delegation of quality* responsibility leads to an *unclear quality definition* and confusion about what is expected.

There is a *lack of awareness of "best."* This is reinforced by a *lack of measures, indicators, or benchmarks* to compare to or track progress. The companies *fail to use all listening posts* available to them. And the information that is collected may not get to the appropriate decision makers due to *weak information systems*.

The quality focus is typically limited to production activities. These quality systems are *reactive systems*—the system does not act to change or improve until a complaint is made or alarm is given. Management may even claim that these are prevention systems and are customer driven. Their logic is that anytime a customer complains, they will initiate corrective action and will set up safeguards to prevent that problem from getting to the customer again. They do not have an understanding of the interrelationships that exist within the total business system. This results in a *partial quality system* and a *lack of alignment* among the various departments and activities.

The Malcolm Baldrige National Quality Improvement Act of 1987 was established to

... help improve quality and productivity by:

A. helping to stimulate American companies to improve quality and productivity for the pride of recognition while obtaining a competitive edge through increased profits;

B. recognizing the achievements of those companies that improve the quality of their goods and services and providing an example to others;

C. establishing guidelines and criteria that can be used by business, industrial, governmental, and other organizations in evaluating their own quality improvement efforts; and

D. providing specific guidance for other American organizations that wish to learn how to manage for high quality by making available detailed information on how winning organizations were able to change their cultures and achieve eminence.

The Award Criteria are the basis for making awards and for giving feedback to the applicants. However, in accord with Public Law 100-107, they were also developed to assist in raising the quality performance of American organizations by facilitating communications and sharing within and between organizations and by serving as a road map and yardstick for quality improvement. The Criteria are built on a set of core values and concepts:

☑ Customer-driven quality

☑ Leadership

☑ Continuous improvement

☑ Employee participation and development

☑ Fast response

☑ Design quality and prevention

☑ Long-range outlook

☑ Management by fact

☑ Partnership development

☑ Corporate responsibility and citizenship

☑ Results orientation.

The Criteria are nonprescriptive. They do not prescribe any specific tools, techniques, technologies, systems, or starting points. The Criteria do not require that there should or should not be a separate quality organization within a company or how the company itself should be organized. An examination of the Award recipients shows a diversity of approaches and implementation strategies. The Criteria do emphasize that all elements, practices, strategies, and processes be regularly evaluated as part of the organization's continual improvement process.

The legislature recognized that the greatest benefit of the MBNQA process was not in the recognition of a few excellent companies, but in the increased understanding of and guidance toward quality excellence. That is, the Award's central purpose is educational.[1] It promotes understanding not only through education and training but also through organizations sharing and networking, and interactions with customers and suppliers. It provides guidance to aid organizations in a self-assessment of their present situation and in planning for continual improvement.

The MBNQA process is itself subject to continual improvement. Since its initial release in 1988, the Criteria have undergone several changes based on the feedback of the users. What has not changed are the basic values and structure (i.e., the seven categories). In recent years, the core values and concepts—the bases of the Criteria—have stabilized, with changes primarily to aid clarity. The theme

---

[1] C. Reimann and H. Hertz, "The Malcolm Baldrige National Quality Award and ISO 9000 Registration: Understanding Their Many Important Differences," NIST, Gaithersburg, MD, 1993.

and the focus of the Criteria continue to evolve toward a more comprehensive road map and yardstick toward a total system performance management.

In the future, the major changes will not be in the Criteria but in the process. Already bills are in the legislature to eliminate the arbitrary two Awards per eligibility Category. Also being developed and piloted is the expansion of the Award process to include health care and educational organizations. However, the number of applicants is not expected to grow dramatically, but the use of the Criteria is expected to grow, with most organizations focusing on self- improvement.

Overall, the MBNQA Criteria provide an integrated, value-oriented framework for the design, implementation, and assessment of a process for managing the performance of all operations. Achievement of world-class status does not simply mean developing responses to the Criteria. The quality journey is long and arduous. The MBNQA Criteria do provide a road map and yardstick to make this journey more efficient and effective.

It is a tremendous undertaking in terms of commitment and change. We have seen it work for others, assisting them in achieving operational and market successes. We know it can work for you.

Appendix

MBNQA PROCESS CONTACT LIST

For more information on the Award process contact:

> Malcolm Baldrige National Quality Award
> National Institute of Standards and Technology
> Route 270 and Quince Orchard Road
> Administration Building, Room A537
> Gaithersburg, MD 20899
>
> 301/975-2036
> 301/948-3716 (FAX)

Note: There are charges for products and services available from the following providers:

For bulk orders of the Award Criteria, videos of the recipients, and additional case studies contact:

> The American Society for Quality Control
> Customer Service Department
> 611 E. Wisconsin Avenue
> P.O. Box 3005
> Milwaukee, WI 53201-3005
>
> 800/248-1946
> 414/272-1734 (FAX)

For Quest for Excellence videotapes contact:

> The National Technological University
> Advanced Technology and Management Programs
> 700 Centre Avenue
> Fort Collins, CO 80526
>
> 303/484-6050
> 303/498-0501 (FAX)

For Quest for Excellence audiotapes contact:

Audio Archives International, Inc.
3043 Foothill Blvd., Suite #2
La Crescenta, CA 91214

800/747-8069
818/957-0876 (FAX)

MBNQA RECIPIENTS CONTACT LIST (in alphabetical order)

Company	Quality Contact	Public Affairs Contact
Ames Rubber Corporation Small Business—1993	Charles A. Roberts Vice President, Total Quality 23–47 Ames Boulevard Hamburg, NJ 07419 201/209-3200 FAX: 201/827-8893	Same as Quality
AT&T Consumer Communications Services Service—1994	Diane Phelps-Feldkamp Director, Quality & Business Improvement 295 N. Maple Avenue Room 6356F3 Basking Ridge, NJ 07920 800/473-5047 FAX: 904/636-3780	Linda Haertlein Public Relations Director 295 N. Maple Avenue Room 2342G2 Basking Ridge, NJ 07920 800/473-5047 FAX: 904/636-3780
AT&T Network Systems Group Transmission Systems Business Unit Manufacturing—1992	Louis E. Monteforte, Director Transmission Quality Planning 475 South Street Room 2W-44 Morristown, NJ 07962-1976 201/606-2488 FAX: 201/606-3363	Same as Quality
AT&T Universal Card Services Service—1992	Greg Swindell Chief Quality Officer 8787 Baypine Rd. Room 3-2-151N Jacksonville, FL 32256 904/954-8875 FAX: 904/954-7118 Tours: 800/682-7759	Bruce Reed Media Relations 8787 Baypine Rd. Room 3-2-197N Jacksonville, FL 32256 904/954-8894 FAX: 904/954-8720

Company	Quality Contact	Public Affairs Contact
Cadillac Motor Car Company Manufacturing—1990	Joseph R. Bransky, Director Quality and Reliability General Motors Building Room 6-162 3044 West Grand Blvd. Detroit, MI 48202 313/556-9050 FAX: 313/974-6899	John F. Maciarz, Manager Media Relations Quality and Reliability NAO Headquarters Building 1-8 30400 Mound Road, MS 2057 Warren, MI 48090 810/986-6132 FAX: 810/986-9253
Eastman Chemical Company Manufacturing—1993	Katherine Watkins Advanced Sales Support Specialist P.O. Box 431 Kingsport, TN 37662-5370 800/695-4322, Ex1150 FAX: 615/229-1195	Rodney D. Irvin, Manager Media Relations P.O. Box 511 Stone East Building 2550 East Stone Drive Kingsport, TN 37662 615/229-4008 FAX: 615/229-1008
Federal Express Corporation Service—1990	Quality Questions & Information Jean Ward-Jones, Manager Corporate Quality P.O. Box 727 Memphis, TN 38194-2142 901/395-4539 FAX: 901/395-4641	Quality Speaker Service Sally Davenport, Senior Specialist Public Relations 2005 Corporate Avenue Memphis,TN 38132 901/395-3466 FAX: 901/395-4928 Reporters & Editors Shirley Finley, Senior Specialist Media Relations 2005 Corporate Avenue Memphis, TN 38132 901/395-3463 FAX: 901/346-1013 or 395-4928

Company	Quality Contact	Public Affairs Contact
Globe Metallurgical Inc. Small Business—1988	Norman Jennings Quality Director P.O. Box 157 Beverly, OH 45715 614/984-2361 FAX: 614/984-8635	Sherryl Hennessey Public Relations P.O. Box 157 Beverly, OH 45715 614/984-8640 FAX: 614/984-8635
Granite Rock Company Small Business—1992	Bruce W. Woolpert President and CEO P.O. Box 50001 Watsonville, CA 95077-5001 408/761-2300 FAX: 408/724-3484	Greg Diehl Manager, Marketing Services P.O. Box 50001 Watsonville, CA 95077-5001 408/724-5611 FAX: 408/724-3484
GTE Directories Corporation Service—1994	Jim Runyan Director, Quality Services GTE Place, West Airfield Drive P.O. Box 619810 D/FW Airport, TX 75261-9810 214/453-7985 FAX: 214/453-6785	Janet P. Stevens, APR Public Relations Director GTE Place, West Airfield Drive P.O. Box 619810 D/FW Airport, TX 75261-9810 214/453-7751 FAX: 214/453-7231
IBM Rochester Manufacturing—1990	IBM Rochester Center for Excellence 3605 Highway 52 North Rochester, MN 55901-7829 507/253-2300 FAX: 507/253-3484	Same as Quality
Marlow Industries Small Business—1991	Tiki Miller Baldrige Activities Coordinator 10451 Vista Park Road Dallas, TX 75238-1645 214/342-4293 FAX: 214/341-5212	Same as Quality

Company	*Quality Contact*	*Public Affairs Contact*
Milliken & Company Manufacturing—1989	Craig Long Vice President, Quality P.O. Box 1926, M-186 Spartanburg, SC 29304 803/573-2003 FAX: 803/573-2505 Tours: Sandra Howell, 803/573-1988	Richard Dillard Director of Public Affairs P.O. Box 1926, MS-285 Spartanburg, SC 29304 803/573-2546 FAX: 803/573-2100
Motorola, Inc. Manufacturing—1988	Richard Buetow Senior Vice President & Director of Quality 1303 East Algonquin Road Schaumburg, IL 60196 708/576-5516 FAX: 708/538-2663	Margot Brown Director, Media Relations 1303 East Algonquin Road 7th Floor Schaumburg, IL 60196 708/576-5304 FAX: 708/576-7653

The Ritz-Carlton Hotel Company
 	
Service—1992
	Patrick Mene
	Corporate Dir. Quality
	3414 Peachtree Road, N.E.,
	   Suite 300
	Atlanta, GA 30326
	
	404/237-5500
	FAX: 404/261-0119
		Karon Cullon, Corporate
		   Dir. Public Relations
		3414 Peachtree Road, N.E.,
		   Suite 300
		Atlanta, GA 30326
		
		404/237-5500
		FAX: 404/365-9643

Solectron Corporation Manufacturing—1991	Margaret Smith Marketing Program Specialist 847 Gibraltar Drive Milpitas, CA 95035 408/956-6768 FAX: 408/956-6056	MEDIA REQUESTS ONLY Jeffrey F. Cox, Director Corporate & Marketing Communications 847 Gibraltar Drive Milpitas, CA 95035 408/956-6688 FAX: 408/956-6056

Texas Instruments, Inc.
 	
Defense Systems & Electronics Group
 	
Manufacturing—1992
	Baldrige Response Center
	P.O. Box 660246, M/S 3124
	Dallas, TX 75266
	
	FAX: 214/480-4880
	(no longer receiving telephone
	   calls for information from
	   persons external to the company)
		Same as Quality

Company	Quality Contact	Public Affairs Contact
Wainwright Industries, Inc. Small Business— 1994	Michael Simms Plant Manager P.O. Box 640 St. Peters, MO 63376 314/278-5850 FAX: 314/278-8806	David A. Robbins Vice President P.O. Box 640 St. Peters, MO 63376 314/278-5850 FAX: 314/278-8806
Wallace Company Small Business— 1990	Assets of the Wallace Company have been acquired by Wilson Industries Wallace Company Wilson Industries 1302 Conti Houston, TX 77002 713/237-3700 FAX: 713/237-5935	
Westinghouse Electric Corp. Commercial Nuclear Fuel Div. Manufacturing— 1988	Carl Arendt Mgr. Total Quality Services Westinghouse Productivity & Quality Center P.O. Box 160 Pittsburgh, PA 15230-0160 412/778-5008 FAX: 412/778-5153	Same as Quality
Xerox Corporation Business Products & Systems Manufacturing— 1989	Donna Bislow or Sandy Frieday Xerox Square 836-01D 100 Clinton Ave., South Rochester, NY 14644 716/423-6487 FAX: 716/423-6041	Samuel M. Malone, Jr., Mgr. Quality Communications 80 Linden Oaks Parkway Rochester, NY 14625 716/383-7534 FAX: 716/264-5379
Zytec Corporation Manufacturing— 1991	Karen Scheldroup Baldrige Office 7575 Market Place Drive Eden Prairie, MN 55344 612/941-1100 Exl04 FAX: 612/829-1837	Same as Quality

MBNQA BOARD OF OVERSEERS

Arden L. Bement, 92, 93, 94
Purdue University (Retired from
TRW, Inc.)

Anthony P. Carnevale, 94
Committee for Economic
Development

William W. Eggleston, 91, **92,**[1] **93**
International Business Machines
Corporation
(Retired)

Mary Jane England, 94
Washington Business Group on
Health

Armand V. Feigenbaum, 88, 89, 90, 91
General Systems Company, Inc.

Meredith M. Fernstrom, 88, 89
American Express Company

Douglas A. Fraser, 88
United Auto Workers (Retired)

Bradley T. Gale, 88, 89, 90
PIMS/Strategic Planning Institute

Robert W. Galvin, **89, 90, 91**
Motorola, Inc.

David Garvin, 88, 89
Harvard Business School

Willian A. Golomski, 89, 90, 91
W. A. Golomski & Associates

John J. Hudiburg, 91, 92, 93
Chairman Emeritus
Florida Power & Light Company

Joseph M. Juran, 88, 89, 90, 91
Juran Institute, Inc.

Rasabeth Moss Kanter, 94
Harvard Business School

William H. Kolberg, 92, 93, 94
National Alliance of Business

Meredith M. Layer, 90

Raymond Marlow, 92, 93, 94
Marlow Industries, Inc.

Thomas J. Murrin, **88**
Carnegie Mellon University

Lionel H. Olmer, 88
Former Undersecretary of Commerce
for International Trade

Frank J. Pipp, 94
Xerox Corporation (Retired)

Elmer B. Staats, 88
Former Comptroller General
of the United States

Ray Stata, 94
Analog Devices

Nancy Harvey Steorts, 91, 92
Nancy Harvey Steots International

Preston Townley, 92, 93, **94**
The Conference Board

---

[1] Bold indicates that the person held the position of chair for that year.

STATE QUALITY AWARDS CONTACTS

State	Contact Name & Address	Telephone/Fax
AZ	Dennis Sowards Executive Director Arizona Quality Alliance 13 N. MacDonald, Suite 535 Mesa, AZ 85201	602/655-1948 FAX: 602/655-8069
CA	C. Lance Barnett President California Center for Quality Education and Development 455 Capitol Mall, Suite 500 Sacramento, CA 95814	916/321-5484
CA	Thomas D. Hinton President California Council for Quality & Service P.O. Box 880774 San Diego, CA 92168	619/491-3050 FAX: 619/491-3055
CT	Ernest J. Nagler Project Facilitator Connecticut Award for Excellence P.O. Box 67 Rocky Hill, CT 06067	800/392-2122 FAX: 203/258-4359
CT	Sheila Carmine Director Connecticut Quality Improvement Award, Inc. P.O. Box 1396 Stamford, CT 06904-1396	203/322-9534 FAX: 203/329-2465
DE	Michael J. Hare Delaware Quality Consortium, Inc. Delaware Economic Development Office 99 Kings Highway P.O. Box 1401 Dover, DE 19903	302/739-4271 FAX: 302/739-5749
FL	John Pieno Chairman Florida Sterling Council Governor's Sterling Award Office Room 313, Carlton Building Tallahassee, FL 32399-0001	904/922-5316 FAX: 904/488-9578

State	Contact Name & Address	Telephone/Fax
MA	Jerold Christen Executive Director Massachusetts State Quality Award 3 Robinson Drive Bedford, MA 01730	617/275-1200 FAX: 617/275-1958
MD	Amit Gupta Maryland Center for Quality & Productivity College of Business & Management University of Maryland College Park, MD 20742-7215	301/405-7099 FAX: 301/314-9119
ME	Nancy Werner Maine Quality Center Margaret Chase Smith Library Norridgewock Ave. P.O. Box 3152 Skowhegan, ME 04976	207/685-3004 FAX: same as telephone call first
MI	William Kalmar Director, Michigan Quality Council Oakland University 525 O'Dowd Hall Rochester, MI 48309-4401	810/370-4552 FAX: 810/370-4202
MN	Carol Gabor Director, Minnesota Quality Award Minnesota Council for Quality 3850 Metro Drive, Suite 633 Bloomington, MN 55425	612/851-3181 FAX: 612/851-3183
MO	John Politi President Excellence in Missouri Foundation Harry S. Truman State Off. Building Room 620, 301 W. High Street P.O. Box 1085 Jefferson City, MO 65102	314/526-1725 FAX: 314/526-1729
NC	Mette Leather Acting Director, Recognition Programs North Carolina Quality Leadership Award 4904 Professional Court, Suite 100 Raleigh, NC 27609	919/872-8198 FAX: 919/872-8199

State	Contact Name & Address	Telephone/Fax
NE	The Edgerton Quality Award Program Nebraska Department of Economic Development Existing Business Assistance Division P.O. Box 94666, 301 Centennial Mall South Lincoln, NE 68509-4666	402/471-4167 or 800/426-6505 FAX: 402/471-3778
NJ	Ed Nelson Executive Director—Operations New Jersey Quality Achievement Award Mary G. Roebling Building, CN 827 Trenton, NJ 08625-0827	609/777-0939 FAX: 609/777-2798
NM	Michael D. Silva & Glenn Walters Award Co-Administrators Quality New Mexico 320 Gold, S.W., Suite 1218 Albuquerque, NM 87102	505/242-7903 FAX: 505/242-7940
NY	Barbara Ann Harms New York State Dept. of Labor Office of Labor Management Affairs Harriman State Office Campus Bldg. 12, Rm. 540A Albany, NY 12240	518/457-6747 FAX: 518/457-0620
OK	Michael Strong Executive Director Oklahoma State Quality Award Foundation, Inc. 6601 N. Broadway, Suite 244 Oklahoma City, OK 73116	405/841-5295 or 800/879-6552 FAX: 405/841-5205
OR	Timothy Dedlow Program Manager Oregon Quality Award One World Trade Center 121 S.W. Salmon, Suite 1140 Portland, OR 97204	503/224-4606 FAX: 503/224-5435

State	*Contact Name & Address*	*Telephone/Fax*
PA	Beverly M. Centini or Sam Garula Office of Executive Director Pennsylvania Quality Leadership Foundation, Inc. P.O. Box 4129 Harrisburg, PA 17111-0129	717/561-7100 FAX: 717/561-7104
RI	Lynne Couture Executive Director Rhode Island Area Coalition for Excellence P.O. Box 6766 Providence, RI 02940 or 18 Imperial Place Providence, RI 02903	401/454-3030 FAX: 401/751-7742
TN	Marie B. Williams Executive Director Tennessee Quality Award Office 2233 Highway 75, Suite 1 Blountville, TN 37617-5840	800/453-6474 or 615/279-0037 FAX: 615/279-0978
TX	Cynthia Wisehart Award Administrator Quality Texas P.O. Box 684157 Austin, TX 78768-4157	512/477-8137 FAX: 512/477-8168
WA	Lois Quinn Dept. of Labor & Industries N. 901 Monroe, Suite 100 Spokane, WA 99201	509/324-2534 FAX: 509/324-2636

BIBLIOGRAPHY

_____, "A Smart Way to Manufacture," *Business Week*, April 1990.

_____, "Copycats," *Industry Week*, November 5, 1990.

_____, "Here Comes Japan's Carmakers—Again," *Fortune*, December 13, 1993.

_____, "Baldrige Winner Gains Market Share," *Quality in Manufacturing*, May/June, 1993.

_____, *ANSI/ASQC ISO 9000—1994 Quality Standards Series*, American Society for Quality Control,1994.

_____, "Quality: Everybody's Job—Many Vacancies," Gallup Survey conducted for the American Society for Quality Control, 1990.

_____, *Management Practices*, GAO/NSIAD-91-190, GAO, Washington, D.C., 1991.

_____, *Handbook for Examiners*, National Institute of Standards and Technology (NIST), Gaithersburg, MD, 1994.

_____, *MBNQA—1995 Award Criteria*, NIST, Gaithersburg, MD, 1995.

_____, *Continuous Improvement—Batavia*, video, Ford Motor Company, Dearborn, MI, 1985.

Albrecht, K., *The Only Thing That Matters*, HarperBusiness, New York, 1992.

ASQC, "Inside the Baldrige Award Guidelines," reprint from seven-part series in the June through December *Quality Progress*, Vol. 25, ASQC Quality Press, Milwaukee, 1992.

Barker, J. A., *Future Edge*, William Morrow & Co., New York, 1992.

Boyett, J. H, *et al.*, *The Quality Journey*, Dutton, New York, 1993.

Breitenberg, M., "Questions and Answers on Quality," NISTIR 4721, NIST, Gaithersburg, MD, 1991.

_____, "More Questions and Answers on the ISO 9000 Standard Series," NISTIR 5122, NIST, Gaithersburg, MD, 1993.

Camp, R. C., "Benchmarking: The Search for Best Practices that Lead to Superior Performance," *Quality Progress*, January 1989.

Covey, S. R., *The 7 Habits of Highly Effective People*, Simon & Schuster, New York, 1990.

_____, *Principle Centered Leadership*, Summit Books, New York, 1991.

Deming, W. E., *Out of Crisis*, MIT Press, Cambridge, MA, 1986.

_____, *The New Economics*, MIT Press, Cambridge, MA, 1993.

Dobyns, L., & Crawford-Mason, C., *Quality or Else*, Houghton Mifflin, Boston, 1991.

Garvin, D. A., "Quality on the Line," *Harvard Business Review*, September–October 1983.

————, "How the Baldrige Award Really Works," *Harvard Business Review*, November–December 1991.

Juran, J. M., *Made in USA: A Break in the Clouds*, Juran Institute, Wilton, CT, 1990.

————, Ed., *Quality Control Handbook*, McGraw-Hill, New York, 1988.

Kohn, A., *No Contest*, Houghton Mifflin, Boston, 1986.

Main, J., "Is the Baldrige Overblown?," *Fortune*, July 1, 1991.

Scherkenbach, W. W., *Deming's Road to Continual Improvement*, SPC Press, Knoxville, TN, 1991.

Senge, P. M., *The Fifth Discipline*, Doubleday, New York, 1990.

Conference:

Quest for Excellence, scheduled in February of each year. This is a $2\frac{1}{2}$-day conference at which the current recipients present their strategies. Contact the Association for Quality and Participation at 800/733-3310 or the MBNQA Office at NIST.

INDEX